Situation Awareness for Smart Distribution Systems

Situation Awareness for Smart Distribution Systems

Editors

Leijiao Ge
Jun Yan
Yonghui Sun
Zhongguan Wang

MDPI • Basel • Beijing • Wuhan • Barcelona • Belgrade • Manchester • Tokyo • Cluj • Tianjin

Editors
Leijiao Ge
Tianjin University
China

Jun Yan
Concordia University
Canada

Yonghui Sun
Hohai University
China

Zhongguan Wang
Tianjin University
China

Editorial Office
MDPI
St. Alban-Anlage 66
4052 Basel, Switzerland

This is a reprint of articles from the Special Issue published online in the open access journal *Energies* (ISSN 1996-1073) (available at: https://www.mdpi.com/journal/energies/special_issues/situation_awareness).

For citation purposes, cite each article independently as indicated on the article page online and as indicated below:

LastName, A.A.; LastName, B.B.; LastName, C.C. Article Title. *Journal Name* **Year**, *Volume Number*, Page Range.

ISBN 978-3-0365-4525-7 (Hbk)
ISBN 978-3-0365-4526-4 (PDF)

Contents

About the Editors

Leijiao Ge

Leijiao Ge received his bachelor's degree in electrical engineering and its automation from Beihua University, Jilin, China, in 2006, and master's degree in electrical engineering from Hebei University of Technology, Tianjin, China, in 2009, and Ph.D. degree in electrical engineering from Tianjin University, Tianjin, China, in 2016. He is currently an Associate Professor with the School of Electrical and Information Engineering, Tianjin University.

Jun Yan

Jun Yan received the B.Eng. degree in information and communication engineering from Zhejiang University, Hangzhou, China, in 2011, and the M.S. and Ph.D. degrees in electrical engineering from University of Rhode Island, Kingston, RI, USA, in 2013 and 2017, respectively. He is currently an Assistant Professor with the Concordia Institute for Information Systems Engineering, Concordia University, Montreal, QC, Canada.

Yonghui Sun

Yonghui Sun received his Ph.D. degree in electrical engineering from the City University of Hong Kong, Hong Kong, China, in 2010. He is currently a Professor with the College of Energy and Electrical Engineering, Hohai University, Nanjing, China.

Zhongguan Wang

Zhongguan Wang received his B.S. and Ph.D. degrees from the Department of Electrical Engineering, Tsinghua University, Beijing, China, in 2014 and 2019, respectively. He is currently an Associate Professor with the School of Electrical and Information Engineering, Tianjin University, Tianjin, China.

Preface to "Situation Awareness for Smart Distribution Systems"

Smart distribution systems are the next-generation large-scale, interconnected electric power grids equipped with numerous widely distributed intelligent nodes. Their operations—ranging from power delivery to voltage regulation, fault response, outage management, and many others—are strongly affected by electricity users, and vice versa. As modern societies accelerate toward a future with massive electrification, the power distribution domain is also embracing a fast-changing landscape, with new systems and technologies for distributed generation, storage, consumption, sensing, control, control, protection, and optimization. Over the last decade, we have witnessed enormous opportunities, challenges, efforts, and progress toward an electrified future with smart distribution systems. Situation awareness, a term originating from military operations, is based on the inclusive view of the environment to provide comprehensive perception, comprehension, and prediction. Situation awareness in SDS transforms complex and intangible situation information into bases for making decisions through signal processing, data mining, knowledge engineering, and, more recently, artificial intelligence, which enables and/or empowers automated decision making in the next-generation power distribution systems.

While it is encouraging to witness extensive efforts in this field, situation awareness remains an emerging research direction for smart distribution systems, with many open yet important questions to answer in the future, making it a timely topic to focus on in this Special Issue. In this Special Issue, we present 10 recent studies on a wide range of topics in situation awareness for smart distribution systems.

Integrated energy systems: Efficient and accurate situation awareness is the key to the effective management of integrated energy systems. However, the traditional situation awareness of power systems cannot fully adapt to the strong nonlinearity and uncertainty. We are pleased to present four papers related to this topic in this Special Issue.

Fault management: The prognosis, diagnosis, and responses of faults also play an important role in smart distribution systems, which have recently attracted significant attention due to the need for a more resilient and reliable grid. We are pleased to present two papers on fault detection of generation systems and secondary equipment.

Non-intrusive load monitoring: As the last mile of the electric power infrastructure, smart distribution system operations critically rely on non-invasive load monitoring to understand the patterns of users' energy consumption in a privacy-preserved manner. We are pleased to present one paper related to this topic.

Load forecasting: Accurate prediction of load demand—considering diversified generation and consumption profiles in the distribution grid—is the basis of efficient, flexible, and reliable operations in smart distribution systems. We are pleased to present two papers on the latest load forecasting techniques in this special issue.

Operation and maintenance: Proactive operation and maintenance is an effective strategy to ensure non-disrupted power distribution and extend the lifecycle of power systems. Our Special Issue is pleased to conclude with a review of operation and maintenance for situation awareness in smart distribution systems.

Leijiao Ge, Jun Yan, Yonghui Sun, and Zhongguan Wang

Editors

Editorial

Situational Awareness for Smart Distribution Systems

Leijiao Ge [1,*], Jun Yan [2], Yonghui Sun [3] and Zhongguan Wang [1]

[1] School of Electrical and Information Engineering, Tianjin University, Tianjin 300072, China; wang_zg@tju.edu.cn

[2] Concordia Institute for Information Systems Engineering, Concordia University, Montréal, QC H3G 1M8, Canada; jun.yan@concordia.ca

[3] College of Energy and Electrical Engineering, Hohai University, Nanjing 210098, China; sunyonghui168@gmail.com

[*] Correspondence: legendglj99@tju.edu.cn; Tel.: +86-013820176750

Citation: Ge, L.; Yan, J.; Sun, Y.; Wang, Z. Situational Awareness for Smart Distribution Systems. *Energies* **2022**, *15*, 4164. https://doi.org/10.3390/en15114164

Received: 30 May 2022
Accepted: 30 May 2022
Published: 6 June 2022

Publisher's Note: MDPI stays neutral with regard to jurisdictional claims in published maps and institutional affiliations.

1. Introduction

In recent years, the accelerating climate change and intensifying natural disasters have called for more renewable, resilient, and reliable energy from more distributed sources to more diversified consumers, resulting in a pressing need for advanced situational awareness of modern smart distribution systems. The continuous connection of distributed generation, energy storage, and renewable energy to the grid also enriches the power supply while introducing new consumption patterns and pressures to the power systems.

Modern situation awareness for the smart distribution systems is based on a holistic, panoramic view of the entire operating environment, including the power supplies and the user behaviors, to provide comprehensive perception, comprehension, and prediction for the system. While advanced situational awareness has been widely used in military, transportation, justice, and other fields, it has also become an enabling technology following the digitization and informatization of society.

In this Special Issue, we present ten recent studies on a wide range of topics in the situation awareness for smart distribution systems.

2. Short Review of Contributions

Situational awareness is essential for the planning and operation of an integrated energy system (IES), which needs to coordinate between different energy sources based on the accurate states of all interconnected systems. In [1], Li et al. proposed a novel situational-awareness-based planning strategy to optimize the system capacity, where a bi-level model optimizes multiple environmental and economic objectives while addressing the system stability requirements. Solved by an improved NGSA-II algorithm and the Cplex solver, their model effectively improves system stability and reduces carbon emission for wind–photovoltaic–thermal power systems.

Microgrids with hydrogen, wind, solar, storage, and other energy sources have become a new norm of IES. In [2], Wang et al. proposed a two-stage IES energy management model for wind–PV–hydrogen–storage microgrids based on receding horizon optimization to tackle the impacts of uncertainties and fluctuations. Their day-ahead optimization in the first stage and intra-day optimization in the second stage have successfully mitigated the uncertainties and maintained the grid stability at low operation costs across the microgrid.

The increasing penetration of renewables such as wind power brings uncertainties with significant challenges to the economic dispatch for IES. To tackle this, [3], Liu et al. proposed a distributed two-stage chance-constrained dispatch model that can optimize the IES operation with robustness against wind uncertainty. Considering practical operation constraints and acceptable risk levels, the new model can be solved efficiently by mixed-integer tractable programming. Its effectiveness is demonstrated on an IEEE electricity–gas–heat test case with reduced operating costs.

The development of IES poses new challenges to traditional demand response (DR) programs in distribution grid energy management and optimization. In [4], Li et al. proposed an integrated DR optimization method based on combined models of responsive electric loads, building thermal dynamics, day-ahead scheduling, and user participation. The final optimal scheduling mechanism can effectively reduce the operation cost of a community while considering users' willingness to participate and the utility's requirement of dispatching, while the robustness is further enhanced based on the conditional value-at-risk (CvaR) theory.

DC series arc faults pose severe challenges to the safety of photovoltaic (PV) systems in a smart distribution grid. In [5], Wang et al. proposed a lightweight convolutional neural network (CNN)-based detector to enable the fast and accurate detection of DC series arc faults on resource-limited embedded sensors in PV systems. As an edge-friendly solution, their computationally efficient model can nonetheless precisely detect most faults in various test conditions on the UL1699B test platform.

While numeric data are the norm of smart distribution system instruction, text data such as event logs and operator reports comprise another crucial information source. In [6], Liu et al. proposed a short-text classifier for secondary distribution equipment based on convolutional neural networks (CNNs). Contextual semantic features are auto-extracted from words to mine the fault information in text descriptions of faults and defects, which demonstrates their effectiveness on the real operation data from a regional power grid.

Non-intrusive load monitoring is a key for informed and flexible energy management in smart distribution systems. In [7], He et al. proposed a new denoising auto-encoder (DAE)-based strategy that can effectively disaggregate the residential load without additional data acquisition. Based on regular active power measurements, their method outperforms traditional hidden Markov model (HMM)-based techniques and accurately monitors household appliance consumption in a non-intrusive manner.

As electric vehicles become the future norm of transportation, their charging demand has also become a focus of short-term load forecast in distribution systems. In [8], Zhang et al. proposed a combined strategy of multi-channel convolutional neural network and temporal convolutional network (MCCNN–TCN) to improve the short-term load forecast of EV-charging demands. By finding temporal characteristics and dependencies in time-series data from urban charging stations and meteorological information, the strategy effectively improved the forecast performance over other state-of-the-art methods.

Thermal load is another important focus of load forecast in distribution systems due to their sensitivity to human preferences and seasonal patterns. In [9], Sun et al. proposed a new load forecast method based on innovative models of thermal comforts and the attention mechanism in long short-term memory (LSTM) networks. Validated on real-world data from Northern China, the new strategy achieved a more accurate forecast of the electric-heating loads to improve the safety and stability of smart distribution systems.

Situational awareness is essential in the high-quality operation and maintenance of smart distribution systems. In [10], Ge et al. provided a brief yet inclusive review of detection, comprehension, and projection technologies to enhance situational awareness in smart distribution systems. The review is expected to provide researchers and utility engineers with insights into technical achievements, barriers, and directions of situational awareness for future smart distribution systems.

3. Conclusions

We sincerely hope the papers included in this Special Issue will inspire future research into situation awareness for smart distribution systems. We strongly believe that there is a need for more work to be carried out, and we hope this issue provides a useful open access platform for the dissemination of new ideas.

Author Contributions: Conceptualization, L.G. and J.Y.; methodology, L.G. and Y.S.; software, Z.W. and L.G.; validation, L.G., J.Y., and Y.S.; formal analysis, L.G.; investigation, L.G.; resources, L.G. and J.Y.; data curation, J.Y. and L.G.; writing—original draft preparation, L.G. and J.Y.; writing—

review and editing, L.G.; visualization, L.G.; supervision, L.G.; project administration, L.G.; funding acquisition, L.G. All authors have read and agreed to the published version of the manuscript.

Funding: This research was funded by Science and Technology project of the Headquarters of State Grid Corporation of China, grant number 5400-202128572A-0-5-SF.

Conflicts of Interest: The authors declare no conflict of interest.

References

1. Li, D.; Cheng, X.; Ge, L.; Huang, W. Multiple Power Supply Capacity Planning Research for New Power System Based on Situation Awareness. *Energies* **2022**, *15*, 3298. [CrossRef]
2. Wang, J.; Li, D.; Lv, X.; Meng, X.; Zhang, J.; Ma, T.; Pei, W.; Xiao, H. Two-Stage Energy Management Strategies of Sustainable Wind-PV-Hydrogen-Storage Microgrid Based on Receding Horizon Optimization. *Energies* **2022**, *15*, 2861. [CrossRef]
3. Liu, H.; Fan, Z.; Xie, H.; Wang, N. Distributionally Robust Joint Chance-Constrained Dispatch for Electricity–Gas–Heat Integrated Energy System Considering Wind Uncertainty. *Energies* **2022**, *15*, 1796. [CrossRef]
4. Li, Y.; Zhang, J.; Ma, Z.; Peng, Y.; Zhao, S. An Energy Management Optimization Method for Community Integrated Energy System Based on User Dominated Demand Side Response. *Energies* **2022**, *14*, 4398. [CrossRef]
5. Wang, Y.; Bai, C.; Qian, X.; Liu, W.; Zhu, C.; Ge, L. A DC Series Arc Fault Detection Method Based on a Lightweight Convolutional Neural Network Used in Photovoltaic System. *Energies* **2022**, *15*, 2877. [CrossRef]
6. Liu, J.; Ma, H.; Xie, X.; Cheng, J. Short Text Classification for Faults Information of Secondary Equipment Based on Convolutional Neural Networks. *Energies* **2022**, *15*, 2400. [CrossRef]
7. He, X.; Dong, H.; Yang, W.; Hong, J. A Novel Denoising Auto-Encoder-Based Approach for Non-Intrusive Residential Load Monitoring. *Energies* **2022**, *15*, 2290. [CrossRef]
8. Zhang, J.; Liu, C.; Ge, L. Short-Term Load Forecasting Model of Electric Vehicle Charging Load Based on MCCNN-TCN. *Energies* **2022**, *15*, 2633. [CrossRef]
9. Sun, J.; Wang, J.; Sun, Y.; Xu, M.; Shi, Y.; Liu, Z.; Wen, X. Electric Heating Load Forecasting Method Based on Improved Thermal Comfort Model and LSTM. *Energies* **2022**, *14*, 4525. [CrossRef]
10. Ge, L.; Li, Y.; Li, Y.; Yan, J.; Sun, Y. Smart Distribution Network Situation Awareness for High-Quality Operation and Maintenance: A Brief Review. *Energies* **2022**, *15*, 828. [CrossRef]

Article

An Energy Management Optimization Method for Community Integrated Energy System Based on User Dominated Demand Side Response

Yiqi Li [1], Jing Zhang [1,*], Zhoujun Ma [2,3], Yang Peng [1] and Shuwen Zhao [1]

[1] School of Electrical Engineering, Southeast University, Nanjing 210096, China; 220192833@seu.edu.cn (Y.L.); 220192711@seu.edu.cn (Y.P.); 220192838@seu.edu.cn (S.Z.)

[2] College of Energy and Electrical Engineering, Hohai University, Nanjing 210098, China; mazhoujunsgcc@163.com

[3] State Grid Jiangsu Electric Power Co., Ltd., Nanjing Power Supply Branch, Nanjing 210019, China

* Correspondence: jzhang@seu.edu.cn; Tel.: +86-137-0145-8973

Abstract: With the development of integrated energy systems (IES), the traditional demand response technologies for single energy that do not take customer satisfaction into account have been unable to meet actual needs. Therefore, it is urgent to study the integrated demand response (IDR) technology for integrated energy, which considers consumers' willingness to participate in IDR. This paper proposes an energy management optimization method for community IES based on user dominated demand side response (UDDSR). Firstly, the responsive power loads and thermal loads are modeled, and aggregated using UDDSR bidding optimization. Next, the community IES is modeled and an aggregated building thermal model is introduced to measure the temperature requirements of the entire community of users for heating. Then, a day-ahead scheduling model is proposed to realize the energy management optimization. Finally, a penalty mechanism is introduced to punish the participants causing imbalance response against the day-ahead IDR bids, and the conditional value-at-risk (CVaR) theory is introduced to enhance the robustness of the scheduling model under different prediction accuracies. The case study demonstrates that the proposed method can reduce the operating cost of the community under the premise of fully considering users' willingness, and can complete the IDR request initiated by the power grid operator or the dispatching department.

Keywords: community integrated energy system; energy management; user dominated demand side response; conditional value-at-risk

Citation: Li, Y.; Zhang, J.; Ma, Z.; Peng, Y.; Zhao, S. An Energy Management Optimization Method for Community Integrated Energy System Based on User Dominated Demand Side Response. *Energies* **2021**, *14*, 4398. https://doi.org/10.3390/en14154398

Academic Editors: Leijiao Ge, Jun Yan and Yonghui Sun

Received: 23 June 2021
Accepted: 19 July 2021
Published: 21 July 2021

Publisher's Note: MDPI stays neutral with regard to jurisdictional claims in published maps and institutional affiliations.

1. Introduction

1.1. Background and Motivation

The development of energy cogeneration and integration technologies as well as renewable energies (e.g., photovoltaic (PV)) has attracted many scholars to undertake research on integrated energy systems (IES). The term IES takes into consideration many kinds of energy subsystems, e.g., electricity supply, gas supply, heating, cooling [1,2]. Different forms of energy are coupled and closely connected through energy conversion equipment (e.g., combined heat and power (CHP) unit, electric heating equipment), and can meet the diverse energy demands of users. However, because of its multienergy coupling characteristic, it is impossible to design, plan and optimize separately the operation of various energy supply systems as the traditional distributed energy supply system does [3]. Therefore, how to efficiently deal with the complementarity and substitution between different energy streams has become a key issue to realize energy cascade utilization and to improve comprehensive energy utilization efficiency. Additionally, the traditional energy management system (EMS) framework cannot adapt to the coexistence and interaction features of centralization and distribution in IES [4,5] (e.g., the energy management policy proposed in [6] only considers the electric energy, and cannot be applied to deal directly with multienergy flow problems). Therefore, it is necessary to study the integrated energy management system (IEMS) technology for multienergy flow.

The control objects of the IEMS can be divided into three layers. The upper layer is the system-level multienergy flow transmission network, which involves the production, transmission and safe operation of energy such as gas, electricity, and heat. The middle layer is a local microenergy unit, with industrial parks, smart communities, and intelligent buildings as typical application scenarios, and it involves the coordinated scheduling and optimized operation of multiple energy sources. The lower layer is the user-level integrated producer and consumer. The multienergy complementarity and alternative features of IEMS not only provide users with more options for energy use, but also bring optimization space for the overall regulation and operation of the system. With the development of IES, existing studies based on the traditional demand response (DR) technologies for a single energy source (electric energy) [7–9] can no longer meet users' actual needs, and there is an urgent need to study integrated demand response (IDR) technology for integrated energy. Reasonable use of user-side responsive resources to participate in the IDR of the system will play an important role in realizing the two-way interaction between the supply and the demand and the win–win situation [10]. On the other hand, information communication and engineering measurement and control technology have developed rapidly recently. Having access to a large number of smart sensors has greatly increased the amount of multienergy flow information that can be collected by the middle layer and user layer of IEMS. The IEMS can adjust in time based on the measurement or user feedback information, and improve energy efficiency and operating economy on the premise of ensuring the user's energy comfort.

Thus far, the IDR strategies and mechanisms have been studied for many purposes. In [11], the concept of IDR was first proposed and gas turbines were introduced to supply power to the power grid during peak time, converting part of the power load into gas load. Additionally, the incentive effect of natural gas prices on IDR was analyzed through Nash game theory. In [12], the physical constraints of the natural gas network and the heating network were processed by piecewise linearization, and an IDR optimal transaction strategy model based on the mixed-integer second-order cone programming algorithm and transaction price incentive was proposed. The authors in [13] summarized the development of IDR from the aspects of system modeling, optimization strategy and power market mechanisms, and affirmed the positive effect of IDR on improving the flexibility of IES load response. In [14], an IDR model based on medium- and long-term time dimensions considering system dynamics was proposed, and taking flexible loads, energy storage, and electric vehicles into account, an IES scheduling model was established in order to simulate the benefits for users participating in IDR. In [15], a day-ahead and intraday optimization scheduling model based on the demand side response was proposed, and the scheduling times for different energy subsystems were considered to perform rolling optimization scheduling.

Current research mostly focuses on the impact of market price mechanisms and the refined modeling of IES equipment and networks on IDR [16,17]. It is assumed that users will continue to participate in IDR events satisfactorily under certain price incentives, or users will maximize their responsive load during IDR events. Additionally, users are assumed to allow their own load equipment to be adjusted by EMS or energy service providers. However, most research ignored users' willingness to participate in DR programs. In fact, users are not necessarily willing to give the control of the equipment to EMS or energy service providers underprice incentives [18]. Users may not provide the maximum responsive load during IDR due to privacy reasons. In [18], a survey was conducted on the willingness of 1499 households from a state in Australia to participate in a direct load control (DLC) plan, and the results showed that only about 13% of customers accepted the DLC plan. For users, the main reason for reluctance to participate in the DLC program is that users have low trust in energy companies. At present, there are very few studies on the relationship between user satisfaction with participating in IDR events and response load capacity. In [19], a user dominated demand side response (UDDSR) scheme that allows energy users to dynamically choose to join or withdraw from DR events was put forward. In this scheme, users can submit flexible DR bids to community EMS for participating in DR events. That is, users can flexibly choose the working hours of each household device. However, this scheme only focuses on electric load, and fails to consider the overall optimization within IES.

1.2. Novelty and Contribution

In this paper, an energy management optimization method for community IES based on UDDSR is put forth, where users can submit the day-ahead IDR bid for load responses that fully meets their own comfort, and respond to the IDR requests issued by the power grid operator or dispatching department according to the planned capacity of the IDR bid on the next day. Additionally, an aggregated buildings thermal model is introduced to establishe the adjustable thermal load model, and the user's power load adjustable time, power load adjustable capacity, thermal load adjustable time and heating temperature are set as optimized parameters to establish a day-ahead scheduling model. Considering the uncertainty of PV output, user load, outdoor temperature, and user actual UDDSR response capacity in the community IES, a penalty mechanism is introduced to punish the participants making imbalanced response against the day-ahead IDR bids, and the conditional value-at-risk (CVaR) theory is introduced to enhance the robustness under different prediction accuracy.

The contributions of this paper are summarized as follows:

(1) The interruptible power load, shiftable power load, and adjustable thermal load are modeled, respectively, and are optimized by UDDSR scheme in order to obtain the aggregated IDR bids.
(2) An aggregated buildings thermal model is introduced to measure the temperature requirements of the entire community of users for heating. The adjustable thermal loads of the IDR bids submitted by users are modeled within the context of air temperature, and can be optimized by regulating the indoor temperature of users.
(3) From the overall perspective of system operation, a day-ahead scheduling optimization model for the community IES based on UDDSR is established, and the CVaR theory is introduced to deal with the uncertainties in IES.

2. Demand Response Load Modeling Based on UDDSR

In this paper, the detailed UDDSR optimization approach is based on the mechanism described in [19]. This mechanism allows users to submit flexible bids for DR events and achieves the optimal aggregation of these bids within the DR events. However, it only considers electric equipment including interruptible appliances (e.g., heating systems) and shiftable appliances (e.g., electric vehicles). In this section, the UDDSR optimization with adjustable thermal loads is further studied within the IDR events.

2.1. UDDSR Optimization with Adjustable Thermal Loads

In this paper, thermal loads of the aggregated buildings are modeled within the context of air temperature, and can be adjusted by regulating the indoor temperature of end users. Regarding the adjustable thermal loads of the IDR bids, the maximum and minimum of the heating temperature, the maximum adjustable temperature for heating, and the adjustable time period for heating can be set by users. Since this paper studies the centralized temperature regulation in the case of central heating, the community energy management system (CEMS) will first classify users according to the maximum adjustable temperature for heating in the IDR bids. For users who have the same adjustable temperature, CEMS will select as many users as possible who are willing to adjust the heating temperature within the IDR request period to participate in the UDDSR thermal load response according to (1). Additionally, CEMS will select the minimum of the highest temperatures and the maximum of the lowest temperatures submitted by all users in the IDR bids as the temperature constraint range of the central heating, as demonstrated in (2).

$$\min_{t \in T} VAR(M_u^t) \tag{1}$$

$$\begin{cases} T_{inmin} = \max\limits_{i=1}^{N_u}\{T_{l,i}\} \\ T_{inmax} = \min\limits_{i=1}^{N_u}\{T_{u,i}\} \end{cases} \tag{2}$$

where M_u^t is the total number of users willing to participate in UDDSR thermal load response at time t; T_{inmin}/T_{inmax} is the minimum/maximum indoor temperature that costumers are willing to accept, respectively; $T_{l,i}/T_{u,i}$ is the minimum/maximum heating temperature submitted by the user i.

2.2. Adjustable Thermal Loads Model Based on UDDSR

According to [20], the thermodynamic model of the aggregated buildings can be formulated as the RC equivalent circuit model, as demonstrated in Figure 1, where R is the equivalent thermal resistance of the house shell; C_{air} is the air specific heat; L_{AC}^t is the adjustable thermal load at time t; T_{in}^t and T_{out}^t are the indoor and outdoor temperature at time t.

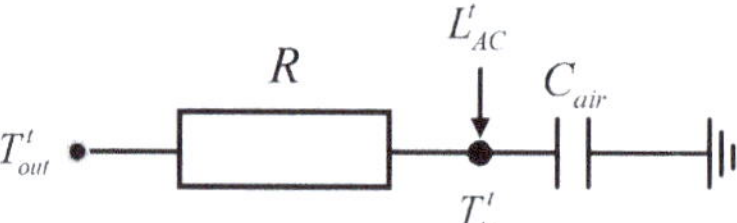

Figure 1. Thermodynamic model of the aggregated buildings.

Therefore, the relation equation between indoor temperature and adjustable thermal load is as follows:

$$\frac{dT_{in}^t}{dt} = -\frac{1}{R \cdot C_{air}} \cdot T_{in}^t + \frac{1}{C_{air}} \cdot \left(L_{AC}^t + \frac{1}{R} \cdot T_{out}^t\right) \tag{3}$$

The discrete model of (3) is

$$T_{in}^t = T_{in}^{t-\Delta t} \cdot e^{-\frac{\Delta t}{R \cdot C_{air}}} + \left(R \cdot L_{AC}^t + T_{out}^t\right) \cdot \left(1 - e^{-\frac{\Delta t}{R \cdot C_{air}}}\right) \tag{4}$$

where e is a constant; Δt is the scheduling interval and is assumed to be 1 h in this paper. Then, the adjustable thermal load L_{AC}^t is calculated from:

$$L_{AC}^t = \frac{1}{R} \cdot \left(\frac{T_{in}^t - T_{in}^{t-\Delta t} \cdot e^{-\frac{\Delta t}{R \cdot C_{air}}}}{1 - e^{-\frac{\Delta t}{R \cdot C_{air}}}} - T_{out}^t\right) \tag{5}$$

$$\begin{cases} T_{inmin} - T_{adj} \cdot T_{DRH}^t \leq T_{in}^t \leq T_{inmax} - T_{adj} \cdot T_{DRH}^t \\ \left|T_{in}^t - T_{in}^{t-\Delta t}\right| \leq \Delta T_{max} \\ T_{inmin}, T_{inmax}, T_{adj} \geq 0 \end{cases} \tag{6}$$

where T_{adj} is the maximum adjustable indoor temperature allowed by end users during IDR event; T_{DRH}^t, determined by IDR bids, is the adjustable time of thermal load allowed by users, and if $T_{DRH}^t = 1/T_{DRH}^t = 0$, the thermal load can/cannot be adjusted; ΔT_{max} is the maximum indoor temperature variation during Δt, and it should be less than 2 °C in order not to affect the comfort of users.

2.3. Electric Loads Model Based on UDDSR

In the community CHP system, the electric loads includes interruptible power loads and shiftable power loads. Based on the aggregated IDR bids obtained from UDDSR optimization in [19], the total response power of the interruptible appliances during the IDR event should be less than the maximum interruptible power at the same time after the aggregated IDR bid. Thus the interruptible power load is expressed as

$$0 \leq L_{DRE,int}^{t} \leq L_{DRE,intmax}^{t} \tag{7}$$

where $L_{DRE,int}^{t}$ is the interruptible power load at time t; $L_{DRE,intmax}^{t}$ is the maximum interruptible power load at time t, which can be obtained from aggregated IDR bid of end users.

The shiftable load model is expressed as

$$L_{DRE,shf}^{t} = L_{DRE,shf,out}^{t} - L_{DRE,shf,in}^{t} \tag{8}$$

$$\sum_{t=1}^{T} L_{DRE,shf,out}^{t} = \sum_{t=1}^{T} \left| L_{DRE,shf,in}^{t} \right| \tag{9}$$

$$\begin{cases} 0 \leq L_{DRE,shf,out}^{t} \leq L_{DRE,shf,outmax}^{t} \\ L_{DRE,shf,inmax}^{t} \leq L_{DRE,shf,in}^{t} \leq 0 \end{cases} \tag{10}$$

where $L_{DRE,shf}^{t}$ is the total shiftable power load at time t; $L_{DRE,shf,out}^{t}$ and $L_{DRE,shf,outmax}^{t}$ are the load and the maximum load shifted from time t to other time; $L_{DRE,shf,in}^{t}$ and $L_{DRE,shf,inmax}^{t}$ are the load and maximum load shifted to time t, respectively; T is the optimized scheduling cycle; $L_{DRE,shf,outmax}^{t}$ and $L_{DRE,shf,inmax}^{t}$ can be obtained from aggregated IDR bid of end users.

3. Distributed Generator and Co-Supply Equipment Model

3.1. PV Model

PV is a common distributed generation device in the community, and can be modeled as:

$$P_{PV}^{t} = P_{stc} \cdot \frac{G^{t}}{G_{stc}} \cdot (1 + \varepsilon (T_{s}^{t} - T_{stc})) \tag{11}$$

where P_{PV}^{t} is the PV output power; P_{stc} is the maximum PV output power under standard test conditions; G^{t} is the light intensity and G_{stc} is that under standard test conditions; ε is the PV power temperature coefficient; T_{s}^{t} is surface temperature of PV and T_{stc} is that under standard test conditions.

3.2. Power Supply Equipment Model

3.2.1. Microgas Turbine (MT) Model

MT is an important CHP equipment in community CHP system, and its model is as follows:

$$\begin{cases} P_{MT}^{t} = V_{MT}^{t} \cdot H_{ng} \cdot \eta_{MT} \\ Q_{MT}^{t} = V_{MT}^{t} \cdot H_{ng} \cdot (1 - \eta_{MT} - \eta_{loss}) \end{cases} \tag{12}$$

where P_{MT}^{t} is the MT output power at time t; V_{MT}^{t} is the MT gas consumption at time t; H_{ng} is the calorific value of natural gas; η_{MT} is the MT power generation efficiency; Q_{MT}^{t} is the MT output heat power at time t; η_{loss} is the MT power loss efficiency.

3.2.2. Gas Boiler (GB) Model

GB burns natural gas to provide heat for community users and can be modeled as:

$$Q_{GB}^{t} = V_{GB}^{t} \cdot H_{ng} \cdot \eta_{GB} \tag{13}$$

where Q^t_{GB} is the GB output heat power at time t; V^t_{GB} is the GB gas consumption at time t; η_{GB} is the GB heat production efficiency.

3.2.3. Waste Heat Recovery (WHR) Device Model

WHR can recover the flue gas waste heat after MT power generation to improve the energy utilization efficiency, and can be modeled as:

$$Q^t_{WHR} = Q^t_{WH} \cdot \eta_{WHR} \tag{14}$$

where Q^t_{WHR} is the WHR recovered heat power at time t; Q^t_{WH} is the MT waste heat at time t; η_{WHR} is the WHR heat recovery efficiency.

3.2.4. Heat Exchanger (HE) Model

HE can convert the heat of hot stream into hot water to provide heating for community end users, and is modeled as:

$$Q^t_{HE} = Q^t_{HE,in} \cdot \eta_{HE} \tag{15}$$

where $Q^t_{HE}/Q^t_{HE,in}$ is the HE heat power output/input at time t; η_{HE} is the HE heat exchange efficiency.

3.3. Energy Storage Equipment Model

3.3.1. Battery (BT) Model

The charging and discharging of BT can greatly improve the utilization rate of the response load on the user side, and the model of BT is:

$$W^t_{BT} = W^{t-\Delta t}_{BT} \cdot (1 - \eta_{BT,loss}) + \left(P^t_{BT,ch} \cdot \eta_{BT,ch} - \frac{P^t_{BT,dis}}{\eta_{BT,dis}} \right) \cdot \Delta t \tag{16}$$

where W^t_{BT} represents the stored energy in BT; $\eta_{BT,loss}$ is the power loss rate of BT; $P^t_{BT,ch}$ and $P^t_{BT,dis}$ are the charging and discharging power of BT, respectively; $\eta_{BT,ch}$ and $\eta_{BT,dis}$ are the charging and discharging efficiency of BT, respectively.

3.3.2. Thermal Storage Tank (TST) Model

When the output thermoelectric power ratio of MT does not match the thermoelectric load ratio of community users, TST can compensate for the difference of thermoelectric ratio through heat storage and release behavior, and improve the utilization efficiency of user-side response heat load. TST can be modeled as:

$$W^t_{TST} = W^{t-\Delta t}_{TST} \cdot (1 - \eta_{TST,loss}) + \left(Q^t_{TST,ch} \cdot \eta_{TST,ch} - \frac{Q^t_{TST,dis}}{\eta_{TST,dis}} \right) \cdot \Delta t \tag{17}$$

where W^t_{TST} is the amount of heat stored in TST at time t; $\eta_{TST,loss}$ is the energy loss rate of TST; $Q^t_{TST,ch}$ is the heat storage power of TST; $\eta_{TST,ch}$ is the heat storage efficiency; $Q^t_{TST,dis}$ is the heat release power; $\eta_{TST,dis}$ is the heat release efficiency.

4. Community CHP System Model Based on UDDSR

In this paper, the community CHP system consists of MT, GB, WHR, HE, PV, BT and TST, and its structure diagram is shown in Figure 2.

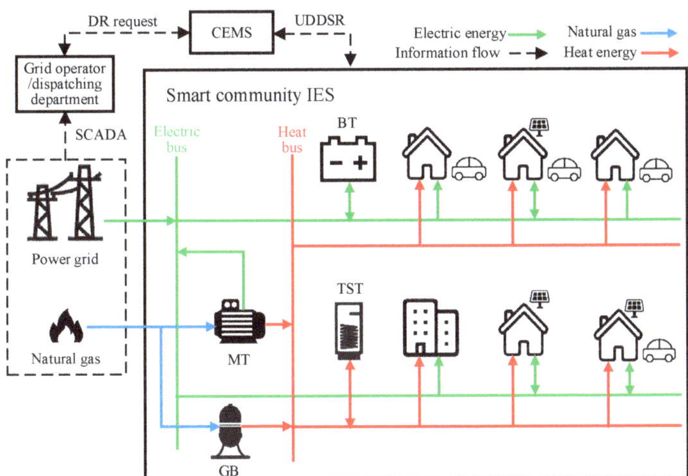

Figure 2. Community CHP system structure diagram based on UDDSR.

Then, an energy hub model based on the bus bar form [21] is adopted to model the community CHP system. The bus bar structure of the community system is shown in Figure 3, and the flow relations of electricity, gas and heat energy are marked by arrows.

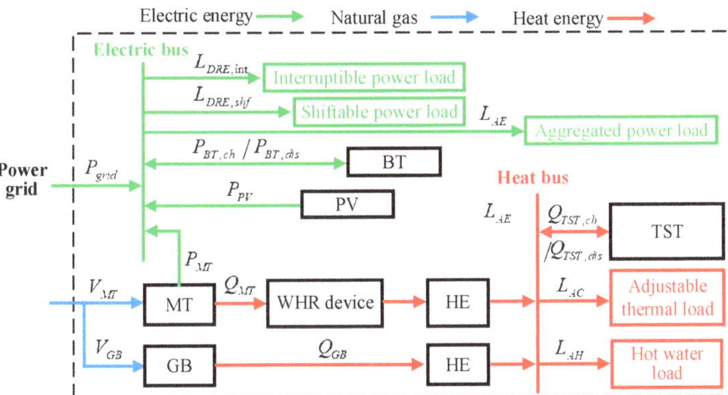

Figure 3. Community CHP system bus bar structure diagram based on UDDSR.

4.1. Day-Ahead Energy Optimization Model

In the community system studied in this paper, by participating in the UDDSR response arranged by CEMS, users can submit the day-ahead IDR bid of load response that fully meets their own comfort, and respond to the IDR request issued by the power grid operator or dispatching department the next day according to the planned capacity of IDR bid. For users, they can reduce or transfer unnecessary loads during the IDR event, and at the same time receive the subsidy of IDR response from the grid operator. For the entire community energy system, CEMS can schedule the user loads to the greatest extent according to the IDR bid plan of users, and thus achieve "peak clipping and valley filling" in energy use. Meanwhile, on the basis of ensuring the stability of the system operation, the overall operation cost of the system can also be reduced, and the economy of the system operation can be improved.

The goal of system optimization is to minimize the total cost of system operation and the temperature change caused by the thermal load response adjustment within the allowable range of users, so as to ensure their satisfaction with energy use as much as possible. This can be described as the objective function below:

$$\min\left\{C_{total} + \sum_{t=1}^{N} \delta^t \cdot \left(T_{in}^t - T_{ref}\right)^2\right\} \tag{18}$$

where C_{total} is the total cost of system operation; T_{ref} is the national standard indoor optimum temperature; δ_t is a time-varying parameter that measure the thermal comfort of users, and during the UDDSR event, δ_t is relaxed to achieve the purpose of temperature regulation and consumption reduction, while in other moments δ_t plays the role of making the indoor temperature close to the optimal temperature; N is the optimal scheduling cycle.

The total cost of system operation is calculated by the following function:

$$C_{total} = C_{grid} + C_{ng} + C_{om} + C_{UDDSR} \tag{19}$$

where C_{grid} is the cost of electricity purchasing from the grid; C_{ng} is the cost of natural gas; C_{om} is the cost of equipment operation and maintenance; C_{UDDSR} is the total subsidy for UDDSR participation given to users by the operator.

C_{grid} and C_{ng} can be calculated as:

$$C_{grid} = \sum_{t=1}^{N} e_P^t \cdot P_{grid}^t \tag{20}$$

$$C_{ng} = \sum_{t=1}^{N} e_{gas} \cdot \left(V_{MT}^t + V_{GB}^t\right) \tag{21}$$

where P_{grid}^t is the power purchased from the grid; e_p^t is the market price; e_{gas} is the price of natural gas.

C_{om} can be calculated as:

$$C_{om} = \sum_{t=1}^{N} \left(C_{om,MT} \cdot P_{MT}^t + C_{om,GB} \cdot Q_{GB}^t + C_{om,PV} \cdot P_{PV}^t\right) \tag{22}$$

where $C_{om,MT}$, $C_{om,GB}$ and $C_{om,PV}$ are the unit power operation and maintenance costs of MT, GB, and PV, respectively.

C_{UDDSR} can be calculated as:

$$C_{UDDSR} = C_u + C_{es} \tag{23}$$

$$C_u = \sum_{t=1}^{N} e_{DRE}^t \cdot \left(L_{DRE,int}^t + \left|L_{DRE,shf}^t\right|\right) + e_{DRH}^t \cdot L_{DRH}^t \tag{24}$$

$$C_{es} = \sum_{t=1}^{N} e_{BT} \cdot \left(P_{BT,ch}^t - P_{BT,dis}^t\right) + e_{TST} \cdot \left(Q_{TST,ch}^t - Q_{TST,dis}^t\right) \tag{25}$$

$$L_{DRH}^t = \frac{1}{R} \cdot \frac{\Delta T^t}{1 - e^{-\frac{\Delta t}{R \cdot C_{air}}}} \tag{26}$$

$$\Delta T^t = \max\left\{T_{in0}^t - T_{in}^t, 0\right\} \tag{27}$$

where C_u is the load response subsidy for users; C_{es} is the energy storage subsidy; e_{DRE}^t is electric load response compensation per unit power; e_{DRH}^t is the thermal load response compensation per unit power; L_{DRH}^t is the change of thermal power caused by lowering the room temperature ΔT^t within the range allowed by users at time t; T_{in0}^t is the indoor temperature before UDDSR event; e_{BT} is the unit power subsidy for the charging and discharging behavior of BT; e_{TST} is the unit power subsidy for heat storage and release behavior of TST.

The operation constraints are described as follows.

1. Energy balancing constraints

$$P^t_{grid} + P^t_{MT} + P^t_{PV} - P^t_{BT,dis} = L^t_{AE} - L^t_{DRE,\text{int}} - L^t_{DRE,shf} + P^t_{BT,ch} \tag{28}$$

$$(Q^t_{GB} + Q^t_{MT} \cdot \eta_{WHR}) \cdot \eta_{HE} - Q^t_{TST,dis} = L^t_{AH} + L^t_{AC} + Q^t_{TST,ch} \tag{29}$$

where L^t_{AE} and L^t_{AH} are the basic electrical load and basic hot water load at time t after load aggregation, which cannot be scheduled during the UDDSR event.

2. Energy supply constraints

$$P^t_{grid} \le P_{grid\max} \tag{30}$$

$$P_{MT\min} \le P^t_{MT} \le P_{MT\max} \tag{31}$$

$$0 \le Q^t_{GB} \le Q_{GB\max} \tag{32}$$

where $P_{grid\max}$ is the maximum interactive power between the community system and the power grid per unit time; $P_{MT\max}$ and $P_{MT\min}$ are the maximum and minimum generating power of MT; $Q_{GB\max}$ is the maximum heating power of GB.

3. Energy storage constraints

For BT, the constraints are:

$$0 \le P^t_{BT,ch} \cdot S^t_{BT,ch} \le P_{BT,ch\max} \tag{33}$$

$$P_{BT,dis\max} \le P^t_{BT,dis} \cdot S^t_{BT,dis} \le 0 \tag{34}$$

$$S^t_{BT,ch} + S^t_{BT,dis} \le 1 \tag{35}$$

$$W_{BT\min} \le W^t_{BT} \le W_{BT\max} \tag{36}$$

where $S^t_{BT,ch}$ and $S^t_{BT,dis}$ are 0–1 variables representing the charging and discharging state of BT; $P_{BT,ch\max}$ and $P_{BT,dis\max}$ are the maximum charging and discharging power of BT; $W_{BT\max}$ and $W_{BT\min}$ are the maximum and minimum energy storage capacity of BT.

For TST, the constraints are:

$$0 \le Q^t_{TST,ch} \cdot S^t_{TST,ch} \le Q_{TST,ch\max} \tag{37}$$

$$Q_{TST,dis\max} \le Q^t_{TST,dis} \cdot S^t_{TST,dis} \le 0 \tag{38}$$

$$S^t_{TST,ch} + S^t_{TST,dis} \le 1 \tag{39}$$

$$W_{TST\min} \le W^t_{TST} \le W_{TST\max} \tag{40}$$

where $S^t_{TST,ch}$ and $S^t_{TST,dis}$ are 0–1 variables representing the heat storing and releasing state of TST; $P_{TST,ch\max}$ and $P_{TST,dis\max}$ are the maximum heat storing and releasing power of TST; $W_{TST\max}$ and $W_{TST\min}$ are the maximum and minimum heat storage capacity of TST.

4.2. CVaR-Based Energy Optimization Model

The day-ahead energy optimization model mentioned in the above section is based on the accurate prediction of the basic electric and heat loads, PV output, and outdoor temperature. It ignores the error between the predicted value and actual value, and assumes that users will maximize the UDDSR response according to the response load capacity of the IDR bid. However, actually, the prediction error may have a significant impact on the optimization results, and users may not respond according to the maximum capacity after UDDSR bid, which must be taken into consideration. In order to solve the above questions, CVaR is applied.

4.2.1. CVaR Model

CVaR theory was firstly used to solve the optimal portfolio problem of investment risk related to financial hedging. It is mainly used to measure the investment loss when the investment loss exceeds the expected maximum loss (i.e., Value-at-Risk (VaR)) under a given confidence level. The CVaR model is shown as follows.

$$CVaR_{con} = E[f(X,\gamma)|f(X,\gamma) > VaR_{con}] \tag{41}$$

where $CVaR_{con}$ is the average excess loss under a given confidence level; *con* is the confidence level; $f(X, \gamma)$ is the loss function; X is the investment portfolio; γ is the risk variable; VaR_{con} is the expected maximum loss under the *con*; $E[.]$ expresses the expect function.

If the probability of γ in different scenarios is known, the formulation of discrete CVaR can be expressed as follows.

$$CVaR_{con} = VaR_{con} + \frac{1}{1 - con}\sum_{t=1}^{N} p_{\gamma}^{t}\max\{f(X,\gamma) - VaR_{con}, 0\} \tag{42}$$

where p_{γ}^{t} is the probability of γ occurring at time t; N is the number of discrete time intervals.

However, (42) needs to obtain VaR at the same confidence level first, which complicates the computing process. To increase the computing speed, the relaxation method in [22] is applied to solve CVaR and VaR simultaneously. The relaxed CVaR discrete function is converted into a common optimization problem, and its calculation formula is expressed as follows.

$$\min g(X,\alpha) = \alpha + \frac{1}{1 - con}\sum_{t=1}^{N} p_{\gamma}^{t}\max\{f(X,\gamma) - \alpha, 0\} \tag{43}$$

where $CVaR_{con}$ is the minimum value of $g(X, \alpha)$; α is the intermediate variable after relaxation of VaR, and when $g(X, \alpha)$ goes to the minimum, α is equal to VaR_{con}.

4.2.2. Day-Ahead Energy Optimization Model Based on CVaR

In the community CHP system, the uncertainties include the prediction errors of electric and heat load, PV output and outdoor temperature, and the response load fluctuation of UDDSR. In this section, the random simulation algorithm is used to generate a set of uncertinty scenarios. It is assumed that the probability distribution of forecast errors and load response fluctuation obeys the normal distribution with the mean value being the forecast, i.e., $\gamma{\sim}N(r_{forecast}, \sigma^2)$, and the probability distribution formula is:

$$h(r) = \frac{1}{\sqrt{2\pi}\sigma} \cdot e^{\frac{-(r - r_{forecast})^2}{2\sigma^2}} \tag{44}$$

where r is the uncertainty variable; σ is the standard deviation of r; $r_{forecast}$ is the forecast value of r.

According to (43), the day-ahead energy optimization model based on CVaR is formulated as follows.

$$CVaR_{con} = \min\alpha + \frac{1}{M(1 - con)}\sum_{i=1}^{M} \phi_i \tag{45}$$

$$\begin{cases} \phi_i \geq C_{total,i} - E[C_{total,i}] - \alpha \\ \phi_i \geq 0(i = 1, 2, \ldots M) \end{cases} \tag{46}$$

where $C_{total,i}$ is the total cost of system operation in scenario i; $E[C_{total,i}]$ is the expected cost of system operation in all simulated uncertinty scenarios; ϕ_i is the middle variable in scenario i; M is the total number of uncertainty scenarios.

Then, after considering the uncertainties of forecast error and response fluctuation, the total cost of system operation can be converted into:

$$C_{total} = C_{grid} + C_{ng} + C_{om} + C_{UDDSR} - C_{punish} \tag{47}$$

$$\begin{cases} C_{punish} = \sum_{t=1}^{N} e_{punish}^t \cdot \left| L_{DRE}^t - L_{DRE0}^t \right| \\ L_{DRE}^t = L_{DRE,int}^t + L_{DRE,shf}^t \end{cases} \tag{48}$$

where C_{punish} is the penalty fee when users do not respond according to the response load optimized by day-ahead UDDSR; L_{DRE}^t is the total actual response load; L_{DRE0}^t is the response load optimized by day-ahead UDDSR.

Additionally, according to (18), the objective function can be converted into

$$\min \left\{ E[C_{total,i}] + \beta \cdot CVaR_{con} + \frac{1}{M} \cdot \sum_{i=1}^{M} \sum_{t=1}^{N} \delta^t \cdot \left(T_{in,i}^t - T_{ref} \right)^2 \right\} \tag{49}$$

where β is the uncertainty factor, i.e., the willingness of the community system to take risks, and $\beta \in [0,1]$.

Meanwhile, the purpose of the energy optimization based on CVaR is to meet the operating conditions in all uncertainty scenarios, thus the bus balancing constraints can be converted into:

$$P_{grid}^t + P_{MT}^t + P_{PV,i}^t - P_{BT,dis}^t \geq L_{AE,i}^t - L_{DRE,int,i}^t - L_{DRE,shf,i}^t + P_{BT,ch}^t \tag{50}$$

$$(Q_{GB}^t + Q_{MT}^t \cdot \eta_{WHR}) \cdot \eta_{HE} - Q_{TST,dis}^t \geq L_{AH,i}^t + L_{AC,i}^t + Q_{TST,ch}^t \tag{51}$$

5. Case Study

The proposed model is conducted on a community IES modified from a central neighborhood in Anhui province in China. The community structure diagram is presented in Figure 2. The forecast curves of electricity load, hot water load, PV output and outdoor temperature of the system on a typical winter day is shown in Figure 4. In the appendix, the peak-valley time-of-use electricity price, the subsidy for users participating in UDDSR and the gas price are shown in Table A1, the equipment operating parameters are shown in Table A2, and the equipment cost and subsidy parameters are shown in Table A3, which are all modified from [23]. The cases were compiled with Python 3.7, and solved by Gurobi solver.

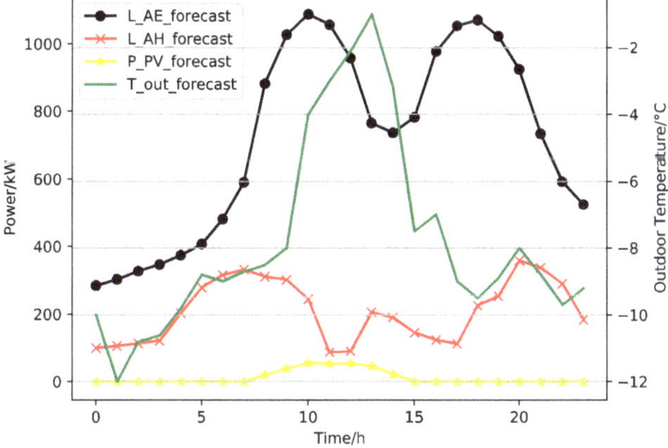

Figure 4. The forecast curves of electric load, hot water load, PV output and outdoor temperature of the system on a typical winter day.

5.1. Day-Ahead Energy Optimization Based on UDDSR

In order to verify the impact of the UDDSR mechanism on the whole community system, the outputs of the system equipment before and after the UDDSR response were analyzed.

5.1.1. Energy Optimization Results without UDDSR Response

When users do not participate in the UDDSR response, the community CHP optimizes energy consumption according to the prediction values of electric and heat loads, PV outputs, and outdoor temperature. The optimization results of equipment outputs are shown in Figure 5. It can be seen that during the valley period of the electricity price, since the cost of purchasing electricity from the grid is lower than that of MT generation, the electrical load is almost entirely satisfied by the power supply from the grid. Meanwhile, since the cost of heat production per unit power of GB is lower than that of MT, and the heat load at this time is higher, GB gives priority to full power to ensure heat supply. During the peak period of the electricity price, the cost of power supply from MT is lower than the electricity price, thus the power supply of MT increases significantly. At this time, the remaining heat load is supplemented by GB.

On the other hand, due to the CHP characteristics of MT, after complementing the heat load, MT has excess power. BT charges at the time of 04:00–05:00 and 15:00–16:00 to dissipate the excess power, and discharges during the peak period of power consumption, which improves the energy utilization rate and operating economy of the system. Similarly, when MT produces too much heat, TST uses the heat storage and release characteristics to meet the thermal load demand.

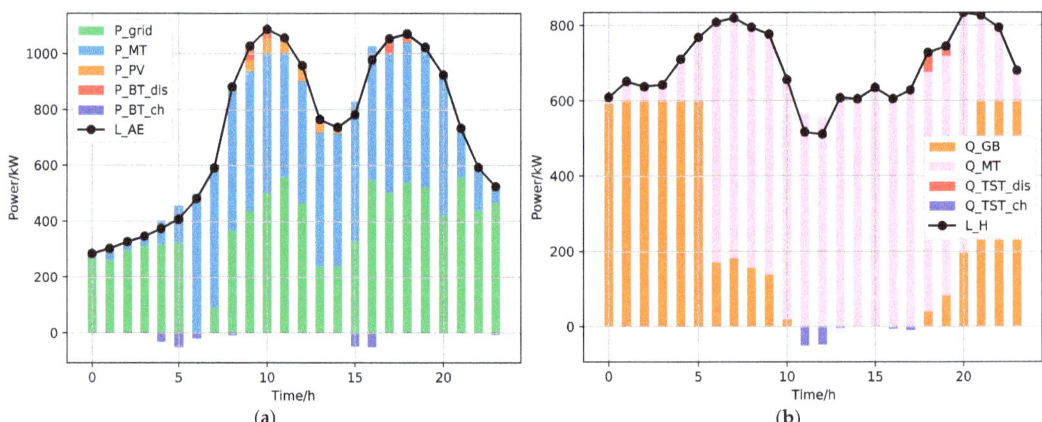

Figure 5. Energy optimization results without UDDSR response. (**a**) Optimization results of electric bus; (**b**) optimization results of heat bus.

The change of indoor heating temperature is depicted in Figure 6. It can be observed that the indoor temperature is always maintained near the optimal room temperature, and the heating needs are met.

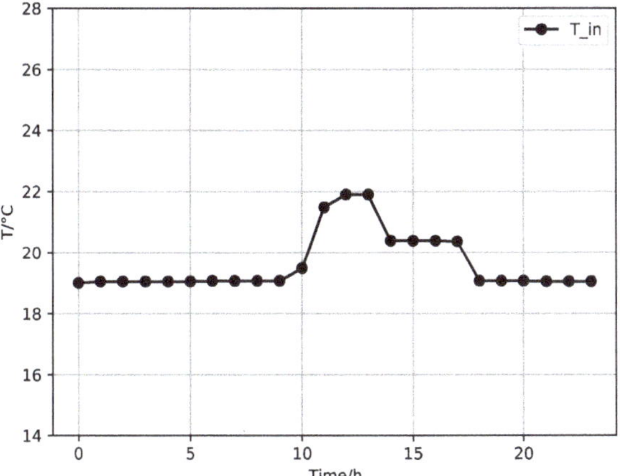

Figure 6. The change of indoor heating temperature without UDDSR response.

5.1.2. Energy Optimization Results with UDDSR Response

When users participate in the UDDSR response, they submit a flexible IDR bid to CEMS according to their own energy demand. The bid content includes the interruptible load, the shiftable load, the time and capacity of the adjustable load and the CEMS aggregates and optimizes the responsive loads of the users. Based on the aggregated results of the responsive loads, the energy use of the community CHP system is optimized. The UDDSR bid results are shown in Figure 7. In this figure, the green curve indicates the adjustable time of the heating temperature allowed by users. When L_DRH State >0, the upper and lower limits of the heating temperature are allowed to be reduced by T_{adj}, i.e., the heating range is changed into $T_{in\min} - T_{adj} \leq Tt\ in \leq T_{in\max} - T_{adj}$; when L_DRH State < 0, the heating temperature cannot be reduced, i.e., the thermal load cannot be adjusted. In this case, it is assumed that $T_{adj} = 1$, $T_{in\min} = 18$, $T_{in\max} = 26$. It can be seen that the operating costs of the community CHP system are lower when users perform UDDSR based on the optimized IDR response load, compared with performing UDDSR according to the maximum response capacity of the aggregated IDR bid.

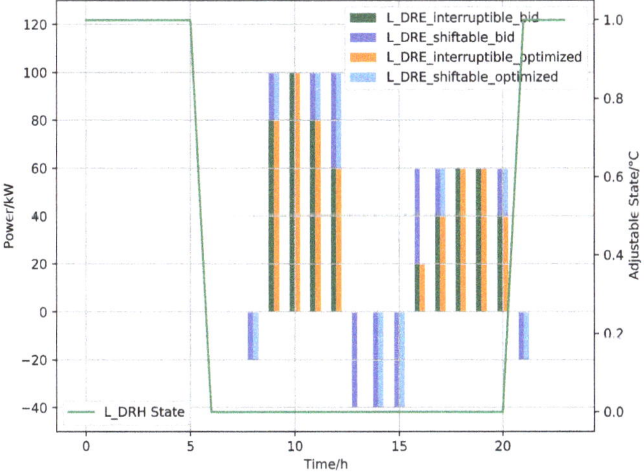

Figure 7. The response loads of users participating in UDDSR.

The optimization results of the equipment output after the UDDSR response are shown in Figure 8. Figure 8a indicates that after the UDDSR response, the power purchasing from the grid during the peak load period is significantly reduced, since part of the unnecessary load is interrupted or shifted. Figure 8b indicates that during the period of 00:00–05:00 and 21:00–23:00, the MT heat supply is significantly reduced, and the heat load of the users has been adjusted.

Figure 9 represents the comparison of the electric heating load before and after the user response. It can be observed that the UDDSR mechanism has an obvious "peak-shaving and valley-filling" effect on the community system, and can successfully complete the demand response events initiated by the grid operator or dispatching department.

Figure 10 displays the indoor heating temperature changes before and after UDDSR response. After the UDDSR response, the heat load during the period of 00:00–05:00 and 21:00–23:00 has been reduced to a certain extent. Although the actual room temperature has been lowered, it is still higher than $T_{in\min}$- T_{adj}. This means the community system does not operate according to the minimum heating temperature, which guarantees the energy satisfaction of users to the greatest extent, and verifies the accuracy and validity of the heating temperature constraint in (18)

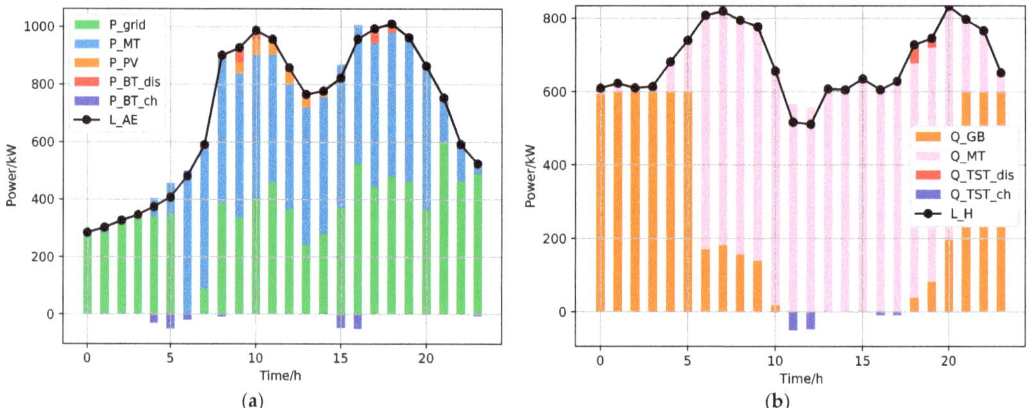

Figure 8. Energy optimization results with UDDSR response. (**a**) Optimization results of electric bus; (**b**) optimization results of heat bus.

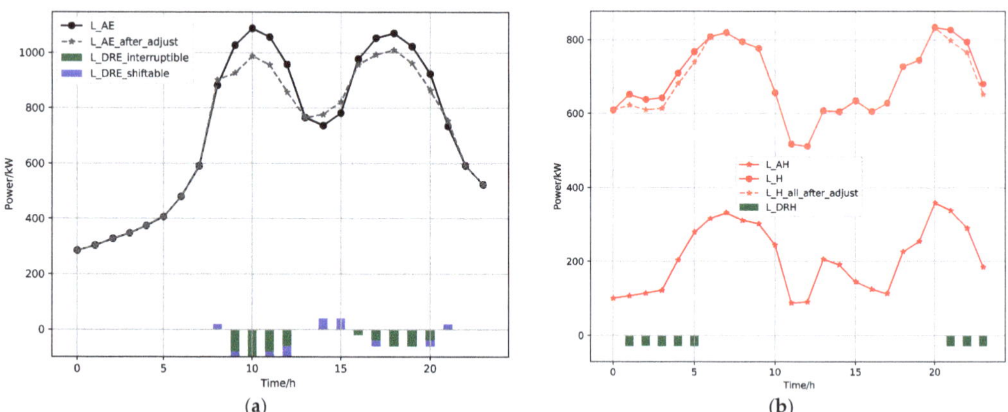

Figure 9. Comparison of electric and heat load before and after UDDSR. (**a**) Comparison of electric load before and after response; (**b**) comparison of heat load before and after response.

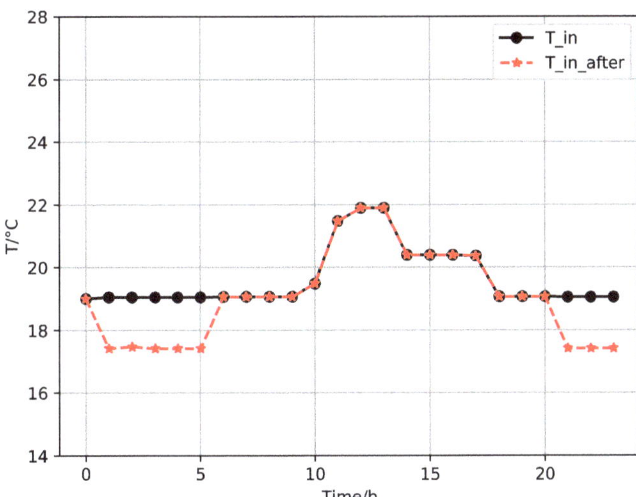

Figure 10. Indoor heating temperature before and after UDDSR.

The comparison of system operating costs before and after UDDSR response is shown in Table 1. It can be seen that after participating in the UDDSR response, users can directly receive a load response compensation of RMB 250.49 (including power load and thermal load response compensation). The total daily operating cost of the community system is reduced by RMB 543.75, and the saving rate can reach 3.09%. The results verify the effectiveness of the proposed UDDSR mechanism.

Table 1. System operation costs before and after UDDSR response.

	Before UDDSR	**After UDDSR**	**Saving (%)**
Electricity purchasing cost (RMB)	7344.90	6715.64	8.57%
Gas purchasing cost (RMB)	9153.34	9001.82	1.66%
Operation and maintenance (RMB)	1122.07	1108.60	1.20%
Power load response compensation (RMB)	0	205	/
Thermal load response compensation (RMB)	0	45.49	/
BT subsidies (RMB)	0	5.70	
Adjustable temperature (°C)	0	1	/
Total cost (RMB)	17,620.31	17,076.56	3.09%

5.2. CVaR-Based Energy Optimization

In this subsection, the random simulation sampling method based on (44) is used to model the uncertainties that the community system may face. Four scenarios where the maximum prediction error and maximum load response fluctuation (maximum uncertainty fluctuations) are not more than 5%, 10%, 15% and more than 15% are set for comparison. Among them the maximum prediction error of outdoor temperature is set to be not more than 2 °C. The number of subscenarios in the uncertainty scenario set for each scenario is 100. The influence of different confidence levels *con* and different uncertainty coefficients β on the system optimization results is analyzed.

5.2.1. Energy Risk Optimization Results Based on CVaR

Scenario 2, where the maximum uncertainty fluctuation does not exceed 10%, is taken as an example to analyze the optimization results when *con* = 0.95, β = 1. A set of uncertainty scenarios is depicted in Figure 11.

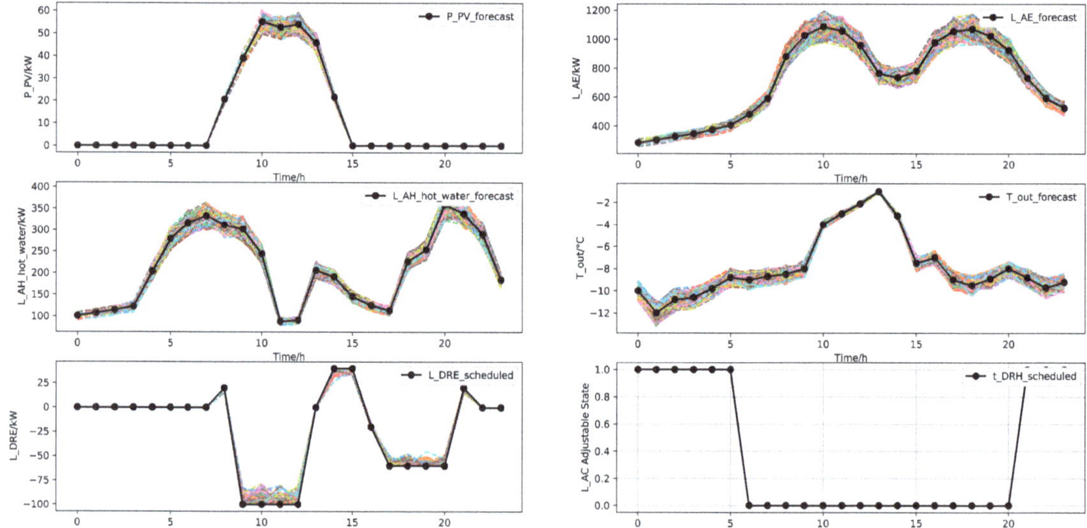

Figure 11. Uncertainty scenarios set with maximum risk fluctuation ≤10%.

The electrical load response is assumed to fluctuate below the optimal response obtained by day-ahead optimization based on UDDSR, i.e., the case only considers the situation where the actual response of users does not meet the standard. The adjustable thermal load is allowed to be regulated at 00:00–05:00 and 21:00–23:00, and this setting has a certain logical consistency with the heating needs of users.

The system energy optimization results of scenario 2 are depicted in Figure 12. From Figure 12a, it can be observed that when $\beta = 1$, the power supply of the community system is greater than the predicted electric load in most periods. In Figure 12b, L_AC0 is the thermal load of users before the UDDSR response, and the heating power of the community system during 00:00 and 06:00–13:00 is greater than the predicted heating load. The community system adopts a completely conservative risk avoidance strategy, i.e., to make the system operate normally under the interference of any risk fluctuations in the second scenario, the system equipment output as much power as possible to meet the electric and heating demand of users.

Figure 13 shows the changes in indoor temperature in the four scenarios. It can be seen that the greater the risk fluctuation, the greater the indoor temperature variation. However, the change of the indoor temperature remains within 2 °C per unit time, and the indoor temperature is kept within the upper and lower limits allowed by users.

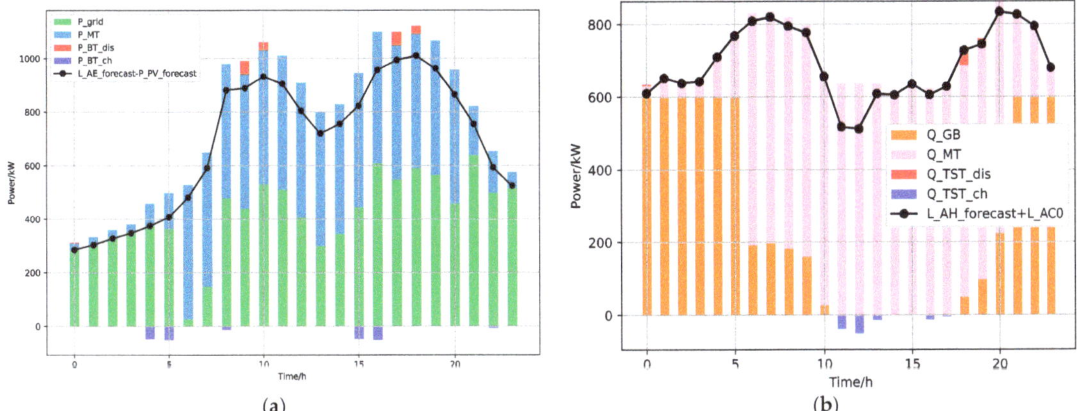

Figure 12. System energy optimization results with maximum uncertainty fluctuation ≤10%. (**a**) Optimization results of electric bus; (**b**) optimization results of heat bus.

The comparison of system operating costs in the four scenarios is shown in Table 2. When the maximum risk fluctuation is less than or equal to 5%, the expected total cost of system operation is reduced by RMB 158.7 compared with the total cost without UDDSR response, that is, a saving of 0.9%. When the maximum risk fluctuation is less than or equal to 10%, the expected cost of the community system operating in the second scenario is RMB 212.33 higher than that without UDDSR response. The system only needs to pay 1.21% more in operating expenses to deal with the impact of 10% risk fluctuation. When the maximum risk fluctuation is larger than 10%, the expected cost of system operation will continue to rise as the risk fluctuation becomes larger. Once the prediction error is large, the system must pay high costs in order to avoid operational risks. On the other hand, the average excess loss of the system increases with the increase in risk fluctuations, indicating that the system needs to increase investment to better deal with risks, which verifies the rationality of the algorithm proposed in this paper.

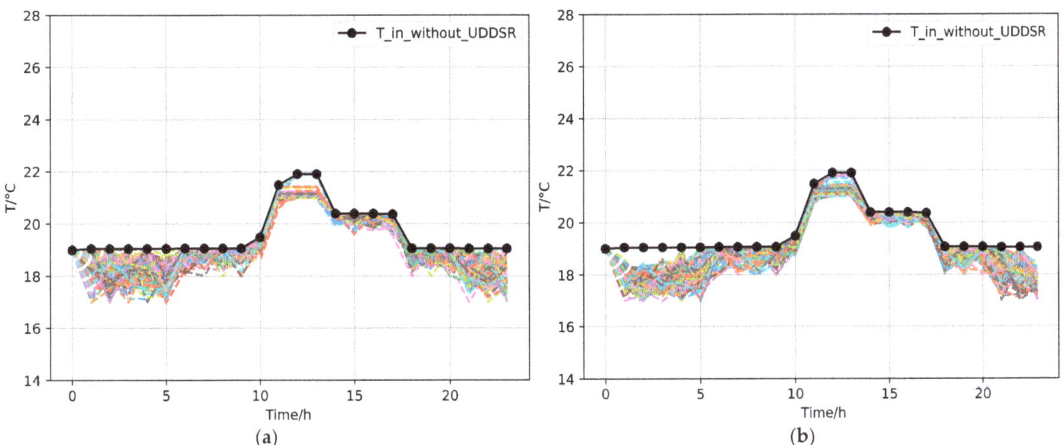

Figure 13. *Cont.*

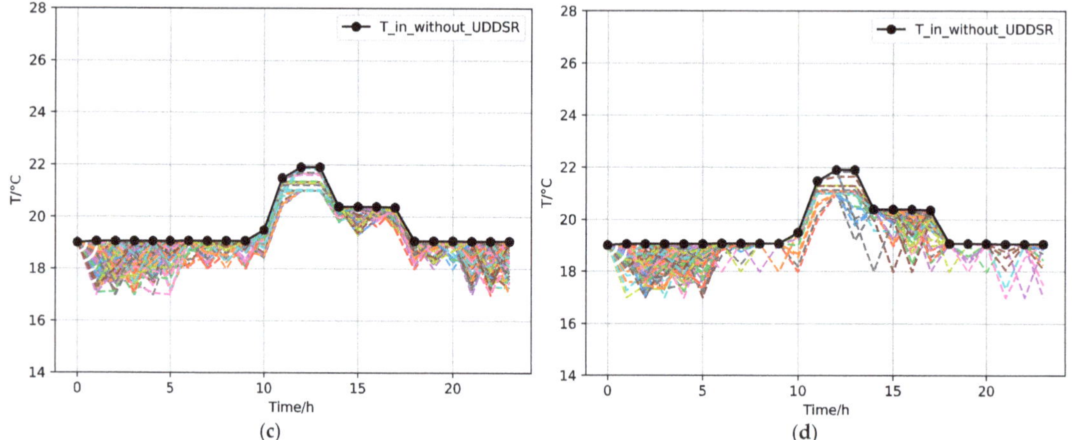

Figure 13. Comparison of indoor heating temperature in different scenarios (con = 0.95, β = 1): (**a**) maximum uncertainty fluctuation $\leq$5%; (**b**) maximum uncertainty fluctuation $\leq$10%; (**c**) maximum uncertainty fluctuation $\leq$15%; (**d**) maximum uncertainty fluctuation >15% (about 50%).

Table 2. System operation costs in different scenarios (con = 0.95, β = 1).

Scenarios	Before UDDSR	1	2	3	4
Maximum risk fluctuation	0%	$\leq$5%	$\leq$10%	$\leq$15%	>15%
Electricity purchasing cost (RMB)	7344.90	6889.18	7293.46	8048.00	11529.32
Gas purchasing cost (RMB)	9153.34	9208.27	9177.42	9338.93	9765.28
Power load response subsidies (RMB)	0	203.25	200.87	196.98	183.88
Thermal load response subsidies (RMB)	0	37.32	45.13	26.54	20.97
BT subsidies (RMB)	0	5.70	5.70	5.71	5.93
Imbalance response penalty (yaun)	0	3.77	8.81	17.21	46.23
Adjustable tmperature (°C)	0	1	1	1	1
Total expected cost of operation (RMB)	17,620.31	17,461.61	17,832.64	18,739.14	22,662.36
CVaR (RMB)	0	4.13	5.21	17.45	41.01
Total cost savings ratio	/	0.90%	−1.21%	−6.34%	−28.61%

5.2.2. Impact of Confidence Level and Uncertainty Coefficient of CVaR on Energy Use Optimization

To further study the impact of confidence level *con* and uncetianty coefficient β (the risk preference of system operators) on the system optimization results, scenario 2 is used as an example to construct the following test set.

$$con = \{0.99, 0.95, 0.9, 0.8, 0.7, 0.6, 0.5, 0.4, 0.3, 0.2, 0.1, 0\} \tag{52}$$

$$\beta = \{1, 0.9, 0.8, 0.7, 0.6, 0.5, 0.4, 0.3, 0.2, 0.1, 0\} \tag{53}$$

The performance of the expected cost of community system operation on the test set under scenario 2 is shown in Figure 14. The expected cost of system operation decreases as the confidence level decreases. This is because the predicted value of the uncertainty variable is used to generate the scenario. The lower the value of *con*, the lower the probability that the predicted value of the uncertainty variable is included in the uncertainty scenario set (i.e., the closer the uncertainty variable is to the predicted value). On the other hand, β represents the weight of CVaR in the optimization objective function. The larger the weight, the more the system tends to avoid uncertainties. Therefore, when *con* is determined, the expected cost of the system increases with the increase of β.

Figure 15 shows the average excess loss CVaR that the community runs under uncertainty on the test set. The average excess loss of the system decreases with the increase in the CVaR weight. The larger the CVaR weight, the more prone the system is to uncertainties, and the system will try to reduce possible uncertainty-induced losses even if this results in of higher operation costs. On the other hand, when β is determined, the average excess loss increases with the rise of *con*. This is because CVaR measures the tail uncertainty outside the confidence interval. The larger *con* is, the greater the deviation between the uncertainty variables and the predicted value in the uncertainty scenario, and the greater the possible uncertainty loss is.

To more clearly show the relationship between CVaR and the expected cost of system operation, a case where *con* = 0.95 is analyzed. In this case, the uncertainty variables contained in the uncertainty scenario set are more volatile, and the system is faced with greater possible uncertainties. The results are shown in Figure 16. When β increases, the expected cost of the system continues to increase, while the average excess loss of the system continues to decrease. This is because the greater β is, the more the system prefers to avoid uncertainties, and the system is willing to pay higher operating costs in exchange for lower excess losses.

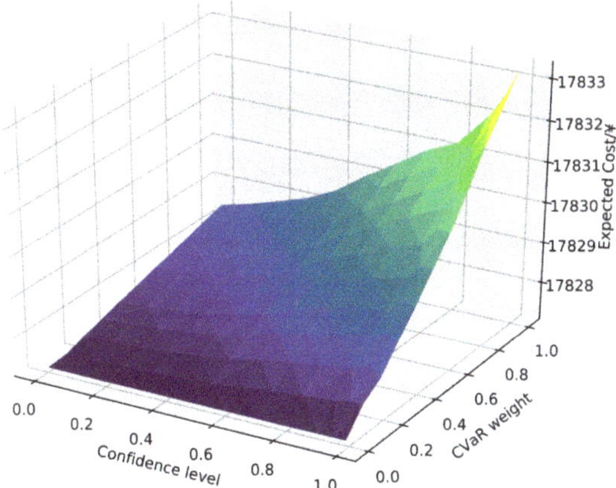

Figure 14. The changes of expected cost of system operation with *con* and β (maximum uncertainty fluctuation $\leq$10%).

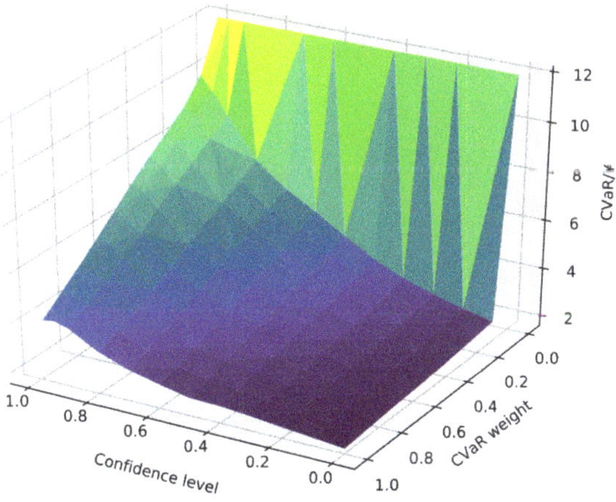

Figure 15. The changes of CVaR with *con* and β (maximum risk fluctuation $\leq$10%).

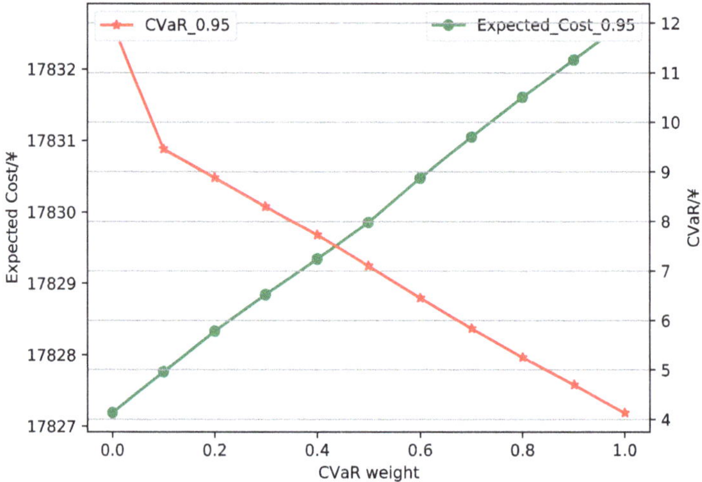

Figure 16. The changes of CVaR and expected cost with β (maximum uncertainty fluctuation $\leq 10\%$).

6. Conclusions

This paper proposes an energy optimization method for community IES based on UDDSR. The thermal model of aggregated buildings is introduced to measure users' adjustable thermal load, and the responsive loads including power loads and thermal loads are aggregated and optimized through UDDSR optimization. Then, a day-ahead scheduling model is proposed to optimize the energy management for the community IES, and CVaR theory is introduced to deal with the volatility of PV output, user load, outdoor temperature, and user actual UDDSR response load. The case study shows that the proposed UDDSR mechanism can effectively reduce the operating costs under the premise of fully considering the willingness of users to participate in IDR events. Additionally, the optimization method based on CVaR enables the community system to pay less than 2% in additional operating costs to deal with the energy deviation caused by the maximum uncertainty of 10%, thus verifying the correctness and effectiveness of the method presented in this paper. For further study, the relationship between user energy consumption behavior and response capacity can be explored, so as to construct a reward and punishment mechanism that is more suitable for the energy needs of users.

Author Contributions: Conceptualization, J.Z.; Formal analysis, Y.L.; Funding acquisition, Z.M.; Investigation, Y.P.; Methodology, J.Z.; Project administration, Z.M.; Resources, Z.M.; Software, Y.L.; Supervision, Y.P. and S.Z.; Validation, J.Z. and Z.M.; Writing—original draft, Y.L.; Writing—review and editing, Y.P. and S.Z. All authors have read and agreed to the published version of the manuscript.

Funding: This research was funded by the Science and Technology Project of Jiangsu State Grid Corporation of China, Grant J20210148.

Institutional Review Board Statement: Not applicable.

Informed Consent Statement: Not applicable.

Data Availability Statement: Not applicable.

Conflicts of Interest: The authors declare no conflict of interest.

Appendix A

Table A1. The electricity price, subsidy for UDDSR participators and gas price.

Time period	Electricity Price (RMB/kWh)	Power Load Response Subsidy (RMB/kWh)	Thermal Load Response Subsidy (RMB/kWh)	Imbalance Response Penalty (RMB/kWh)	Gas Price (RMB/m^3)
Peak time ((09:00–13:00], [17:00–20:00))	1.19	0.3	0.2	0.6	3
Normal time ((06:00–08:00], [14:00–16:00))	0.75	0.1	0.2	0.4	3
Valley time ((00:00 05:00], [21:00–23:00))	0.36	0.05	0.2	0.18	3

Table A2. The equipment operating parameters.

Parameter	Value	Parameter	Value
MT generating efficiency	0.36	TST heat releasing efficiency	0.95
MT maximum output power	500 kW	TST self-loss rate of thermal energy	0.04
MT minimum output power	10 kW	TST maximum capacity	100 kWh
GB heat production efficiency	0.85	TST minimum capacity	0 kWh
GB maximum thermal output power	600 kW	TST maximum heat storage/release power	50 kW
GB minimum thermal output power	0 kW	Maximum power purchased from the grid	1000 kW
BT charging efficiency	0.95	Minimum power purchased from the grid	0 kW
BT discharging efficiency	0.95	Maximum power of interruptible power load	$L^t_{DRE,intmax}$
BT self-loss rate of electrical energy	0.04	Maximum power of shiftable power load	$L^t_{DRE,shf,outmax} / L^t_{DRE,shf,inmax}$
BT maximum capacity	100 kWh	Maximum indoor temperature	26
BT minimum capacity	0 kWh	Minimum indoor temperature	18
BT maximum charging/discharging power	50 kW	Optimum indoor temperature	21
TST heat storing efficiency	0.95	Maximum adjustable temperature	T_{adj}

Table A3. The operation and maintenance cost of equipment and subsidy parameters.

Equipment	Operation and Maintenance Cost (RMB/kWh)	Equipment	Subsidy (RMB/kWh)
MT	0.075	BT	0.01
GB	0.08	TST	0.01
PV	0.01		

References

1. Jia, B.; Li, F.; Sun, B. Design of energy management platform for integrated energy system. In Proceedings of the 2019 14th IEEE Conference on Industrial Electronics and Applications (ICIEA), Xi'an, China, 19–21 June 2019; pp. 1593–1597.
2. Tao, H.; Fang, L.; YaLou, L.; Xiaoxin, Z. The Design of the Simulation System for the Integrated Energy System. In Proceedings of the 2018 International Conference on Power System Technology (POWERCON), Guangzhou, China, 6–9 November 2018; pp. 410–416.
3. Zhou, X.; Bao, Y.-Q.; Chen, P.-P.; Ji, T.; Deng, W.; Jin, W. Integrated Electricity and Natural Gas Energy System Scheduling Model Considering Multi-Energy Storage Systems. In Proceedings of the 2018 IEEE PES Asia-Pacific Power and Energy Engineering Conference (APPEEC), Sabah, Malaysia, 7–10 October 2018; pp. 178–182.
4. Miao, B.; Lin, J.; Li, H.; Liu, C.; Li, B.; Zhu, X.; Yang, J. Day-Ahead Energy Trading Strategy of Regional Integrated Energy System Considering Energy Cascade Uti-lization. *IEEE Access* **2020**, *8*, 138021–138035. [CrossRef]

5. Chen, Y.; Wei, W.; Liu, F.; Sauma, E.E.; Mei, S. Energy Trading and Market Equilibrium in Integrated Heat-Power Distri-bution Systems. *IEEE Trans. Smart Grid* **2019**, *10*, 4080–4094. [CrossRef]
6. Tedesco, F.; Mariam, L.; Basu, M.; Casavola, A.; Conlon, M.F. Economic Model Predictive Control-Based Strategies for Cost-Effective Supervision of Community Microgrids Considering Battery Lifetime. *IEEE J. Emerg. Sel. Top. Power Electron.* **2015**, *3*, 1067–1077. [CrossRef]
7. Qinwei, D. A price-based demand response scheduling model in day-ahead electricity market. In Proceedings of the 2016 IEEE Power and Energy Society General Meeting (PESGM), Boston, MA, USA, 17–21 July 2016; pp. 1–5.
8. Junhui, L.; Hongkun, B.; Hujun, L.; Qinchen, Y.; Fangzhao, D.; Wenjie, Z. Optimal Decomposition of Flexible Load Resource Reserve Target Based on Power Demand Response. In Proceedings of the 2021 IEEE International Conference on Power Electronics, Computer Applications (ICPECA), Shenyang, China, 22–24 January 2021; pp. 1046–1049.
9. Parisio, A.; Rikos, E.; Glielmo, L. A Model Predictive Control Approach to Microgrid Operation Optimization. *IEEE Trans. Control Syst. Technol.* **2014**, *22*, 1813–1827. [CrossRef]
10. Du, Y.; Zheng, N.; Cai, Q.; Li, Y.; Li, Y.; Shi, P. Research on Key Technologies of User Side Integrated Demand Response for Multi-Energy Coordination. In Proceedings of the 2020 IEEE 4th Conference on Energy Internet and Energy System Integration (EI2), Wuhan, China, 30 October–1 November 2020; pp. 2461–2464.
11. Bahrami, S.; Sheikhi, A. From Demand Response in Smart Grid toward Integrated Demand Response in Smart Energy Hub. *IEEE Trans. Smart Grid* **2015**, *7*, 650–658. [CrossRef]
12. Liu, P.; Ding, T.; Zou, Z.; Yang, Y. Integrated demand response for a load serving entity in multi-energy market considering network constraints. *Appl. Energy* **2019**, *250*, 512–529. [CrossRef]
13. Wang, J.; Zhong, H.; Ma, Z.; Xia, Q.; Kang, C. Review and prospect of integrated demand response in the multi-energy system. *Appl. Energy* **2017**, *202*, 772–782. [CrossRef]
14. Ren, S.; Dou, X.; Wang, Z.; Wang, J.; Wang, X. Medium- and Long-Term Integrated Demand Response of Integrated Energy System Based on System Dynamics. *Energies* **2020**, *13*, 710. [CrossRef]
15. Yang, H.; Li, M.; Jiang, Z.; Zhang, P. Multi-Time Scale Optimal Scheduling of Regional Integrated Energy Systems Con-sidering Integrated Demand Response. *IEEE Access* **2020**, *8*, 5080–5090. [CrossRef]
16. Tahir, M.F.; Haoyong, C.; Mehmood, K.; Ali, N.; Bhutto, J.A. Integrated Energy System Modeling of China for 2020 by Incorporat-ing Demand Response, Heat Pump and Thermal Storage. *IEEE Access* **2019**, *7*, 40095–40108. [CrossRef]
17. Jiang, Z.; Ai, Q.; Hao, R. Integrated Demand Response Mechanism for Industrial Energy System Based on Multi-Energy Interaction. *IEEE Access* **2019**, *7*, 66336–66346. [CrossRef]
18. Stenner, K.; Frederiks, E.R.; Hobman, E.V.; Cook, S. Willingness to participate in direct load control: The role of consumer distrust. *Appl. Energy* **2017**, *189*, 76–88. [CrossRef]
19. Zhou, S.; Zou, F.; Wu, Z.; Gu, W.; Hong, Q.; Booth, C. A smart community energy management scheme considering user dominated demand side response and P2P trading. *Int. J. Electr. Power Energy Syst.* **2020**, *114*, 105378. [CrossRef]
20. Brahman, F.; Honarmand, M.; Jadid, S. Optimal electrical and thermal energy management of a residential energy hub, integrating demand response and energy storage system. *Energy Build.* **2015**, *90*, 65–75. [CrossRef]
21. Wang, C.S.; Hong, B.W.; Guo, L.; Zhang, D.J.; Liu, W.J. A General Modeling Method for Optimal Dispatch of Combined Cooling, Heating and Power Microgrid. *Proc. CSEE* **2013**, *33*, 26–33.
22. Zhang, Q.; Wang, X. Hedge Contract Characterization and Risk-Constrained Electricity Procurement. *IEEE Trans. Power Syst.* **2009**, *24*, 1547–1558. [CrossRef]
23. Sheikhi, A.; Rayati, M.; Bahrami, S.; Ranjbar, A.M. Integrated Demand Side Management Game in Smart Energy Hubs. *IEEE Trans. Smart Grid* **2015**, *6*, 675–683. [CrossRef]

Article

Electric Heating Load Forecasting Method Based on Improved Thermal Comfort Model and LSTM

Jie Sun [1], Jiao Wang [1], Yonghui Sun [1], Mingxin Xu [1], Yong Shi [1], Zifa Liu [2] and Xingya Wen [2,*]

[1] State Grid Inner Mongolia Eastern Electric Power Co., Ltd., Economic and Technical Research Institute, Hohhot 010011, China; candalf@126.com (J.S.); wj_18686056760@163.com (J.W.); 15771346563@163.com (Y.S.); 18547175858@163.com (M.X.); shiyong5787@163.com (Y.S.)

[2] School of Electrical and Electronic Engineering, North China Electric Power University, Beijing 102206, China; tjubluesky@163.com

[*] Correspondence: wenxingya@ncepu.edu.cn; Tel.: +86-18810119275

Abstract: The accuracy of the electric heating load forecast in a new load has a close relationship with the safety and stability of distribution network in normal operation. It also has enormous implications on the architecture of a distribution network. Firstly, the thermal comfort model of the human body was established to analyze the comfortable body temperature of a main crowd under different temperatures and levels of humidity. Secondly, it analyzed the influence factors of electric heating load, and from the perspective of meteorological factors, it selected the difference between human thermal comfort temperature and actual temperature and humidity by gray correlation analysis. Finally, the attention mechanism was utilized to promote the precision of combined adjunction model, and then the data results of the predicted electric heating load were obtained. In the verification, the measured data of electric heating load in a certain area of eastern Inner Mongolia were used. The results showed that after considering the input vector with most relative factors such as temperature and human thermal comfort, the LSTM network can realize the accurate prediction of the electric heating load.

Keywords: electric heating; load forecasting; thermal comfort; attention mechanism; LSTM neural network

Citation: Sun, J.; Wang, J.; Sun, Y.; Xu, M.; Shi, Y.; Liu, Z.; Wen, X. Electric Heating Load Forecasting Method Based on Improved Thermal Comfort Model and LSTM. *Energies* **2021**, *14*, 4525. https://doi.org/10.3390/en14154525

Academic Editor: Leijiao Ge

Received: 14 July 2021
Accepted: 23 July 2021
Published: 27 July 2021

Publisher's Note: MDPI stays neutral with regard to jurisdictional claims in published maps and institutional affiliations.

1. Introduction

Electric heating is a clean, efficient, and flexible form of heating equipment. In recent years, coal-fired heating has been gradually replaced by electric heating in northern China. In order to control urban haze pollution and improve the quality of life of residents, in recent years, the relevant departments of the state have launched the policies of "electricity instead of coal" and "electricity instead of oil" [1]. These policies promote the process of clean energy gradually replacing polluting energy and greatly improve the effect of reducing pollutant emissions. With the continuous improvement of residents' requirements for indoor comfort, the scale of electric heating in winter is increasing year by year, and electric heating is used more and more frequently. Meanwhile, the daily maximum load in winter is also increasing.

Electric heating equipment can be divided into centralized (direct heating electric boiler, regenerative electric boiler, etc.) and distributed (heating cable, electric heating film, carbon crystal heating, etc.). Because electric heating in operation will not produce pollution gas and noise, it is very clean and environmentally protective. The typical characteristics of electric heating are high power, concentrated load, easy-to-produce peak load, and large peak valley difference, and thus it has a great impact on distribution lines [2]. Therefore, accurate load forecasting of electric heating load has great practical significance.

The influence of meteorological factors on short-term load forecasting cannot be ignored. The relevant literature mainly analyzes the factors such as temperature, humidity, wind, and precipitation. The article [3] studies the influence of meteorological time series characteristics on urban power consumption and proposes a prediction method different from traditional methods. Articles [4,5] analyzed the prediction model of meteorological sensitive load under the influence of temperature, humidity, snowfall, and other meteorological factors, and put forward the strategy of data processing.

Electric heating load is a kind of temperature control load [6]. In recent years, scholars from all over the world have carried out research work on temperature-controlled load characteristics. The authors of [7,8] predicted the dispatchable capacity and the ability to respond to grid dispatching from the perspective of temperature-controlled load providing auxiliary services for the power system. In the study of [9], the characteristic law of typical microgrid temperature-controlled load is analyzed, and a physical model and a rough scheme for optimal scheduling is established. The authors of [10,11] analyzed and modeled the typical temperature control load characteristics in the centralized area, evaluated the load more accurately in the multi-state situation, and proposed a real-time management and control scheme for the temperature control load.

Load forecasting is based on historical load and weather data in order to analyze the possible influence of historical load data on future load changes, so as to achieve accurate load forecasting in a certain period of time in the future [12]. Short-term load forecasting only forecasts the data of each period in the next few days. The classical load forecasting algorithms generally include artificial neural network (ANN) [13], support vector machine (SVM) [14], and gray neural network [15]. For the learning of time series data, the long-short term memory (LSTM) network algorithm is more mature. In the study of [16], the convergent cross mapping (CCM) method was used to study the internal relationship between power consumption and temperature, wind speed, and other factors. The LSTM neural network model was established, and the urban power consumption was predicted. The results show that the accuracy was good. In [17], different training steps of electric heating load forecasting are compared on the basis of the LSTM network. The results show that LSTM network can achieve accurate electric heating load forecasting in different time scales.

The research on the influence of absolute temperature on power load forecasting has been relatively mature. Few studies have considered the influence of users' thermal comfort temperature in different environments, taking the difference between users' thermal comfort temperature and air temperature as the input of load forecasting model.

On the basis of the analysis of electric heating load characteristics in distribution network, this paper focused on the analysis of meteorological factors and the comfort temperature of a main crowd. Firstly, the interfering factors of electric heating load were studied by gray relational analysis method. Then, the thermal comfort temperature model of residents was constructed. Finally, the historical data of electric heating load were connected with the traditional influencing factors and the difference between thermal comfort temperature and air temperature, and the electric heating load was predicted by the improved LSTM network. Meanwhile, the proposed model was compared with other models. The results showed that the prediction effect of the proposed method was better.

This paper proposes an electric heating load forecasting method based on improved human thermal comfort model and improved LSTM neural network. The main contributions of this paper are as follows:

1. Modeling the thermal comfort of the human body.
2. The difference between the user's thermal comfort temperature and the temperature is introduced, rather than the absolute temperature value as the input in the network model.
3. On the basis of LSTM network, we added attention mechanism and dropout layer.

2. Thermal Comfort Model of the Human Body

The use of electric heating devices in heating areas in China (such as eastern Inner Mongolia) has gradually become mature, and its comfort is very important to the user experience. In the use of decentralized electric heating, human thermal comfort will affect the heating time, heating temperature, and other factors, thus affecting the electric heating load data. As the most important driving force of user response, thermal comfort should be considered in load forecasting.

Indoor environment quality will directly affect the physical and mental health and work efficiency of human body. It is very important and fundamental for people in a heated area to achieve a comfortable indoor temperature. Thermal comfort is used to indicate that most people are satisfied with the objective thermal environment, both physically and psychologically. It is mainly affected by physical conditions, physiological conditions, and psychological conditions [18]. The physical conditions include the heat transfer performance and shading coefficient of the walls and windows of the building where people live, the internal disturbance of lighting and equipment, the growth rate of indoor microorganism, and so on, which are not affected by the human body's own activities. Physiological conditions include the change of perspiration rate caused by the roughness or cracking of human skin, the intensity of exercise when carrying out routine activities, and the regulation of local or overall sensation of radiation temperature. Psychological conditions refer to the deviation between the factors and psychological expectation in the thermal environment, which are closely related to subjective feeling.

At present, the thermal comfort of people's environment is usually analyzed according to the ISO 7730 thermal comfort model [19], which is proposed by the international standards organization. The calculation results are expressed by predicted mean vote (PMV), and the formula is as follows [20,21]:

$$
\begin{aligned}
PMV = &\left[0.303 \times e^{-0.036M} + 0.028\right] \times \left\{(M - W) - 3.05 \times 10^{-3} \times \right.\\
&\left[5733 - 6.99 \times (M - W) - P_a\right] - 0.42 \times \left[(M - W) - 58.15\right]\\
&-1.7 \times 10^{-5} \times M \times (5867 - P_a) - 0.0014 \times M \times (34 - t_a) -\\
&3.96 \times 10^{-8} \times f_{cl} \times \left[(t_{cl} + 273)^4 - (t_r + 273)^4\right] - f_{cl} \times h_c \times\\
&\left. (t_{cl} - t_a)\right\}
\end{aligned} \tag{1}
$$

where M is metabolic rate of human body, W/m^2; W is the mechanical power consumed by the human body, W/m^2; P_a is partial pressure of water vapor in ambient air around human body, Pa; t_a is air temperature around human body, °C; t_r is average radiation temperature, °C; f_{cl} is the ratio of clothing area covered by human body to bare area; t_{cl} is the temperature of outer surface for clothing, °C; and h_c is the heat transfer coefficient, $W/(m^2 \cdot K)$.

ISO 7730 thermal comfort model has a high accuracy in obtaining the user comfort temperature range, but it is difficult to obtain the real-time environmental data required by the model. Therefore, the ISO 7730 model can be simplified properly without affecting the accuracy. In [22], the Rohles simplified model was improved, and the results were extended to a wider range of clothing insulations. Only the indoor air temperature and relative humidity in the test environment were used as the input parameters, and therefore the thermal comfort parameters can be easily evaluated. The results show that the method is very close to ISO 7730 thermal comfort model and is easy to operate and greatly enhanced. The simplified and improved model is as follows:

$$
I_{PMV} = aT_a + bP_v - c \tag{2}
$$

where I_{PMV} is index value of PMV; T_a is indoor temperature; P_v is relative humidity, %; and a, b, and c are known parameters.

When the indoor temperature and relative humidity are on the high side or on the low side, they will interfere with people's core temperature. At present, people's heating temperature is increasing day by day, and therefore the temperature of people's thermal comfort zone will also rise as a whole, and the regulation ability of cold and heat stimulation of people who stay in the thermal comfort zone for a long time will be weakened. In the end, peoples' sensitivity and reaction time to adjust the temperature will become longer.

When the indoor temperature is not the expected thermal comfort temperature, people will adjust the temperature setting to achieve the expected value. Therefore, in order to consider the impact of users' thermal comfort temperature, we used the difference between the air temperature and human thermal comfort temperature to improve the input data of LSTM neural network prediction model.

3. Analysis of Factors Affecting Electric Heating Load

3.1. Load Characteristics of Electric Heating

Electric heating load is different from general electric load, and it has obvious seasonal climate characteristics. Taking the electric heating data of a certain year in eastern Inner Mongolia as an example, from the change trend of annual load curve, we found that the electric heating load in northern region is more intensive in winter (December to March of the next year), in which December to January are the months with the lowest average temperature. From the daily load curve, we found that electric heating load also has obvious characteristics of daily type. From Monday to Friday, the load of office buildings is higher, while the load of weekends and holidays is lower, but the load of commercial and residential electric heating is higher, and the overall trend of daily change is not large. It can be seen from Figure 1 that the typical daily load curve of electric heating in eastern Inner Mongolia presents the characteristics of morning peak, afternoon trough, and evening peak. In terms of electricity consumption, this is mainly due to the start-up of industrial and commercial electric heating in the morning, the general rise of temperature in the afternoon, and the start-up of residential load gathering in the evening.

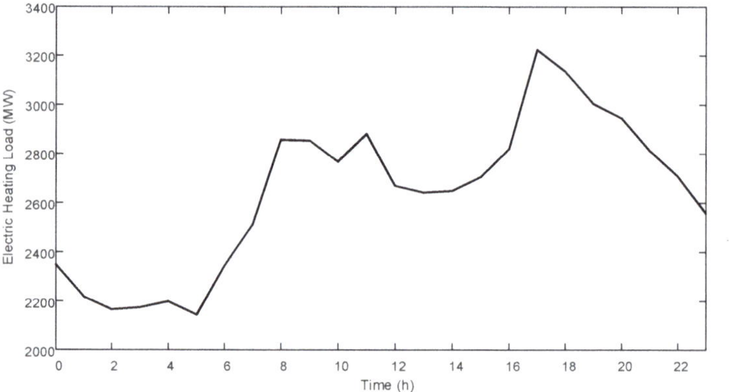

Figure 1. Typical electric heating load curve of a coal to electricity area in eastern Inner Mongolia.

The key areas of electric energy substitution in eastern Inner Mongolia are distributed electric heating and centralized electric heating, and electric heating accounts for more than 50% of the proportion of electric energy substitution in eastern Inner Mongolia. With the increasing application of electric heating and large-scale access to the power grid, the impact on the operation of the power system is to further narrow the gap between the winter and summer load.

3.2. Correlation Analysis of Electric Heating Load and Influencing Factors

The idea of association analysis is to compare the similarity degree of data series, so as to clarify the association degree and regular pattern between each series. It belongs to an effective and practical method of gray system theory to analyze the correlation degree of various factors in the research object system [5,23]. In order for the variation characteristics of electric heating load in winter in eastern Inner Mongolia to be studied, the relationship between the meteorological factors such as temperature difference (the difference between human thermal comfort temperature and actual temperature), relative humidity, wind speed and snow falling, and electric heating load should be analyzed. The calculation steps of correlation analysis method are as follows:

Step 1: Construct electric heating load characteristic sequence and influence factor sequence. The electric heating load sequence is expressed as X_0, and the related influencing factor sequence is expressed as X_i; the complete sequence is as follows:

$$X_0 = (x_0(1) \quad x_0(2) \cdots x_0(k) \cdots x_0(n)) \tag{3}$$

$$X_i = (x_i(1) \quad x_i(2) \cdots x_i(k) \cdots x_i(n)) \tag{4}$$

where k is serial number; n is number of samples, $k = 1, 2, \cdots, n$; and i is the number of related factors, $i = 1, 2, \cdots, m$.

Step 2: Obtain the correlation degree.

(a) Each sequence is dimensionless as the initial value, as shown in the following formula:

$$X'_i = \frac{X_i}{x_i(1)} = \left(x'_i(1) x'_i(2) \cdots x'_i(n) \right) \tag{5}$$

where $i = 1, 2, \cdots, m$, and X'_i is initial value after processing.

(b) Determine the difference between electric heating load sequence and each influencing factor Δ_i.

$$\Delta_i(k) = \left| x'_0(k) - x'_i(k) \right| \tag{6}$$

$$\Delta_i = (\Delta_i(1) \quad \Delta_i(2) \cdots \Delta_i(k) \cdots \Delta_i(n)) \tag{7}$$

Record the minimum value of all sequence differences as a, the minimum range is b.

$$\begin{cases} a = \min\{\Delta_i(1), \Delta_i(2), \cdots, \Delta_i(k), \cdots, \Delta_i(n)\} \\ b = \max\{\Delta_i(1), \Delta_i(2), \cdots, \Delta_i(k), \cdots, \Delta_i(n)\} \\ \quad i = 1, 2, \cdots, m \end{cases} \tag{8}$$

(c) Find the correlation coefficient of each sample in the sequence $\gamma_i(k)$.

$$\gamma_i(k) = \frac{a + \varepsilon b}{\Delta_i(k) + \varepsilon b} \tag{9}$$

where $\gamma_i(k)$ is the correlation coefficient between the k-th parameter of the i-th subsequence and the k-th parameter of the electric heating load sequence, and ε is the resolution coefficient, usually 0.5.

(d) Calculate the average correlation coefficient as the following:

$$\gamma_i = \frac{1}{n} \sum_{k=1}^{n} \gamma_i(k) \tag{10}$$

where $i = 1, 2, \cdots, m$.

Step 3: Analyze the correlation coefficient.

Obtain the correlation coefficient between the electric heating load data series X_0 and each related factor series X_i. The larger the correlation coefficient, the greater the influence of the factor series on the electric heating load data series. Therefore, the correlation coefficient between electric heating load and various factors can be calculated, as shown in Table 1.

It can be seen from the data in Table 1 that temperature difference and humidity are the most influential factors on electric heating load data, while snowfall and wind speed are relatively less influential. This is mainly because temperature difference and humidity will affect human thermal comfort to a greater extent. Although snowfall and wind speed will also affect people's psychological expectation and feeling of temperature and humidity, their influence is relatively small relative to temperature difference and humidity.

Table 1. Coefficient of correlation between electric heating load and meteorological factors.

Influence Factor	Correlation Coefficient	Influence Factor	Correlation Coefficient
Temperature difference [1]	0.9601	Snowfall	0.8326
Humidity	0.9416	Wind speed	0.7952

[1] Table notes: In this paper and Table 1, "temperature difference" refers to the difference between human thermal comfort temperature and actual temperature.

After the most relevant factors of electric heating load are analyzed, in addition to the historical electric heating load data, temperature difference data and humidity also become the main source data of electric heating load prediction.

4. Improved LSTM Neural Network Prediction Model

4.1. Long Short-Term Memory Network

Due to the inherent time series of load data, the selected forecasting model must have a good ability to express the time series characteristics. In this paper, the long short-term memory network (LSTM) was taken as the main body and improved as the model to study its applicability for short-term load forecasting modeling of electric heating load in eastern Inner Mongolia.

LSTM is a kind of special recurrent neural network (RNN). It can use the information learned at the last moment to learn at the current moment and can set gradient threshold to prevent the gradient disappearing or exploding in RNN training. LSTM algorithm adds cell state C to the original RNN hidden layer to keep the long-term state, thus solving the long-term dependence problem of RNN. Therefore, LSTM is superior to other neural network models. Figure 2 is the schematic diagram of LSTM expansion structure.

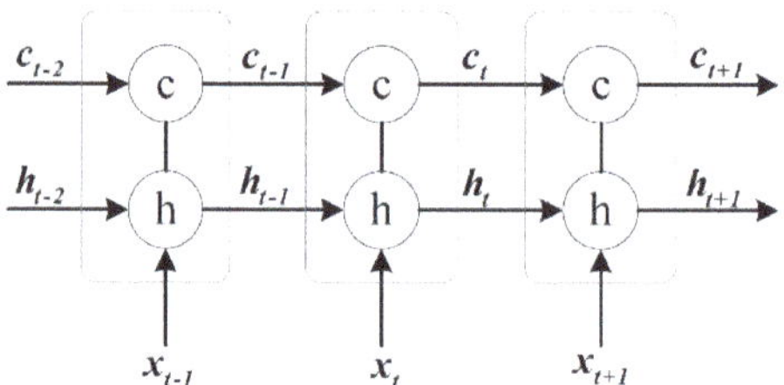

Figure 2. Network deployment structure of LSTM.

In Figure 2, the input of LSTM consists of three parts: the input value at the current time x_t, the output value at the previous time h_{t-1}, and the cell state at the previous time c_{t-1}. The output of LSTM consists of the output unit state c_t and the output value of hidden layer h_t.

Compared with RNN, LSTM redesigns the internal memory unit while maintaining its basic structure. The architecture diagram of each unit of LSTM is shown in Figure 3. The key of every LSTM cell is the control of cell state c. There are three control gates in the unit state, which are forgetting gate f_t, input gate i_t, and output gate o_t. Through these gates, information can be filtered or added to achieve a new unit state.

According to Figure 3, from left to right, it can be seen that the unit state of the previous time c_{t-1} and the output value of the hidden layer of the previous time h_{t-1} together memorize the historical information of the sequence data. Step-by-step analysis of LSTM architecture can be divided into three parts.

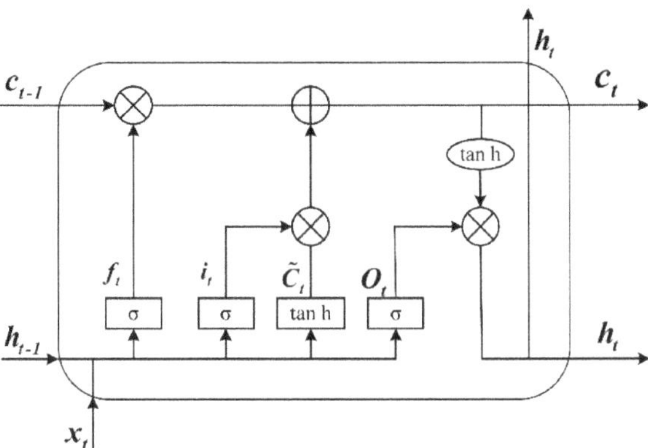

Figure 3. Network deployment structure of LSTM.

The first step is to filter the information selectively. The forgetting gate removes the information in the last unit according to h_{t-1} and x_t, that is, it removes the useless part of the information learned at the last moment. The forgetting gate is as follows:

$$f_t = \sigma\left(w_f\cdot[h_{t-1}, x_t] + b_f\right) \tag{11}$$

where $\sigma(\cdot)$ is Sigmoid activation function, w_f is the weight of forgetting gate, and b_f is the bias of forgetting gate.

The second step is to generate new information that needs to be updated. This part is combined by input gate i_t and candidate value $\tilde{C}_t$. h_{t-1} and x_t use sigmoid function to obtain the data that need to be input into the cell state (i.e., input gate) and create a new candidate state through **tanh** layer. The formula is as follows:

$$i_t = \sigma(w_i\cdot[h_{t-1}, x_t] + b_i) \tag{12}$$

$$\tilde{C}_t = \tanh(w_c\cdot[h_{t-1}, x_t] + b_c) \tag{13}$$

where i_t is information to memorize, that is, input gate; $\tilde{C}_t$ is the candidate value to update the original cell state; w_i and w_c represent the weight of input gate and candidate value, respectively; and b_i and b_c represent the bias of input gate and candidate value, respectively.

The third step is to generate new cell state c_t and hidden layer outputs h_t. By multiplying the input gate i_t and the candidate value $\tilde{C}_t$ and adding them to the forgetting gate f_t, one can obtain the updated cell state value c_t, as shown in the following formula:

$$c_t = f\cdot c_{t-1} + i_t\cdot k_t \tag{14}$$

The new cell state c_t is processed by a **tanh** function, and then multiplied by the output gate o_t to obtain the output value of the hidden layer h_t:

$$o_t = \sigma(w_o\cdot[h_{t-1}, x_t] + b_o) \tag{15}$$

$$h_t = o_t\cdot\tanh(c_t) \tag{16}$$

where w_o is the weight of output gate, and b_o is the bias of output gate.

Through the analysis of LSTM structure system, we can see that using LSTM to replace neurons in RNN to build load forecasting model can solve the problem of long-term dependence and we can learn the hidden historical operation law in power load forecasting.

4.2. Improved LSTM with Attention Mechanism

For different times, the brain will focus on the areas that need to be focused on and reduce or ignore the attention to other areas. This kind of attention allocation mechanism can help people to obtain important and detailed information and reduce the influence of other irrelevant information.

Attention mechanism refers to the idea of human brain attention resource allocation [24]. By assigning different probabilities to generate different attention distribution coefficients, the model can better learn the information in the input sequence and improve the accuracy of the model.

The attention structure is shown in Figure 4, where $x_t(t \in [1, n])$ is the input to the hidden layer of the LSTM model, $h_t(t \in [1, n])$ is the hidden layer output through the LSTM corresponding to each input, $\alpha_t(t \in [1, n])$ is the probability distribution value of the attention mechanism output to hidden layer, and y is the LSTM output value with attention mechanism.

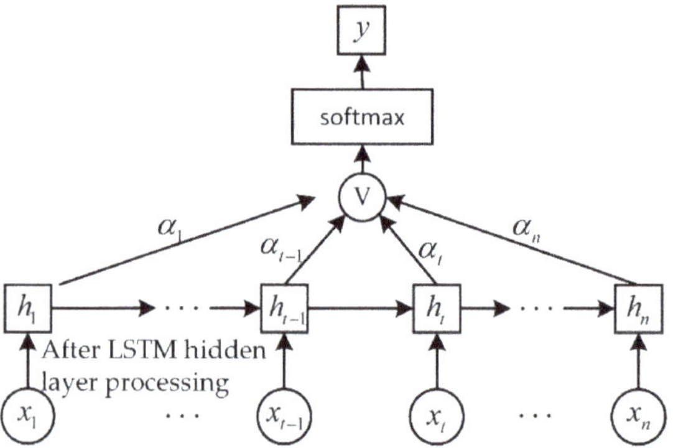

Figure 4. Structure of attention mechanism.

The formulas of attention weight matrix and eigenvector in attention mechanism are as follows:

$$e_t = u_s \tanh(w_s h_t + b_s) \tag{17}$$

$$\alpha_t = \frac{\exp(e_t)}{\sum\limits_{n=1}^{t} e_n} \tag{18}$$

$$V = \sum_{t=1}^{n} \alpha_t h_t \tag{19}$$

where e_t is the non-normalized weight matrix, and w_s, b_s, and u_s represent randomly initialized attention mechanism weight matrix, bias vector, and time series matrix, respectively.

To sum up, the structure of the improved LSTM electric heating load forecasting model designed in this paper is shown in Figure 5, which is mainly composed of input layer, LSTM layer, attention layer, dropout layer, and output layer. The function of dropout layer is to prevent over learning and set the discard rate, so that some neurons extracted from the model can be "discarded" (do not participate in network training).

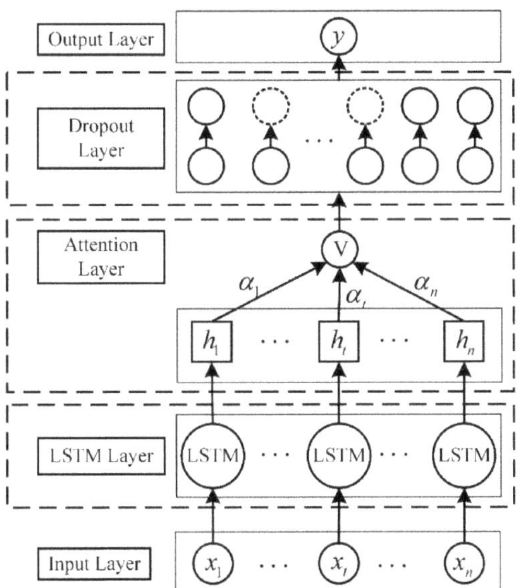

Figure 5. Improved LSTM short term forecast model of electric heating load.

Considering the climate characteristics of northern China, according to the results of correlation analysis, we took the historical electric heating load data from January to March l, the difference between human thermal comfort temperature and air temperature Δt, and relative humidity p_v as the original sample set of the prediction model. The sample data were standardized by 0-1 as the input matrix X_s of the model. The data of temperature and relative humidity were from the National Meteorological Data Center.

The input data X_s of the input layer was simply extracted with feature vectors, and the neural network unit was controlled by three "gates" structures. The output data of LSTM layer was the matrix $H = [h_1 \cdots h_i \cdots h_n]$, which represents the output value of electric heating load of this layer. The input of attention mechanism was the output matrix H of LSTM layer, and the feature vectors V were obtained by different attention weights.

5. Case Study

5.1. Date Preprocessing

The data used in this paper are the historical data of 66 days of electric heating load from January to the first week of March in 2018 in an area of eastern Inner Mongolia. At the same time, the thermal comfort of 300 individuals of different ages was investigated, and the model parameters were fitted by Equation (2), and the thermal comfort temperature of the main population was obtained. Among the 300 individuals, there were 150 men and 150 women, mainly young people aged about 20 years old and middle-aged and old people aged about 60 or 70 years old.

The thermal comfort questionnaire survey was conducted on the subjects, and the temperature and relative humidity during the survey were investigated. The model parameters of the same user under different clothing and activity intensity were obtained by fitting (see Table 2).

It can be seen from Table 2 that users had different adaptability to temperature under different clothes and different activity intensities. In order to make the model more universal, we took the average value of 23.275 °C as the thermal comfort temperature of the human body.

Table 2. Thermal comfort model parameters of users.

Clothing Fever/clo	Activity Intensity/met	a	b	c	Thermal Comfort Temperature/°C
0.5	Weak 0.6	0.263	0.456	6.576	26.5
	Strong 1.2	0.267	0.378	6.243	23.7
1	Weak 0.6	0.145	−0.127	2.823	22.6
	Strong 1.2	0.114	−0.135	2.211	20.3

5.2. Parameter Setting and Analysis

The input data were divided into training set and test set. The first 90% of the input samples were taken as the training set for the data samples of model fitting; the last 10% of the input sample was taken as the test set to evaluate the accuracy of the final model, that is, the training prediction of the prediction day. We set the initial learning rate as 0.05, learning decay rate as 0.6, and data training cycle as 250. In addition, the dropout layer discard rate was set to 0.25.

The number of hidden layers of the LSTM network and the number of LSTM units in each hidden layer had an impact on the accuracy of electric heating load forecasting. Under-learning or over-learning will affect the accuracy of the model. The enumeration method was used to record the training effect of different hidden layers and different number of neurons in each layer, so as to determine the optimal network structure. Firstly, the number of hidden layers was set to 1, and different numbers of neurons were set one by one to train and record MAPE; then, we kept the optimal number of neurons in the first layer, set the number of hidden layers to 2, continued to set the number of different units one by one for training, and so on. In this paper, the maximum number of hidden layers was set to 3, and the performance of each training is shown in Table 3.

Table 3. Forecasting performance of different LSTM network structures.

Number of Hidden Units	1 Hidden Layer e_{MAPE}/%	2 Hidden Layers e_{MAPE}/%	3 Hidden Layers e_{MAPE}/%
5	4.2486	8.5961	7.3803
10	8.5121	7.607	5.0794
15	6.7676	6.4442	10.6352
20	4.4361	5.0683	9.1701
25	7.5099	7.4918	9.3492
30	6.7286	9.7442	9.7862
35	8.7444	5.6755	6.7193
40	10.7386	10.2154	8.7017

According to the results in Table 3, when the number of hidden layers was 1 and the number of neurons in each layer was 5, the minimum e_{MAPE} was 4.2486%; when the number of hidden layers was 2, the number of neurons in the first layer was fixed to 5, and the number of neurons in the second layer was set to 20, and the minimum e_{MAPE} was 5.0683%. When the hidden layer was 3, the first two layers were fixed with the optimal number. When the number of neurons in the third layer was 10, the minimum e_{MAPE} was 5.0794%.

5.3. Test Results and Analysis

In order to verify the performance of the thermal comfort model and the improved LSTM neural network method proposed in this paper, we selected the optimal prediction model (one hidden layer, five neurons per layer). In addition, the hourly load from January to early March 2018 was used as the dataset to test the prediction performance of the model, which was compared with the other three cases.

Figure 6 shows the mean absolute percentage error (MAPE) of the prediction results of the proposed method. Figure 7 shows the comparison curve between the actual electric heating load and the load predicted by each method. The curve LSTM-T-A represents the prediction result of the LSTM model with thermal comfort temperature and attention mechanism added, the curve LSTM-T represents the prediction result with thermal comfort temperature added only, and the curve LSTM-A represents the prediction result with attention mechanism added only. The curve LSTM represents the LSTM prediction results without thermal comfort temperature and attention mechanism. It can be seen from Figure 7 that compared with the other three methods, LSTM-T-A had little change in amplitude compared with the real value, and the curve characteristics were closest to the real value.

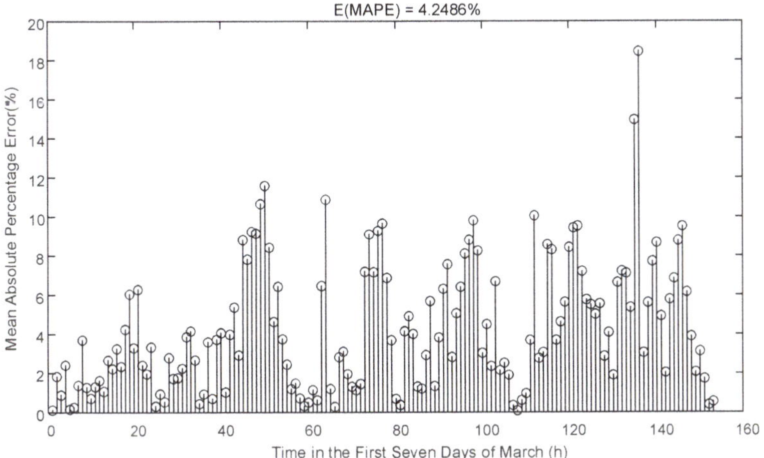

Figure 6. MAPE of the LSTM-T-A prediction results.

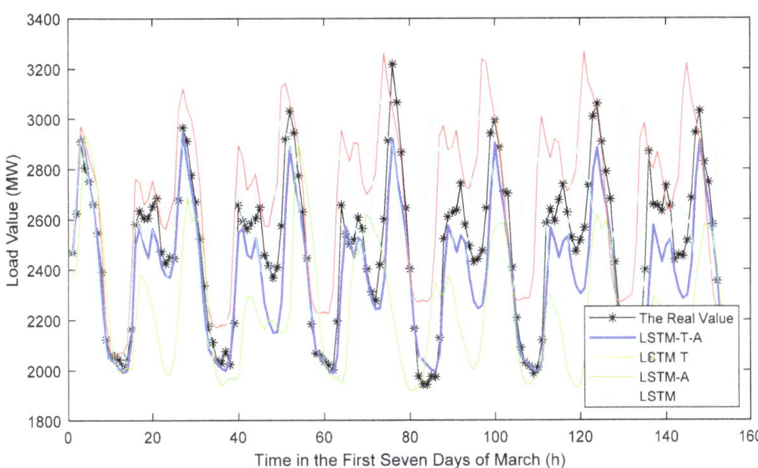

Figure 7. Forecast results of electric heating load.

The MAPE (Mean Absolute Percentage Error), MAE (Mean Absolute Error), and RMSE (Root Mean Square Error) of the above four models are shown in Table 4. In addition to comparing the improved part of LSTM, the errors of SVM and ANN are also compared.

Table 4. Prediction performance comparison of different neural network algorithms.

Models	e_{MAPE}/%	e_{MAE}/MW	e_{RMSE}/MW
LSTM-T-A	4.2486	109.3525	141.2577
LSTM-T	9.5517	228.4801	297.6025
LSTM-A	11.3527	293.5961	358.2558
LSTM	12.7182	311.0858	399.6952
SVM	13.6543	346.7190	424.6283
ANN	14.7216	384.1764	457.4381

It can be seen from Figure 7 and Table 4 that for the LSTM model, the improvement after adding human thermal comfort temperature and attention mechanism will significantly improve the prediction accuracy of electric heating load. LSTM-T-A prediction curve fitted the real value best, and the selected error index values were the smallest, which showed a better prediction effect.

6. Conclusions

According to the load of electric heating in northern China, we analyzed the load characteristics of electric heating in winter and constructed the thermal comfort temperature model of the human body. The main meteorological factors affecting electric heating load were screened out by the gray correlation analysis method. Meanwhile, the difference between thermal comfort temperature and actual temperature of main users was analyzed and considered. Attention mechanism and dropout layer were added to improve the LSTM neural network, and the optimal number of hidden layers and hidden neurons were obtained.

The actual electric heating load data were used to verify the model and were compared with several models. The results show that:

1. Comprehensive historical data showed that the shape of the typical daily load curve of electric heating load fluctuated greatly, and the peak valley difference was large. Moreover, the electric heating load had a strong time correlation, which was closely related to temperature, relative humidity, and thermal comfort temperature.
2. It is necessary to find the optimal number of hidden layers and neurons in order to mine more data information and improve the prediction accuracy of improved LSTM network.
3. As far as the improvement of LSTM prediction method is concerned, considering human thermal comfort temperature and attention mechanism accuracy, the training effect is the best. When considering the difference between thermal comfort temperature and air temperature in the model input, we found that the conclusion was more accurate and performed better than SVM, ANN, and other algorithms, and thus it is a more suitable electric heating load forecasting method.

Author Contributions: Methodology, J.S.; investigation and data curation, J.W.; resources, Y.S. (Yonghui Sun); project administration, M.X.; supervision, Y.S. (Yong Shi); writing—original draft preparation, X.W.; writing—review and editing, Z.L. All authors have read and agreed to the published version of the manuscript.

Funding: This work was supported in part by State Grid Inner Mongolia East Electric Power Co., Ltd.; the project number is KH20010229.

Informed Consent Statement: Informed consent was obtained from all subjects involved in the study.

Data Availability Statement: Data are contained within the article. Data sharing is not applicable to this article.

Conflicts of Interest: The authors declare no conflict of interest.

References

1. China Development and Reform Energy No. 2100. Winter Clean Heating Planning in Northern China (2017–2021). 2017. Available online: http://www.gov.cn/xinwen/2017-12/20/5248855/files/7ed7d7cda8984ae39a4e9620a4660c7f.pdf (accessed on 25 February 2021).
2. Fan, S.; Jia, K.; Guo, B.; Jiang, L.; Wang, Z.; He, G. Collaborative Optimal Operation Strategy for Decentralized Electric Heating Loads. *Autom. Electr. Power Syst.* **2017**, *41*, 20–29.
3. Wei, X.; Wang, X. Prediction of Urban Electricity Load Based on Meteorological Big Data. *Electr. Meas. Instrum.* **2021**, *58*, 90–95.
4. Shi, J.; Zhang, J. Load Forecasting Based on Multi-model by Stacking Ensemble Learning. *Proc. CSEE* **2019**, *39*, 4032–4041.
5. Guo, Z.; Zhang, Z.; Zhou, X.; Hu, S.; Ma, G.; He, C. Study on Forecasting Method of Electric Heating Load Considering Meteorological Factors. *Electr. Meas. Instrum.* **2021**. Available online: http://kns.cnki.net/kcms/detail/23.1202.TH.20200818.1113.010.html (accessed on 30 March 2021).
6. Huang, Y.; Zhu, Y.; Mu, G.; Ding, D.; Cui, Y. Evaluation of Adjustable Capacity of Household Electrical Heating Load Based on Temperature Forecast. *Power Syst. Technol.* **2018**, *42*, 2487–2493.
7. Yu, Y.; Quan, L.; Jia, Y.; Mi, Z. Improved model predictive control of aggregated thermostatically controlled load for power fluctuation suppression of new energy. *Electr. Power Autom. Equip.* **2021**, *41*, 92–99.
8. Wang, X. Forecast and Application of Schedulable Capacity of TCLs Based on Big Data Analysis Method. Master's Thesis, Hefei University of Technology, Hefei, China, 2018.
9. Liu, Q. Research on Control Strategy of Thermostatically Controlled Loads in Microgrid. Master's Thesis, Beijing Jiaotong University, Beijing, China, 2018.
10. Kevin, P.S.; Sortomme, E.; Venkata, S. Evaluating the Magnitude and Duration of Cold Load Pick-up on Residential Distribution Feeders Using Multi-State Load Models. *IEEE Trans. Power Syst.* **2016**, *31*, 3765–3774.
11. Saeid, B.; Hosam, K. Modeling and Control of Aggregate Air Conditioning Loads for Robust Renewable Power Management. *IEEE Trans. Control Syst. Technol.* **2013**, *21*, 1318–1327.
12. Xiao, B.; Zhou, C.; Mu, G. Review and Prospect of the Spatial Load Forecasting Methods. *Proc. CSEE* **2013**, *33*, 78–92.
13. Wang, L.; Wang, Y.; Wei, Z. Short Term Load Forecasting Based on Regression Analysis and Neural Network. *Electrotech. Appl.* **2007**, *26*, 37–39.
14. Lu, N.; Wu, B.; Liu, Y. Application of Support Vector Machine Model in Load Forecasting Based on Adaptive Particle Swarm Optimization. *Power Syst. Prot. Control* **2011**, *39*, 43–46.
15. Jiao, R.; Su, C.; Lin, B. Short term load forecasting by grey model with weather factor based correction. *Power Syst. Technol.* **2013**, *37*, 721–724.
16. Xu, Y.; Fang, L.; Zhao, D. Electricity Consumption Prediction Based on LSTM Beural Betworks. *Power Syst. Big Data* **2017**, *20*, 25–27.
17. Li, X.; Ma, L.; Zhao, X.; Zhu, J.; Xu, Z. Research on Electric Heating Load Forecasting Method Based on Multi-Scale Time and Long-term Memory Network. *Proc. CSU-EPSA* **2020**. Available online: http://kns.cnki.net/kcms/detail/12.1251.TM.20200526.0944.001.html (accessed on 13 April 2021).
18. Peng, W.; Chen, L.; Lei, J.; Hu, X. Research on Human Thermal Comfort Prediction Model for Machining Workshop in Natural Ventilation. *Eng. J. Wuhan Univ.* **2020**, *53*, 261–267.
19. ISO. *ISO 7730. Ergonomics of the Thermal Environment-Analytical Determination and Interpretation of Thermal Comfort Using Calculation of the PMV and PPD Indices and Local Thermal Comfort Criteria*; ISO: Geneva, Switzerland, 2005.
20. Zhu, Y. *Building Environment Science*; China Construction Industry Press: Beijing, China, 2005.
21. Fanger, P. *Thermal Comfort*; McGraw-Hill Company: New York, NY, USA, 1970.
22. Buratti, C.; Ricciardi, P.; Vergoni, M. HVAC systems testing and check: A Simplified Model to Predict Thermal Comfort Conditions in Moderate Environments. *Appl. Energy* **2013**, *104*, 117–127. [CrossRef]
23. Zhao, P.; Dai, Y. Power Load Forecasting of SVM Based on Real-time Price and Weighted Grey Relational Projection Algorithm. *Power Syst. Technol.* **2020**, *44*, 1325–1332.
24. Zhao, B.; Wang, Z.; Ji, W.; Gao, X.; Li, X. A Short-term Power Load Forecasting Method Based on Attention Mechanism of CNN-GRU. *Power Syst. Technol.* **2019**, *43*, 4370–4376.

 *energies*

Review

Smart Distribution Network Situation Awareness for High-Quality Operation and Maintenance: A Brief Review

Leijiao Ge [1,*], Yuanliang Li [1], Yuanliang Li [2], Jun Yan [2] and Yonghui Sun [3]

[1]   School of Electrical and Information Engineering, Tianjin University, Tianjin 300072, China; tjlyliang@foxmail.com
[2]   Concordia Institute for Information Systems Engineering, Concordia University, Montreal, QC H3G 1M8, Canada; yuanliang.li@concordia.ca (Y.L.); jun.yan@concordia.ca (J.Y.)
[3]   College of Energy and Electrical Engineering, Hohai University, Nanjing 210098, China; sunyonghui168@163.com
*   Correspondence. legendglj99@tju.edu.cn; Tel.: +86 013820176750

Abstract: In order to meet the requirements of high-tech enterprises for high power quality, high-quality operation and maintenance (O&M) in smart distribution networks (SDN) is becoming increasingly important. As a significant element in enhancing the high-quality O&M of SDN, situation awareness (SA) began to excite the significant interest of scholars and managers, especially after the integration of intermittent renewable energy into SDN. Specific to high-quality O&M, the paper decomposes SA into three stages: detection, comprehension, and projection. In this paper, the state-of-the-art knowledge of SND SA is discussed, a review of critical technologies is presented, and a five-layer visualization framework of the SDN SA is constructed. SA detection aims to improve the SDN observability, SA comprehension is associated with the SDN operating status, and SA projection pertains to the analysis of the future SDN situation. The paper can provide researchers and utility engineers with insights into the technical achievements, barriers, and future research directions of SDN SA.

Keywords: smart distribution network; situation awareness; high-quality operation and maintenance; critical technology; comprehensive framework

Citation: Ge, L.; Li, Y.; Li, Y.; Yan, J.; Sun, Y. Smart Distribution Network Situation Awareness for High-Quality Operation and Maintenance: A Brief Review. *Energies* **2022**, *15*, 828. https://doi.org/10.3390/en15030828

Academic Editor: Alberto Geri

Received: 30 December 2021
Accepted: 20 January 2022
Published: 24 January 2022

Publisher's Note: MDPI stays neutral with regard to jurisdictional claims in published maps and institutional affiliations.

1. Introduction

1.1. Motivation

Due to the rapid development of emerging information and communication technologies (ICT) and advanced metering infrastructure (AMI), distribution networks are in an evolvement from passive to active distribution networks (ADN), also called smart distribution networks (SDN) [1]. In addition, with the rapidly increasing penetration of distributed generations (DGs) inspired by the smart grid (SG) concept [2], the SDN integrates multiple renewable energy sources (RES) and focuses on reliable operation [3]. To achieve the environmental objective for gas emission reduction and accommodate the high penetration of DGs, supervisory control and data acquisition (SCADA) systems are employed to monitor the SDN, and distribution management systems (DMS) and energy management systems (EMS) act as decision-support information systems for the coordination of remote SDN equipment. Additionally, the widespread application of devices such as distribution transformer terminal unit (TTU), feeder terminal unit (FTU), remote terminal unit (RTU), and distribution automation terminal (DTU) contributes to the maturity of SDN [4,5].

Operation and maintenance (O&M) cost is an economic factor that the SDN management must consider. Mansor et al. [6] presented operational planning of SDN based on utility planning concepts, considering the cost minimization of O&M, switching, losses, and reliability. Based on the volatilities of wind speed and demand load, ref. [7] presented advanced real-time dispatching strategies to minimize long-run expected cost instead of immediate myopic cost. In addition, the quality of O&M technology directly affects the operating status of SDN. To prevent persistent faults in distribution transformers (DTs), Al Mhdawi et al. [8] proposed a remote condition internet of things (IoT) monitoring and fault prediction system using customized software-defined networking technology. In [9], a multi-status simulation based on event-driven for the SDN O&M was investigated, which can simulate the specific events in the SDN with different time constants within the same simulation framework. To improve the reliability of SDN O&M, Kiaei et al. [10] proposed a hybrid fault location for SDN using available multi-source data, which can precisely calculate the fault location in distribution networks with many sub-laterals. The O&M level of multi-terminal SDN directly connected to each user determines the power quality of end-users. Among multiple O&M technologies, situation awareness (SA) emerges and is gradually integrated into the SDN. Facing a high proportion of RES, adequate monitoring, analysis, and prediction of the SDN operating status are urgent. Therefore, comprehensive SA, which contains detection, comprehension, and projection, becomes a significant guarantee for the optimal operation of SDN [11]. Due to the strong adaptability, SA can dynamically evolve with the future SDN technology development to provide higher quality O&M of SDN.

The concept of SA means to percept elements in the environment within a volume of time and space, comprehend their meaning, and project their future status [12]. In general, the process of SA can be divided into three stages: situation detection, situation comprehension, and situation projection [13]. To visualize the concept of SA, SA can be analogous to human psychology. In psychology, the sensory, perception, and behavioral habits can be expressed as follows:

1. The sensation is the brain's reflection of various attributes in objective things that directly act on the human sensory organs [14]. Human cognition of objective things starts with sensation. It is the initial detection of complex things and the basis of complex cognitive activities such as perception and behavior. That is similar to the concept of situation detection.
2. Based on sensory information, perception processes multiple sensory information in a specific way, interprets the sensory information on individual experience, and taps the deep meaning of sensory information. That is similar to the concept of situation comprehension.
3. Based on sensory and perception, behavior refers to human activities after receiving internal and external stimuli. The theory of planned behavior [15] can explain human decision-making behaviors from the perspective of perceptual information processing and predict the future behavioral tendency based on the expectation value theory [16]. That is similar to the concept of situation projection.

Therefore, the human collects multiple sensory information and relies on perception to process the sensory information. The following behaviors can be explained and predicted by the theory of planned behavior [17]. The human situation refers to the comprehensive integration of mental activity, physiological state, and environmental information. Similarly, the basic principle of the SA corresponds to the above psychological terms, which represents detecting, comprehending, and projecting various elements with specific spatial–temporal properties [18]. In general, three SA stages can be defined as follows:

1. Situation detection. The task of the stage is to detect essential features in the environment. Multi-dimensional data can be collected and completed in this stage. In addition, situation detection is the data basis of situation comprehension and projection.
2. Situation comprehension. The essence of the stage is to understand the environment through data analysis. Specifically, the data obtained in the situation detection are integrated, and the connection and potential information between multi-source data are explored.
3. Situation projection. The core of situation projection is to achieve the practical application of SA knowledge. Based on the information gained from situation detection and comprehension, this stage can predict the future environmental situation in time.

1.2. Related Work

Although it initially appeared as a tool in the military [19], SA has been researched across a wide range of domains for individual and team activities. For example, [20] presented the distributed swarm SA of unmanned aerial vehicles based on the representative SA model. A convolutional neural network (NN) has been proposed for road traffic SA in [21]. For telecommunication, network SA becomes a security priority to perceive the network threat globally [22]. For robotics, Anjaria et al. [23] investigated the relationship between the SA theory and cybernetics and adopted this relationship to validate the feasibility of implementing SA-based information security risk management (ISRM) in organizational scenarios. SA has also been identified as a critical skill in maintaining safety in high-risk industries. For example, the influence of some variables on safety performance was investigated, and the mediating effects of SA were examined in [24]. In agriculture, Irwin et al. [25] explored SA among farmers in the United Kingdom when operating heavy agricultural machinery. In navigation, considering existing models of SA and ontology-based approach for maritime SA, seaborne SA was applied to navigation safety control in [26]. For healthcare, SA has been recognized as a critical technology for making effective and quick decisions for emergency response [27].

For the SDN, the situation represents the operating status of the SDN's equipment, structure, status, security, and environment. SDN SA is also composed of the same three SA stages. In the situation detection stage, the information related to critical elements of the SDN is captured and completed. In the situation comprehension stage, the operating status of SDN and the potential information of the perceived data are analyzed. In the situation projection stage, the future behavior of SDN components based on their operating status and the perceived information is predicted [28]. Compared with the past, the architecture of SDN has undergone tremendous changes. The traditional distribution network is passive where the operation, control, and management modes are all determined by the power of the transmission network. In the developing SDN, AC/DC hybrid [29], multi-energy complementarity [30], energy internet [31], and other distribution network forms emerge. In addition, the higher proportion of RES and the disorderly access of DGs also lead to a significant increase in the SDN uncertainty. For example, the outputs of wind turbines and photovoltaic generators are greatly affected by meteorology rather than produced entirely based on the plan. These changes make SDN have more complex operating conditions and fault types. Moreover, there is a variety of system states that should be monitored for SA detection, which cannot be fully covered by remote measurement devices. Meanwhile, with the increase of regional electrical loads, power electronic devices become diverse, and the requirements for power quality increase. Therefore, it is urgent to explore the SDN SA from the perspective of high-quality O&M.

In the modern SDN, it is challenging to operate SA efficiently as the SDN has diversified characteristics in network topology, equipment types, energy types, and system configurations. Many studies have been trying to tackle the challenge from different aspects. For example, a security SA of the SDN was conducted by the random deletion of network nodes to simulate the network attack, which can meet the requirements of energy internet and is highly compatible with the RES [32]. Facing the power uncertainty brought by a high proportion of RES, a hybrid factor analysis (FA), gray wolf optimization (GWO), and generalized regression neural network (GRNN) was proposed for short-term load forecasting [33]. Due to the lack of definitions of a generic indicator framework that can uniformly characterize the critical operating states of the SDN, limited work has been done to evaluate the effectiveness of the SDN SA. To quantify the SDN SA performance effectively, ref. [18] proposed an improved interval-based analytic hierarchy process-based subjective weighting and a multi-objective programming-based objective weighting. To transfer more knowledge of the real-time SDN situation to the control center operator, [34] proposed two design strategies for SDN SA in real-time distribution operations. One strategy is for the preparation of standardized data acquisition networks. The other is a real-time security analysis for SDN. Diez et al. [35] presented a graphical user interface for a power grid based on SA-oriented design principles, where the control room operators can achieve an appropriate SA level.

1.3. Contributions

Although SA has become a significant element in enhancing the O&M of SDN, there are very few studies about SDN SA. For the early stage of SDN SA, ref. [18] presented a candidate SA framework for SDN, consisting of situation perception, situation comprehension, situation projection, and communication networks over the physical SDN elements. It is a pity that the background and functions of the critical technologies have not been explained in detail. To this end, this paper constructs a five-layer comprehensive framework to introduce the critical technologies of SDN SA, which can be regarded as a solid base for high-quality O&M in SDN. To the best of our knowledge, only [13] initially explored critical technologies of situation perception, comprehension, and projection prospect from the perspective of system access. However, its preliminary exploration of SA for SDN is merely a vision. Modern SDN technology is constantly updated, and high-quality O&M has become the core demand. Traditional SA cannot meet the O&M requirements of the existing SDN. To this end, this paper provides a more detailed and appropriate description of SDN SA from the perspective of O&M. The critical technologies of different SA stages are selected based on their significance to O&M, their relevance to SA, and their practicality to SDN. Based on the presented technical framework of SDN SA, distribution network researchers and utility engineers can be provided with insights into the technical achievements, barriers, and future research directions of SDN SA.

The purpose of this paper is to provide an updated picture of the SDN SA and contribute to the high-quality O&M of SDN. In order to promote the development of SA technology in the power distribution field, the research background and concept of SDN SA are clearly explained in Section 1. The challenges and objectives of future SDN SA are analyzed, which indicate the exploration directions of SDN SA. In addition, a five-layer comprehensive framework is presented to help the researchers understand the SDN SA in Section 3. Specifically, this paper constructs a virtuous circle between SDN and SA to improve the O&M quality of SDN, where SA transmits the SDN situation information to the management team, who formulated measures to guide the SDN to a better situation. To adapt to the evolving SDN, the critical technologies of SA are updated and analyzed based on the O&M requirements. Ultimately, we believe this paper can provide positive guidance for the future research and application of SDN SA.

1.4. Organization

The present paper is structured as follows: an overview of the objectives and challenges of SDN SA is discussed in Section 2. A five-layer comprehensive framework of SDN SA is conducted in Section 3. From the O&M perspective, the analysis of the critical technologies for situation detection, situation comprehension, and situation projection is proposed in Sections 4–6, respectively. Finally, the paper is concluded and prospected in Section 7. The brief review aims to address the challenges faced in the deployment of SDN SA and provide helpful information and guidance in selecting suitable technologies for specific SDN applications.

2. Description of Situation Awareness for Smart Distribution Networks

2.1. Objectives of Situation Awareness for Smart Distribution Networks

1. The primary goal is to achieve real-time or quasi-real-time SA for SDN, which can accurately obtain the critical information of SDN, quickly determine the operating status of the distribution networks, and predict the development trend of SDN at the same time [11]. Based on the historical records of SDN data, SA provides a comprehensive SDN situation to ensure high-quality O&M.

2. Observability is a significant technical indicator of SA. High-level SA can provide SDN with a highly visual situation and solve the shortcomings of insufficient measurement devices in the SDN [35].

3. SA has a significant contribution to SDN reliability. Specifically, conduct the SDN self-healing technology, detect potential SDN risks, and predict security situations in advance. Finally, a scientific basis for the SDN active defense can be provided [13].

4. Through continuous innovation of intelligent algorithms, SA is cultivating SDN self-adaptive capabilities [36]. Based on the information obtained by SA, SDN can independently recognize and improve the situation in an informed way.

2.2. Challenges of Situation Awareness for Smart Distribution Networks

Due to SDN's diverse scenarios with more equipment and complex operating status, traditional SA cannot adapt to the modern SDN environment. The O&M challenges for modern SA are as follows:

1. Situational detection challenges. New measurement technologies such as AMI [37] and phasor measurement units (PMUs) [38] are gradually deployed in SDN. Therefore, the data dimensions collected by SDN scale rapidly, which inevitably increases the computational pressure of SA. Due to the insufficient measurement devices, the collected data are challenging to recognize the poor operating status of the SDN. Therefore, the input data of the SA system are asymmetric, and some missing data are necessary to be accurately completed by calculation. How to comprehensively detect SDN status remains a challenging point in high-quality O&M.

2. Situational comprehension challenges. Large-scale DGs lead the traditional dispatch mode to unsuitable. As a result, the phenomenon of reverse power transmission at the distribution network terminals is prominent, and the risk of voltage fluctuations and power loss increases [39]. In addition, different SDN topologies, operation modes, energy types, and automation levels have higher requirements for the compatibility of situational comprehension in different regions. Traditional situation comprehension technology is challenging to adapt to the current SDN. As the decision center of SDN, situation comprehension should assist the high-quality O&M of multi-form SDN. How to accurately understand the operating situation of the SDN is the focus of research.

3. Situation projection challenges. Unlike passive distribution networks, SDN has a higher proportion of DGs and electric vehicles (EVs) and more diverse operating modes [40]. The uncertain outputs of DGs and EVs lead to an imbalance between power supply and consumption. Although the SDN flexibility is improved, the RES outputs, three-phase unbalanced load, EV charging, inspection schedule, and stability margin are challenging to determine in the situation projection. Additionally, situation projection for complex scenarios requires sufficient mathematical analysis, computational capability, and robustness capability. How to effectively predict the operational trend of SDN needs to be solved urgently.

3. Comprehensive Framework of Situation Awareness

A five-layer comprehensive framework of SDN SA is shown in Figure 1, which includes distribution network equipment, communication network, situation detection, situation comprehension, and situation projection. In addition, SCADA systems [41], 5G communications [42], distribution automation systems [43], distribution network equipment [44], SA systems, and communication networks [45] are integrated into Figure 1. First, the distribution network equipment at the bottom layer transmits measurement information, equipment status, and network topology to the communication network at the second layer. Then, the communication network summarizes the SDN data and transmits it to situation detection at the third layer. After situation detection collects the data, it completes the pre-processing, completion, and visualization of multi-source data through various critical technologies. Meanwhile, the processed information is transmitted to the management team and situation comprehension at the fourth layer. Situation comprehension combines various critical technologies to explore the detected data, analyze the operating status of SDN, and provide information support for the high-quality O&M. An intelligent O&M mode can be realized based on the operating status of SDN. In addition, SDN historical data is transmitted to the situation projection at the top layer. Next, the situation projection combines meteorological, economic, social, resource, and other factors to predict the developing situation of SDN. After experiencing the forward cycle, the predicted information is fed back to the situation comprehension at the fourth layer. Next, situation comprehension can summarize and analyze all the information and then transmit a more comprehensive SDN situation to the management team. As a result, the management team and the SA system can coordinate to operate an optimal SDN based on the exact situation. A virtuous circle of SA is constructed for the high-quality O&M of SDN.

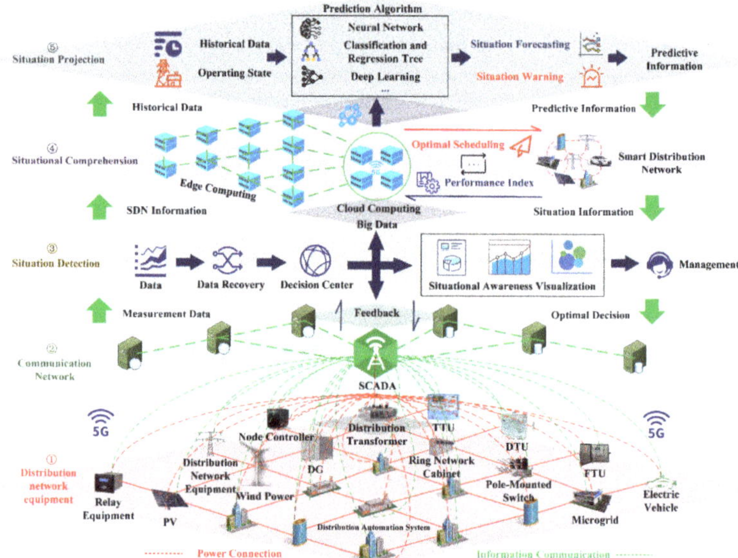

Figure 1. A 5-layer comprehensive framework of SDN SA.

4. Critical Technologies of Situation Detection

Situation detection includes data acquisition, processing, completion, and visualization, which is the prerequisite of situation comprehension and projection [11]. To improve the SDN visibility, the comprehensive perception of the SDN is realized in both breadth and depth, whose implementation framework is shown in Figure 2. First, multi-source SDN data are collected by smart meters, terminal equipment, PMU, TTU, FTU, DTU, and other equipment. Then, the data are preliminarily processed through pre-processing technologies such as data storage, data fusion, and data cleaning. Next, the critical technologies of situation detection are used in data completion and data presentation to improve the observability of SDN, including big data analytics, 5G communication, virtual acquisition, and optimal configuration of measurement. Finally, the completed data are sent to the situation comprehension and projection. To our knowledge, the four critical technologies can synergistically contribute to situation detection effects.

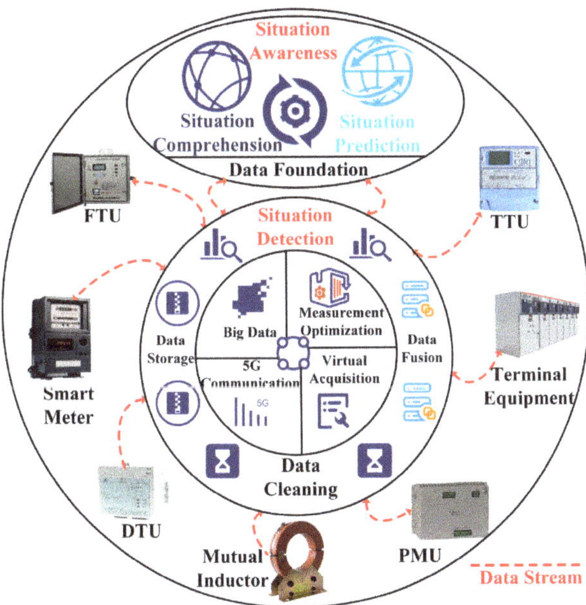

Figure 2. The implementation framework of situation detection.

When facing the core O&M goals, enough collected data are significant for situation comprehension to analyze the operating status of SDN. To deal with the uncertainties, it is necessary to have enough data for situation projection to predict the future SDN situation. Otherwise, inaccurate or incomplete SDN data might mislead O&M in a worse direction. Thereby, data construction is the foundation of SDN O&M. With the rapid development of the SDN construction, the power data stored in the SDN enterprise database show explosive growth with the O&M [46]. These data are usually stored in the form of unstructured data, such as images and text, which contain vital information about the operating status of SDN equipment. Through SDN situation detection technology, the O&M data can be collected, mined, and completed, where the data abundance can provide the possibility for high-quality SDN O&M.

4.1. Big Data Analytics

In the data-intensive era of SDN, SA data are large-scale, multi-source, changeable, and heterogeneous. Recently, studies have been looking into SDN situation detection, and big data analytics technology has gradually been applied to SA. Most of the existing methods employ different ways to store different data types, which leads to the inefficiency of data queries and analyses. To this end, Tao et al. proposed a graph database-based hierarchical multi-domain SA data storage to store the situation information, combining multi-dimensional data to improve the SDN visibility after data pre-processing [47]. An innovative data-fusion method was proposed in [48] to detect incipient faults by integrating data collected from multiple sources instead of a single data source. Using the status information of SDN equipment, a defect texts mining model for a secondary device in a smart substation was proposed in [49] to achieve the accurate classification of secondary device defect texts. In addition, power equipment data mining is a rapidly growing area of big data analytics, contributing to more O&M data. As a use case, the H-mine algorithm was adopted in [50] to quickly mine fault data of the secondary system of smart substations.

4.2. 5G Communication

Communication technology is the core factor that affects SDN observability. Wireless communication systems were preferred over wired for various reasons and various applications with reliable costs at lower speeds [51], which expands the infrastructure and provides easily accessible connections even in remote areas. Due to the characteristics of low power consumption, low cost, high capacity, low latency, high bandwidth, and multiple functions, flexible 5G communication technology has begun to be invested in the SCADA system [52]. Basnet et al. [53] simulated the false data injection (FDI) attack and the syn flood distributed denial of services (DDoS) attacks on 5G-enabled remote SCADA systems, which can detect the stealthy cyber-attacks that bypass the cyber layer and go unnoticed in the monitoring system with more than 99.9999% detection accuracy for both training and validation data.

The IoT enables all energy consumption and production components to be connected, improves O&M visibility, and provides real leverage at every stage of energy flow from use to supply and end-user [54]. Due to 5G's higher data transmission speed and lower transmission delay than the existing 4G networks, 5G would ensure the convergence of widespread broadband, perception, and intelligence and then promote the development of IoT. A comprehensive review of the role of 5G cellular networks in the growth of IoT technology was presented in [51]. For example, the implementation of IoT based on the smart inverters can be achieved such as a solar-charged inverter that employs WiFi technology to engage in two-way communication with the user, informing the user of both the battery voltage of the inverter and run time of the loads which the user chooses to run. The deployment of advanced wireless networks in SDN would allow faster data transmission and processing [55]. 5G communication technology might become the future road of sustainable energy systems paving to state-of-the-art technologies and networks. In [55], 5G was employed to optimize demand-side response management in integrated energy systems. Combining the 5G and measurement equipment, such as PMU and AMI, can enhance the distribution network O&M [56]. Moreover, 5G-based SA provides the possibility of precise load control at the millisecond level [57]. The energy consumption reduction of 5G networks in SDN will become a vital research direction.

4.3. Virtual Acquisition

To improve the completeness of O&M data, SDN virtual acquisition technology is becoming a research hotspot. The technology is independent of the full coverage of the SDN measurement equipment installed, such as sensors, collectors, and concentrators. For areas that cannot be equipped with monitoring systems to collect real-time data, the virtual acquisition uses machine learning techniques based on data from similar areas to generate data for the objective areas [58]. Similar areas and dates can be selected based on data clustering results. By mining the inherent mapping relationship between the objective distribution network and similar areas, the anonymous data can be supplemented by historical data in similar areas and existing real-time data. The data supplement method can be based on machine learning such as NN [59]. Currently, the virtual acquisition technology remains in its infancy. The authors of [58] presented a virtual acquisition of distributed PV data based on the combination of bat algorithm and wavelet NN, which realizes the acquisition of O&M data of nine distributed PV stations when only one station is equipped with complete measurement equipment. In addition, a virtual acquisition based on a mixture of grey relational degree and back-propagation NN was proposed in [60], which can accurately acquire unknown O&M data of distributed PV without complete measurement equipment. In the future, virtual acquisition technology is worthy of research.

4.4. Optimal Configuration of Measurement

SDN SA strongly relies on various digital measurement devices and well-designed monitoring systems. The AMI is a typically configured infrastructure that integrates many technologies to achieve its objective, including meter data management systems, communication networks in different levels of the infrastructure hierarchy, smart meters, and ways to integrate the acquired data into software application platforms and interfaces [61]. To ensure data observability, the AMI adopts measurement equipment configuration optimization, PMU configuration optimization, and data analysis technology. Dua et al. [62] proposed a novel method to detect the configuration of the distribution network by collecting and processing real-time measurement from the optimally placed micro-phasor measurement unit. The authors of [63] presented a data-driven method based on the measurements of micro-phasor measurement units to deal with the optimal hourly configuration of the distribution network in a real-time manner. PV intelligent edge terminal (IET) is one of the notable devices to achieve high-quality O&M with a high proportion of distributed PV [64]. A mathematical model and improved coyote optimization were proposed in [64] to optimize the configuration of PV IETs, which acquires the optimal number, location, and connection way of PV IETs.

5. Critical Technologies of Situation Comprehension

Situation comprehension is the data analysis stage, which explores the potential information of the data collected in the situation detection. Many key operational performance indicators need to be correctly evaluated in SDN, such as reliability [65], flexibility [66], stability [67], and power quality [68], which are integrated into the analysis of the SDN situation. As the foundation of high-quality O&M, the implementation framework of situation comprehension is shown in Figure 3. First, SDN data are collected and completed by situation detection. Then, the data are transferred to the situation comprehension system to explore potential information. By conducting critical technologies of situation comprehension, many key operational performance indicators can be acquired and used as the data basis for O&M technologies. Then, the technologies contribute to high-quality O&M based on the situation comprehension results and return the calculation results to the situation guidance. Ultimately, the intelligent O&M combined with situation comprehension and management can be realized. The critical technologies of situation comprehension include uncertain power flow calculation, hybrid state estimation (HSE), reliability analysis, voltage stability analysis, flexibility evaluation, and power quality evaluation technology.

Energy equipment such as wind power, photovoltaic, DC electrolysis of water into hydrogen, hydrogen storage, AC ice storage, and water storage equipment has been increasingly connected to SDN. The introduction of various energy equipment increases the need for real-time scalable and reliable monitoring, control, and protection of SDN. Situation comprehension establishes the mathematical model compatible with multiple types of SDN terminal equipment, adopts the SDN information provided by situation detection to evaluate the SDN key operational indicators, and then realizes the flexible correction of SDN operating status. Based on the critical technologies of the situation comprehension above, the management team can take more specific measures to improve the quality of O&M. For example, the configuration optimization of DGs based on the results of situation comprehension can be applied to improve the economy of SDN O&M. Meanwhile, many uncertainties and power data in SDN can be determined through situational understanding to reduce the blindness of O&M decision making. In addition, self-learning evaluation technology can achieve dynamic evaluation and the weight balance of multiple indicators to effectively evaluate the key operational indicators of SDN [69]. To coordinate different DGs and energy storages, coordinated dispatch technology can be adopted to build an integrated energy system based on the results of situation comprehension and contribute to high-quality O&M [70]. In addition, the popularization of electric IoT gives SDN more powerful computing capabilities, which promotes the miniaturization and intellectualization of IoT terminals. As IoT has found its way to SDN, demand-side management can be more efficient in the presence of IoT [71]. Edge computing technology [72] enables flexible collaboration between smart terminals and improves the response speed of SDN O&M. In sum, situation comprehension can provide O&M with richer information through various technologies and help the management team make the optimal decision.

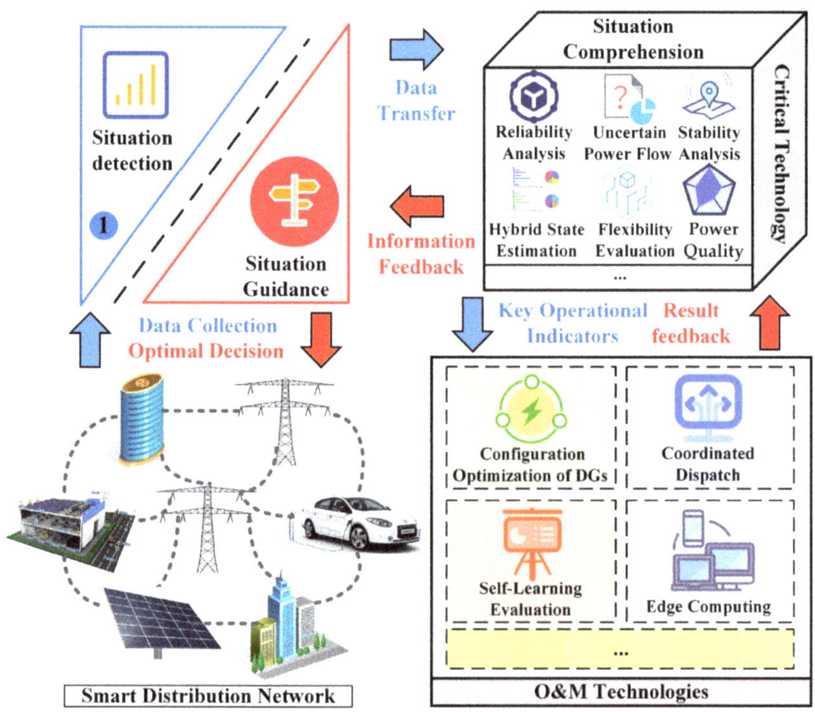

Figure 3. The implementation framework of situation comprehension.

5.1. Uncertain Power Flow Calculation

Uncertain power flow calculation (PFC) technology involves interval PFC [73], fuzzy PFC [74], and probabilistic PFC [75], which estimate the influence of uncertain factors on the SDN. Unlike deterministic PFC, only a single uncertain PFC can provide SDN with more information of power flow within a volume of time and space, reducing the number of repeated PFC caused by uncertain SDN parameter changes. The known and to-be-calculated quantities in deterministic PFC are considered as random variables. The SDN's uncertain PFC model can be established based on affine arithmetic, fuzzy numbers, or probability statistics theory. Liu et al. [76] presented an interval PFC method for multi-terminal DC distribution networks to deal with the uncertainties of DG output powers and loads. The power flow of DC distribution network in affine arithmetic is explained by the following equation:

$$P_k = -U_k \sum_{j=1}^{n} g_{kj} U_j \quad k = 1, 2, \ldots, n \tag{1}$$

where P_k is the nodal power of the kth load node in affine form, g_{kj} is the admittance of the positive line from the kth node to the jth node, U_k is the positive voltage of the kth node in affine form, U_j is the positive voltage of the jth node in affine form, and n is the total number of nodes. The interval PFC algorithm provides an essential tool for SDN SA to solve the uncertainties of loads and RES outputs.

Due to the uncertainties of the DGs and loads, Yang et al. [77] presented a random fuzzy PFC model, which adopts cumulant technology in the random stage and the fuzzy simulation in the fuzzy stage. The normal distribution can usually represent the load power, and their random fuzzy models are explained by the following equation:

$$f(P_{\text{Load}}) = \frac{1}{\sqrt{2\pi}\xi_{\sigma P}} \exp\left(-\frac{\left(P_{\text{Load}} - \xi_{\mu P}\right)^2}{2\xi_{\sigma P}{}^2}\right)$$
$$f(Q_{\text{Load}}) = \frac{1}{\sqrt{2\pi}\xi_{\sigma Q}} \exp\left(-\frac{\left(Q_{\text{Load}} - \xi_{\mu Q}\right)^2}{2\xi_{\sigma Q}^2}\right) \tag{2}$$

where P_{Load} and Q_{Load} are the active and reactive load powers, $\xi_{\mu P}$ and $\xi_{\mu Q}$ are the means of the active power and reactive power, and $\xi_{\sigma P}$ and $\xi_{\sigma Q}$ are standard deviations of the active power and reactive power.

Liu et al. [78] presented an improved dependent probabilistic sequence algorithm based on the traditional linear PFC to obtain the probability distribution information of power flow, which can achieve more accurate results and computational efficiency of probabilistic PFC. The following equation explains the nth node's voltage probability distribution:

$$P\{X_n = X_{n0} + i \cdot (\Delta S \cdot \Delta P)\} = \Delta X_n(i) \tag{3}$$

where ΔP is discrete step length of power, ΔS is discrete step length of sensitivity factor, X_n is n^{th} node's voltage, X_{n0} is reference state of nth node's voltage, $\Delta X_n(i)$ is a variety of nth node's voltage, and i is the number of a corresponding expansion sequence group. Simultaneously, the lth branch flow's probability distribution can be expressed by the following equation:

$$P\{Z_1 = Z_{10} + i \cdot (\Delta T \cdot \Delta P)\} = \Delta Z_1(i) \tag{4}$$

where ΔT is discrete step length of sensitivity factor, Z_1 is l^{th} branch's power flow, Z_{10} is reference state of lth branch's power flow, and $\Delta Z_1(i)$ is a variety of lth branch's power flow. Because of the low demand for the sample size, this method is suitable for SA to analyze the power flow uncertainties of SDN with incomplete measurement information.

5.2. Hybrid State Estimation

The current distribution network O&M data mainly come from the SCADA system. To improve the estimation accuracy in the distribution network, PMU with more comprehensive measurement information has gradually become popular in SDN [79]. A PMU delivers time-synchronized values of voltage and current phasors and other system-related quantities [80]. However, the current SDN remains in a state where many traditional and new measurement devices coexist. The main challenge in the HSE is the mismatch of the refresh rates between the SCADA and PMU measurements [81]. Therefore, there is an urgent need for PMU/SCADA HSE technology to improve the accuracy and breadth of SA.

A novel HSE method was presented in [82], which decouples the SCADA and PMU measurements to deal with different accuracy levels between them. The novel HSE model, based on weighted least-squares formulation including both SCADA and PMU measurements, is as:

$$\min_{x=(x_{\text{PMU}},x_{n-\text{PMU}})} J(x) - [z - h(x)]^T \cdot R^{-1} \cdot [z - h(x)]$$
$$\text{s.t. } c(x) = 0 : \lambda \tag{5}$$
$$x_{\text{PMU}} - x_{\text{st-PMU}} = 0 : \mu$$

where x is the vector of system states including voltage angles and magnitudes, λ is the Lagrange multiplier vector of the equality constraints of zero injection busses, $x_{\text{st-PMU}}$ is the PMU states estimated, μ is the Lagrange multiplier vector, x_{PMU} and $x_{n-\text{PMU}}$ indicate the PMU and non-PMU states, R is the covariance matrix, z is a vector consisting the system measurements, vector $h(x)$ includes nonlinear functions which relate the states with the measurements through power flow equations, $J(x)$ is the Jacobian matrix, and $c(x)$ is constraint condition. The condition number, as well as the run time of the HSE method, are significantly better than those of conventional state estimation, which can effectively improve the efficiency of the situation comprehension.

Considering the fast applications of intelligent electronic equipment in the SDN, Kong et al. [83] presented an HSE method based on SCADA and PMU measurements, which can help situation comprehension effectively converge and quickly track the system states while ensuring the estimation accuracy. To comprehensively utilize multi-source measurement data, future research should explore suitable data processing methods for the differences between different measurement devices regarding frequency, time scale, structure, and delay.

5.3. Reliability Analysis

As a significant part of the SG, DG penetration in the SDN becomes an ever-increasing problem, and the protection system has significant influences on SDN reliability. Therefore, the comprehensive reliability evaluation of SDN consists of primary distribution networks and a protection system. As the traditional reliability assessment of distribution networks ignores the influence of relay protection and the complex configuration mode of the area-centralized distribution protection system, Xiao et al. [84] proposed an improved failure mode and effect analysis method to evaluate the comprehensive reliability of SDN based on fault location and protection system. Alves et al. [85] presented a reliability assessment methodology to evaluate instantaneous and average measurements of reliability and availability, which is validated in a low-voltage distribution network. Aiming to evaluate the potential rate of exposure to the failure of system components, smart monitoring systems (SMSs) are applied in SG to improve the component reliability. Honarmand et al. [86] presented a new mathematical model to evaluate the reliability of a distribution network equipped with the process-oriented SMSs using the Markov method, which shows SMSs increase the reliability of the distribution network by 90%. The uncertainty of EV charging load challenges the distribution network, especially SDN with a higher proportion of DGs. The objective of [87] is to conduct a comprehensive analysis of spatial–temporal EV charging from the perspective of both system reliability and EV charging service reliability.

The least erroneous knowledge on fault detection and location in SDN helps with the restoration process, expedites maintenance, and reduces power outage duration. Khavari et al. [88] presented a novel framework for fault detection and location for SDN equipped with data loggers, including faulty section identification, area detection, and high impedance fault location. Gilanifar et al. [89] presented a multi-task logistic low-ranked dirty model for fault detection in SDN utilizing the distribution PMU data, which improves the fault detection accuracy by the similarities in the fault data streams among multiple locations across an SDN. Automatic and accurate fault detection and location are critical components of effective situation comprehension. In addition, low voltage direct current (LVDC) distribution systems have recently been considered an alternative to power system infrastructure. Mohanty et al. [90] proposed a fault location based on the offline connection of external discharge equipment using the probe power unit. However, the offline method relies on isolating the faulty section first, while extra operating time is required. To tackle this, Jia et al. [91] presented an online fault location technology for the DC distribution network, which calculates the fault distance based on voltage resonance. Wang et al. [92] proposed a new fault let-through energy-based DC fault location working strategy to facilitate post-fault network maintenance.

5.4. Voltage Stability Analysis

With the development of existing SDN structures, the probability of a voltage collapse in distribution networks has increased. Voltage stability represents the ability to keep node voltages within an acceptable range after a disturbance [93]. A stable SDN can maintain the voltage near an acceptable value after the disturbance occurs. Otherwise, voltage collapse will occur. To prevent potential risk, it is necessary to predict the voltage collapse. The voltage drop caused by overload causes most of the voltage instability problems. Therefore, finding the network nodes prone to voltage collapse becomes a research hotspot.

Sadeghi et al. [93] presented a novel approach for static voltage stability evaluation in distribution networks, introducing a new indicator to assess the voltage stability of distribution networks. The voltage stability indicator *VSI* is as follows:

$$VSI = V_1^2 - 4\left(|V_2||V_1|\cos(\delta_1 - \delta_2) - |V_2|^2\right) \tag{6}$$

where V_2 is the receiving end bus voltage and V_1 is the sending end bus voltage. δ_1 and δ_2 are voltage angles at the sending and receiving buses, respectively. The voltage stability indicator includes only the bus voltage and voltage angle, which is suitable for SDN SA with high response speed requirements.

The penetration level of DGs is increasing and has a significant impact on voltage stability. Hu et al. [94] presented a relatively available transmission capacity indicator (RATCI) based on the power transfer margin of the power–voltage curve considering the distribution network resistance, which is defined as follows:

$$RATCI = (P_{cri} - P_0)/P_{cri} \tag{7}$$

where P_0 is an initial operational point of the system and P_{cri} is the critical point of the system. The novel RATCI assesses the voltage stability by combining DGs and the defined reactive power types, helping SA achieve the optimal penetration rate of the RES while still maintaining voltage security.

In some scenarios, voltage stability can be evaluated accurately by separate static modeling of the distribution network. Nevertheless, simultaneous dynamic modeling of distribution networks is needed in other cases [95]. Song et al. [96] proposed a novel voltage stability indicator using the network-load admittance ratio, where simulation results verify that the indicator has satisfactory linearity with load increase and acceptable estimation accuracy of the voltage stability margin.

5.5. Flexibility Evaluation

As a vital operation indicator of situation comprehension, the flexibility evaluation of distribution networks is gradually being paid attention to by scholars with the increasing penetration of RES. Meanwhile, the SDN faces challenges from decentralizing DGs and the electrification of heating and transportation. To this end, Fonteijn et al. [97] proposed four theoretical possibilities for flexibility as a solution for congestion management based on four pilot projects on congestion management in the Netherlands. However, limited attention has been paid to the probabilistic characteristics of uncertain regions. Ge et al. [98] presented a new sequential flexibility assessment based on the feasibility analysis of the uncertain region of PV active power and load demand, which explores the influence of probabilistic characteristics of uncertain variables on flexibility assessment. To tackle random disturbances and improve O&M quality, a large number of power electronic devices such as soft normally open point (SNOP) are integrated into SDN. The authors of [99] presented a new node flexibility assessment model of distribution systems for SNOP integration. As a new variable load, EVs can increase the system flexibility through interactions with the grid and promote RES consumption. Liu [100] proposed a flexibility evaluation method considering the interaction between distribution networks and EVs.

5.6. Power Quality Evaluation

One of the significant purposes of situation comprehension is to analyze the power quality of SDN. With the gradual deployment of sensitive loads in frequency converters and relays, voltage sag has become a significant power quality issue of SDN. To improve the comprehension of voltage sag severity in SDN, Guo et al. [101] proposed a comprehensive weight-based severity evaluation of voltage sag. In most practical distribution networks, there is insufficient information available about harmonic contents of customers for SA.

Therefore, Amini et al. [102] proposed a novel assessment model of harmonic distortion level emphasizing the impedance characteristics of the network buses, which can also be employed as a valuable tool in SDN, where harmonic contents of nonlinear loads are not available. The acceptable value of impedance characteristic Z_{acc} is determined based on voltage and current of network buses as follows:

$$Z_{\mathrm{acc}} = \frac{V_{\mathrm{h}}}{I_{\mathrm{h}}} \tag{8}$$

where V_{h} and I_{h} are acceptable harmonic voltage and current of ith buses, respectively. If the impedance characteristic is less than the acceptable value, it can be ensured that harmonic voltage limits will be satisfied if harmonic currents are within the standard range.

Time-varying nonlinear loads in SDN frequently interfere with the judgment of the SA system. To this end, Lamedica et al. [103] presented a novel model of time-varying nonlinear loads in SDN based on demand conditions, which achieves a pre-evaluation of harmonic disturbances under variable conditions using normal and uniform distribution to randomize the electrical values of the nonlinear loads. In addition, Bajaj et al. [104] presented an analytic hierarchy process-based approach for evaluating and benchmarking the power quality performance of grid-integrated renewable energy systems, which includes voltage harmonic distortion, current harmonic distortion, voltage and frequency fluctuations, and voltage imbalances. For example, power quality indicators of voltage and current harmonic distortion [104] can be expressed as follows:

$$TVHD = \frac{100 \times \sqrt{V_{\mathrm{rms}}^2 - V_{\mathrm{f\_rms}}^2}}{V_{\mathrm{f\_rms}}} \tag{9}$$

$$TCHD = \frac{100 \times \sqrt{I_{\mathrm{rms}}^2 - I_{\mathrm{f\_rms}}^2}}{I_{\mathrm{f\_rms}}} \tag{10}$$

where $TVHD$ is total voltage harmonic distortion, $TCHD$ is total current harmonic distortion, V_{rms} is RMS value of overall voltage, $V_{\mathrm{f\_rms}}$ is RMS value of fundamental frequency voltage, I_{rms} is RMS value of overall current, and $I_{\mathrm{f\_rms}}$ is RMS value of fundamental frequency current. Power quality indicators of voltage and frequency fluctuations [104] can be expressed as follows:

$$VSS = 1 - \left(\frac{V_{\mathrm{a}} + V_{\mathrm{b}} + V_{\mathrm{c}}}{3} \right) \tag{11}$$

$$FRR = 100 \times \frac{f_{\mathrm{m}} - f_{\mathrm{r}}}{f_{\mathrm{r}}} \tag{12}$$

where VSS is voltage sag score, FRR is frequency regulation ratio, f_{m} is the measured value of frequency, and f_{r} is the rated frequency. V_{a}, V_{b}, and V_{c} are post-sag RMS voltages of phases A, B, and C, respectively. Power quality indicator of voltage imbalance VIF [104] can be expressed as follows:

$$VIF = \frac{82 \cdot \sqrt{V_{\mathrm{abe}}^2 + V_{\mathrm{bce}}^2 + V_{\mathrm{cae}}^2}}{\text{average line voltage}} \tag{13}$$

where V_{ab}, V_{bc}, and V_{ca} are three-phase imbalanced line voltages. V_{abe} is the difference between the line voltage V_{ab} and the average line voltage, V_{bce} is the difference between the line voltage V_{bc} and the average line voltage, and V_{cae} is the difference between the line voltage V_{ca} and the average line voltage.

6. Critical Technologies of Situation Projection

Situation projection is the stage of state prediction to predict the SDN development, evaluate the operational risks, and provide predicted information for SDN management. With the intelligent O&M, the self-adaptation of SDN relies on accurate situation projection. The implementation framework of the situation projection is shown in Figure 4. First, a large amount of processed data from situation detection and situation comprehension is transferred to the situation projection system. Then, multiple factors such as meteorology, economy, society, resources, and load are comprehensively considered. In addition, state-of-the-art intelligent algorithms such as deep learning [105] and Adaboost [106] are applied to situation projection. Finally, critical technologies of situation projection are conducted to simulate and predict the SDN developing trend in different aspects. Meanwhile, the predicted information is sent back to SDN to provide theoretical support for optimal decision making. The critical technologies of situation projection include three-phase unbalanced load prediction technology, renewable energy output prediction technology considering uncertainty, state of energy estimation technology, fault prediction and inspection management technology, and security situation projection technology.

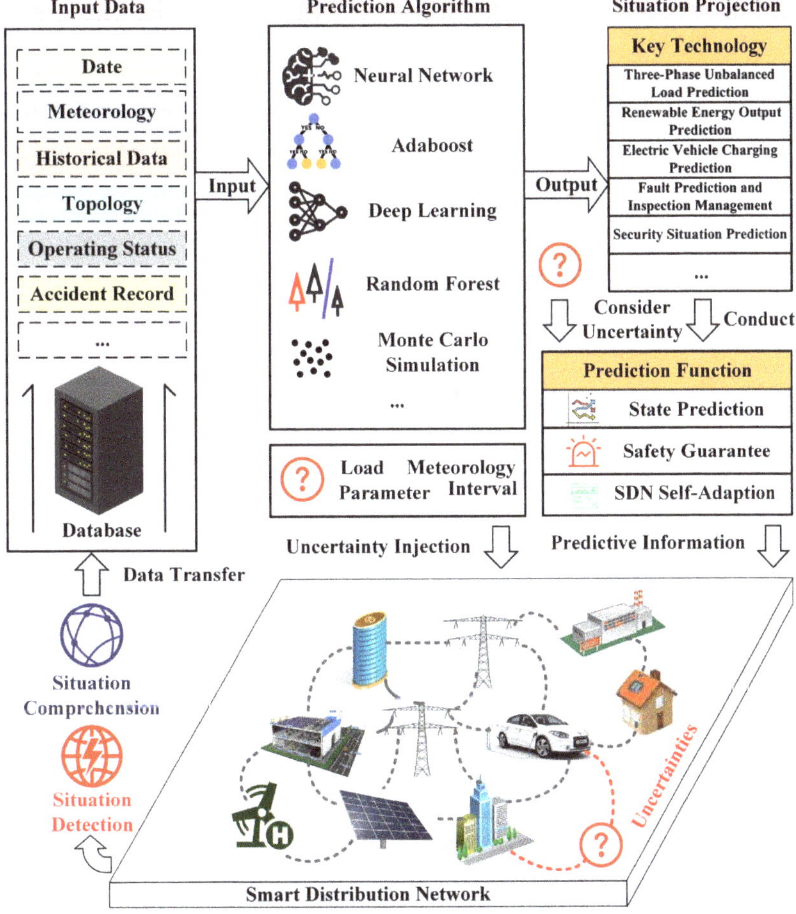

Figure 4. The implementation framework of situation projection.

With the rapid development of new SDN equipment, the O&M of SDN is facing many urgent issues. The integration of high-penetration RES [107] and EVs [108] into the distribution systems increases the uncertainty of SDN operations. In addition, various equipment faults [109] and three-phase unbalance problems [110] can frequently occur in SDN. The security situation is also a vital challenge in establishing secure communication networks for SDN [28]. To this end, situation projection is employed to simulate the behaviors and predict the future development of SDN. The critical technologies of situation projection are related to the security, stability, reliability, and other aspects of the SDN. The goal of situation projection is multifaceted, including reducing the occurrence of three-phase unbalance, assessing the operating risks, evaluating the state-of-energy of EVs, addressing the uncertainty of RES output, assuring information security, providing information support, and guiding SDN management to achieve high-quality O&M [11]. To sum up, situation projection plays a role in SDN in the energy transformation and the upgrade toward future smart cities.

6.1. Three-Phase Unbalanced Load Prediction

Three-phase unbalance means that the amplitude of the three-phase currents or voltage in the power system is inconsistent, and the amplitude difference is beyond the prescribed range [111]. The difference in electricity and electricity usage time between the three phases leads to an unbalanced current [112]. The problem of power three-phase unbalance is closely related to the O&M quality of SDN.

To this end, some studies have investigated three-phase unbalanced predictions. Based on the hierarchical temporal memory, a three-phase unbalanced forecasting model was proposed in [112], where the encoder was adopted for binary coding, the spatial pooler was used for frequency pattern learning, the temporal pooler was employed for pattern sequence learning, and the sparse distributed representations classifier was conducted for unbalance forecasting. Based on the historical data, Han et al. [113] adopted the Elman NN to forecast the daily power consumption of each user and three-phase outlet current in the distribution networks on the day of phase modulation. Therefore, the line loss and three-phase load unbalance can be effectively reduced by changing the access phase sequence of the load. For the unbalanced three-phase SDN, Zhou et al. [114] developed regression analysis for PFC and adopted recurrent NN to predict the load demands. The model that requires fewer distribution-level PMU than nodes is more suitable for existing distribution networks.

6.2. Renewable Energy Output Prediction Considering Uncertainty

Despite the transformation of the SDN energy structure, the intermittency of RES affects the stable operation of SDN. In order to solve the uncertainty issue of RES output, many scholars study the prediction of RES output. The renewable energy output prediction technology quantifies the impact of the RES uncertainty, which can provide a comprehensive RES situation, offer theoretical support for SDN scheduling and configuration, and ensure high-quality O&M. In general, the prediction methods can be divided into (a) physical model prediction and (b) data-driven prediction.

The physical model prediction refers to modeling the physical characteristics of RES [115]. Cui et al. [116] established mathematical models of PV cells and inverters to calculate PV output under different conditions. However, the physical model prediction involves multiple links and has high requirements on the parameters of PV power station components. Therefore, the method may suffer complex modeling, poor robustness, and poor prediction accuracy [117].

Meanwhile, RES output prediction based on the data-driven method mainly considers historical output and meteorological data, which can overcome the shortcomings of the physical model prediction. To deal with the short-term PV output uncertainty characteristics, Ge et al. [118] proposed a PV output prediction technology based on a GRNN. The GWO was adopted to optimize the network parameters of GRNN and achieved a high precision in day-ahead short-term PV output forecasting. In addition, Wang et al. [119] proposed a two-stage attention mechanism prediction model based on long short-term memory (LSTM) for the problem of wind power output prediction.

The above research is deterministic renewable energy forecasting. In recent years, the uncertain method for forecasting RES output has attracted widespread attention from scholars. Algorithms such as probability and statistics laws, interval estimation, and probability theory were employed to predict the RES output [120]. Peng et al. [121] proposed an interval prediction based on the gated recurrent unit for wind power forecasting. Yang et al. [122] proposed a probability prediction for wind power output, which is compatible with SDN areas containing various uncertain parameters.

6.3. State-of-Energy Estimation

The state-of-energy is a vital evaluation index for energy optimization and management of power battery systems in EVs. Unlike the state-of-charge, state-of-energy is the residual energy of the battery in traditional applications, represents the integral result of battery power, and refers to the product of current and terminal voltage. Additionally, the state-of-energy affects the terminal voltage like the state-of-charge. Based on NN, Zhao et al. [123] combined fault and defect diagnosis results with big data statistical regulation to construct a comprehensive EV battery system fault diagnosis. The charging energy of EVs changes based on different actual operating conditions, and the complexity of these changes increases the difficulty of prediction.

To tackle this challenge, Dong et al. [124] presented an online model-based estimation approach against uncertain dynamic load currents and environmental temperatures, which simulates battery dynamics robustly with high accuracy. As a result, the estimates of the dual filters can converge to the real state-of-energy with an error no greater than 4%. To accurately estimate the state-of-charge and state-of-energy for a lithium-ion battery pack, Zhang et al. [125] estimated the battery's energy state online using an adaptive H infinity filter, which can estimate the battery states in real-time with the higher accuracy compared with an extended Kalman filter and an H-infinity filter.

6.4. Fault Prediction and Inspection Management

With the increasingly complex SDN structure, there are many types of faults in the distribution network. Additionally, the redundancy of influencing factors increases. According to the configuration of maintenance personnel, constructing a dynamic inspection strategy can provide reliable decision support for high-quality O&M and reduce the risk of accidents. The main challenges of inspection management include extracting fault features and decoupling fault location layers [126]. Fu et al. [127] proposed a short-term preventive inspection scheduling for SDN, considering the support potential of the DGs and batteries; the results show that the supporting potential of DGs and batteries in preventive maintenance scheduling contributes to a significant reduction of load losses. Liu et al. [128] established various constraints between lines based on the network topology and proposed an optimization model for the inspection plan of distribution network equipment. The results show that the proposed inspection scheduling effectively reduces outage power loss. Moreover, accurate and fast fault prediction in SDN is significant for increasing reliability, fast restoration, optimal electrical energy consumption, and customer satisfaction [129]. Due to the causal ambiguity of written fault records, [130] demonstrated using natural language processing techniques to disambiguate the free text in maintenance tickets to achieve supervised learning of fault prediction technologies. Tsioumpri et al. [131] demonstrated that localized weather data could support fault prediction on distribution networks, taking evasive behaviors for imminent events over short timescales.

6.5. Security Situation Projection

Existing security measures are insufficient to avert attackers' infringement into wireless SDN communication networks [132]. The security situation projection becomes significant to build a secure and resilient SDN. It remains challenging to rapidly extract SDN security situation elements and identify abnormal situations [28]. To hide personal power consumption data from the adversary, Shakila et al. [132] presented the concept of time-variant key generation along with lightweight encryption and device verification technique. To address the security issues of the wireless, private time-division long-term evolution (TD-LTE) network in SDN, Chen et al. [133] proposed a systematic security protection architecture. Considering the security of wireless public network access, Liu et al. [134] proposed a wireless public network access control based on the Bayesian classification, which realized the intelligent distribution of communication networks and improved the operating efficiency of SDN. Although the introduction of smart meters improves measurement and control functions of SDN, cyber-attacks such as electricity theft are constantly emerging, where the attackers increase the power consumption record of other users while reducing their own records. To this end, Tao et al. [135] presented a statistical strategy using the information on higher-order statistics of power consumption, which can detect electricity theft attacks and identify the attackers and victims.

7. Conclusions

With the development of distribution network automation, SA has gradually been popularized and applied in SDN. As more SDN operating technologies and energy forms appear, critical technologies of SA need to be adjusted to adapt to the evolving SDN. Consolidating the critical technologies of SDN SA, promoting the organic integration of various technologies, and improving them based on the implementation effect of SA will be the future research directions. To provide technical support for high-quality O&M of SDN, this paper explains the background of SDN SA, introduces the SA concept, establishes a five-layer integrated framework for SA, and finally analyzes the critical technologies of SA. Especially in SDN SA, the situation detection guarantees the SDN observability by completing the information related to critical elements of the SDN, the situation comprehension facilitates the O&M quality by exploring the operating status and the potential information of SDN, and the situation projection assists O&M personnel in decision making by forecasting the future behavior of SDN components based on their operating status and the perceived information.

For the future perspectives in SDN SA, the scope of SA will be extended from SDN to underdeveloped distribution networks. Future studies will focus on the synergetic effect of personnel, equipment, events, and networks. With the advancement of intelligent algorithms, the improvement of SA operational efficiency will be one of the key research directions. Only a fast-response SA can assist in realizing the intelligent O&M of SDN. In addition, the proposed virtuous circle of SA and SDN is a significant element in the high-quality O&M, while proposing a proper SA effect evaluation method can prevent SDN from falling into a vicious circle. The critical techniques of SA will continue to expand as power demands change and SDN technology advances. We believe this paper can support the development and application of the future SDN SA system.

Author Contributions: Conceptualization, L.G. and Y.L. (Yuanliang Li, tjlyliang@foxmail.com); methodology, L.G. and Y.L. (Yuanliang Li, tjlyliang@foxmail.com); software, L.G. and J.Y.; validation, Y.S., J.Y. and L.G.; formal analysis, L.G. and Y.L. (Yuanliang Li, yuanliang.li@concordia.ca); investigation, J.Y.; resources, L.G.; data curation, Y.S. and Y.L. (Yuanliang Li, yuanliang.li@concordia.ca); writing—original draft preparation, Y.L. (Yuanliang Li, tjlyliang@foxmail.com); writing—review and editing, Y.L. (Yuanliang Li, tjlyliang@foxmail.com) and Y.L. (Yuanliang Li, yuanliang.li@concordia.ca); visualization, L.G.; supervision, J.Y. and Y.L. (Yuanliang Li, yuanliang.li@concordia.ca); project administration, Y.S.; funding acquisition, L.G. All authors have read and agreed to the published version of the manuscript.

Funding: This research was funded by the Science and Technology Project of State Grid Corporation of China, grant number 5100-202155018A-0-0-00, the National Natural Science Foundation of China, grant number 51807134, and the State Key Laboratory of Power System and Generation Equipment, grant number SKLD21KM10.

Institutional Review Board Statement: Not applicable.

Informed Consent Statement: Not applicable.

Data Availability Statement: Not applicable.

Conflicts of Interest: The authors declare no conflict of interest.

Abbreviations

ICT	Information and communication technologies
AMI	Advanced metering infrastructure
ADN	Active distribution networks
SDN	Smart distribution networks
DGs	Distributed generations
SG	Smart grid
RES	Renewable energy sources
SCADA	Supervisory control and data acquisition
DMS	Distribution management systems
EMS	Energy management systems
TTU	Transformer terminal unit
FTU	Feeder terminal unit
RTU	Remote terminal unit
DTU	Distribution automation terminal
O&M	Operation and maintenance
DTs	Distribution transformers
IoT	Internet of things
SA	Situation awareness
NN	Neural network
ISRM	Information security risk management
FA	Factor analysis
GWO	Gray wolf optimization
GRNN	Generalized regression neural network
PMUs	Phasor measurement units
EVs	Electric vehicles
FDI	False data injection
DDoS	Distributed denial of services
IET	Intelligent edge terminal
PFC	Power flow calculation
HSE	Hybrid state estimation
SMSs	Smart monitoring systems
LVDC	Low voltage direct current
RATCI	Relatively available transmission capacity indicator
SNOP	Soft normally open point
LSTM	Long short-term memory
TD-LTE	Time-division long-term evolution

References

1. Evangelopoulos, V.A.; Georgilakis, P.S.; Hatziargyriou, N.D. Optimal operation of smart distribution networks: A review of models, methods and future research. *Electr. Power Syst. Res.* **2016**, *140*, 95–106. [CrossRef]
2. Gill, S.; Kockar, I.; Ault, G.W. Dynamic Optimal Power Flow for Active Distribution Networks. *IEEE Trans. Power Syst.* **2014**, *29*, 121–131. [CrossRef]
3. Antoniadou-Plytaria, K.E.; Kouveliotis-Lysikatos, N.; Georgilakis, P.S.; Hatziargyriou, N.D. Distributed and Decentralized Voltage Control of Smart Distribution Networks: Models, Methods, and Future Research. *IEEE Trans. Smart Grid* **2017**, *8*, 2999–3008. [CrossRef]
4. Shicong, M.A.; Bingyin, X.U.; Houlei, G.A.O.; Yongduan, X.U.E.; Jinghua, W. An Earth Fault Locating Method in Feeder Automation System by Examining Correlation of Transient Zero Mode Currents. *Autom. Electr. Power Syst.* **2008**, *32*, 48–52.
5. Ruj, S.; Nayak, A. A Decentralized Security Framework for Data Aggregation and Access Control in Smart Grids. *IEEE Trans. Smart Grid* **2013**, *4*, 196–205. [CrossRef]

6. Mansor, N.N.; Levi, V. Operational Planning of Distribution Networks Based on Utility Planning Concepts. *IEEE Trans. Power Syst.* **2019**, *34*, 2114–2127. [CrossRef]
7. Meng, F.Y.; Bai, Y.; Jin, J.L. An advanced real-time dispatching strategy for a distributed energy system based on the reinforcement learning algorithm. *Renew. Energy* **2021**, *178*, 13–24. [CrossRef]
8. Al Mhdawi, A.K.; Al-Raweshidy, H.S. A Smart Optimization of Fault Diagnosis in Electrical Grid Using Distributed Software-Defined IoT System. *IEEE Syst. J.* **2020**, *14*, 2780–2790. [CrossRef]
9. Lv, C.; Sheng, W.X.; Liu, K.Y.; Dong, X.Z.; Deng, P. Multi-status modelling and event simulation in smart distribution network based on finite state machine. *IET Gener. Transm. Distrib.* **2019**, *13*, 2846–2855. [CrossRef]
10. Kiaei, I.; Lotfifard, S. Fault Section Identification in Smart Distribution Systems Using Multi-Source Data Based on Fuzzy Petri Nets. *IEEE Trans. Smart Grid* **2020**, *11*, 74–83. [CrossRef]
11. Ge, L.; Li, Y.; Chen, Y.; Cui, Q.; Ma, C. Key Technologies of Situation Awareness and Implementation Effectiveness Evaluation in Smart Distribution Network. *Gaodianya Jishu/High Volt. Eng.* **2021**, *47*, 2269–2280.
12. Yin, J.; Lampert, A.; Cameron, M.; Robinson, B.; Power, R. Using Social Media to Enhance Emergency Situation Awareness. *IEEE Intell. Syst.* **2012**, *27*, 52–59. [CrossRef]
13. Wang, S.; Liang, D.; Ge, L. Key Technologies of Situation Awareness and Orientation for Smart Distribution Systems. *Autom. Electr. Power Syst.* **2016**, *40*, 2–8.
14. Dundes, A. On the psychology of legend. In *American Folk Legend*; University of California Press: Oakland, CA, USA, 2020; pp. 21–36.
15. Lim, H.-R.; An, S. Intention to purchase wellbeing food among Korean consumers: An application of the Theory of Planned Behavior. *Food Qual. Prefer.* **2021**, *88*, 104101. [CrossRef]
16. Ajzen, I. The theory of planned behavior. *Organ. Behav. Hum. Decis. Processes* **1991**, *50*, 179–211. [CrossRef]
17. Luiza Neto, I.; Matsunaga, L.H.; Machado, C.C.; Gunther, H.; Hillesheim, D.; Pimentel, C.E.; Vargas, J.C.; D'Orsi, E. Psychological determinants of walking in a Brazilian sample: An application of the Theory of Planned Behavior. *Transp. Res. Part F-Traffic Psychol. Behav.* **2020**, *73*, 391–398. [CrossRef]
18. Ge, L.J.; Li, Y.L.; Li, S.X.; Zhu, J.B.; Yan, J. Evaluation of the situational awareness effects for smart distribution networks under the novel design of indicator framework and hybrid weighting method. *Front. Energy* **2021**, *15*, 143–158. [CrossRef]
19. McKeown, C. Designing for Situation Awareness: An Approach to User-Centered Design. *Ergonomics* **2013**, *56*, 727–728. [CrossRef]
20. Gao, Y.; Li, D.S.; Cheng, Z.X. UAV Distributed Swarm Situation Awareness Model. *J. Electron. Inf. Technol.* **2018**, *40*, 1271–1278.
21. Zhu, Q. Research on Road Traffic Situation Awareness System Based on Image Big Data. *IEEE Intell. Syst.* **2020**, *35*, 18–26. [CrossRef]
22. Liu, X.W.; Yu, J.G.; Lv, W.F.; Yu, D.X.; Wang, Y.L.; Wu, Y. Network security situation: From awareness to awareness-control. *J. Netw. Comput. Appl.* **2019**, *139*, 15–30. [CrossRef]
23. Anjaria, K.; Mishra, A. Relating Wiener's cybernetics aspects and a situation awareness model implementation for information security risk management. *Kybernetes* **2018**, *47*, 58–79. [CrossRef]
24. Mohammadfam, I.; Mahdinia, M.; Soltanzadeh, A.; Aliabadi, M.M.; Soltanian, A.R. A path analysis model of individual variables predicting safety behavior and human error: The mediating effect of situation awareness. *Int. J. Ind. Ergon.* **2021**, *84*, 103144. [CrossRef]
25. Irwin, A.; Caruso, L.; Tone, I. Thinking Ahead of the Tractor: Driver Safety and Situation Awareness. *J. Agromed.* **2019**, *24*, 288–297. [CrossRef]
26. Smirnova, O.V. Situation Awareness for Navigation Safety Control. *Transnav-Int. J. Mar. Navig. Saf. Sea Transp.* **2018**, *12*, 383–388. [CrossRef]
27. Blandford, A.; Wong, B.L.W. Situation awareness in emergency medical dispatch. *Int. J. Hum.-Comput. Stud.* **2004**, *61*, 421–452. [CrossRef]
28. Xiao, J.; Zhang, B.; Luo, F. Distribution Network Security Situation Awareness Method Based on Security Distance. *IEEE Access* **2019**, *7*, 37855–37864. [CrossRef]
29. Espina, E.; Cardenas-Dobson, R.; Simpson-Porco, J.W.; Saez, D.; Kazerani, M. A Consensus-Based Secondary Control Strategy for Hybrid AC/DC Microgrids With Experimental Validation. *IEEE Trans. Power Electron.* **2021**, *36*, 5971–5984. [CrossRef]
30. Wu, X.; Li, Y.; Tan, Y.; Cao, Y.; Rehtanz, C. Optimal energy management for the residential MES. *IET Gener. Transm. Distrib.* **2019**, *13*, 1786–1793. [CrossRef]
31. Lv, Z.; Kong, W.; Zhang, X.; Jiang, D.; Lv, H.; Lu, X. Intelligent Security Planning for Regional Distributed Energy Internet. *IEEE Trans. Ind. Inform.* **2020**, *16*, 3540–3547. [CrossRef]
32. Xu, C.; Liang, R.; Cheng, Z.; Xu, D. Security situation awareness of smart distribution grid for future energy internet. *Electr. Power Autom. Equip.* **2016**, *36*, 13–18.
33. Ge, L.; Li, Y.; Xian, Y.; Wang, Y.; Liang, D.; Yan, J. A FA-GWO-GRNN Method for Short-Term Photovoltaic Output Prediction. In Proceedings of the 2020 IEEE Power and Energy Society General Meeting (PESGM), Montreal, QC, Canada, 2–6 August 2020.
34. Song, I.K.; Yun, S.Y.; Kwon, S.C.; Kwak, N.H. Design of Smart Distribution Management System for Obtaining Real-Time Security Analysis and Predictive Operation in Korea. *IEEE Trans. Smart Grid* **2013**, *4*, 375–382. [CrossRef]
35. Diez, D.; Romero, R.; Tena, S. Designing a Human Supervisory Control System for Smart Grid. *IEEE Lat. Am. Trans.* **2016**, *14*, 1899–1905. [CrossRef]

36. Ge, L.; Li, Y.; Yan, J.; Wang, Y.; Zhang, N. Short-term Load Prediction of Integrated Energy System with Wavelet Neural Network Model Based on Improved Particle Swarm Optimization and Chaos Optimization Algorithm. *J. Mod. Power Syst. Clean Energy* **2021**, *9*, 1490–1499. [CrossRef]
37. Choi, J.S.; Lee, S.; Chun, S.J. A Queueing Network Analysis of a Hierarchical Communication Architecture for Advanced Metering Infrastructure. *IEEE Trans. Smart Grid* **2021**, *12*, 4318–4326. [CrossRef]
38. De Oliveira-De Jesus, P.M.; Rodriguez, N.A.; Celeita, D.F.; Ramos, G.A. PMU-Based System State Estimation for Multigrounded Distribution Systems. *IEEE Trans. Power Syst.* **2021**, *36*, 1071–1081. [CrossRef]
39. Zhou, X.; Farivar, M.; Liu, Z.; Chen, L.; Low, S.H. Reverse and Forward Engineering of Local Voltage Control in Distribution Networks. *IEEE Trans. Autom. Control* **2021**, *66*, 1116–1128. [CrossRef]
40. Mohammadrezaee, R.; Ghaisari, J.; Yousefi, G.; Kamali, M. Dynamic State Estimation of Smart Distribution Grids Using Compressed Measurements. *IEEE Trans. Smart Grid* **2021**, *12*, 4535–4542. [CrossRef]
41. Taylor, Z.; Akhavan-Hejazi, H.; Cortez, E.; Alvarez, L.; Ula, S.; Barth, M.; Mohsenian-Rad, H. Customer-Side SCADA-Assisted Large Battery Operation Optimization for Distribution Feeder Peak Load Shaving. *IEEE Trans. Smart Grid* **2019**, *10*, 992–1004. [CrossRef]
42. Ye, J.; Ge, X.; Mao, G.; Zhong, Y. 5G Ultradense Networks With Nonuniform Distributed Users. *IEEE Trans. Veh. Technol.* **2018**, *67*, 2660–2670. [CrossRef]
43. Siirto, O.K.; Safdarian, A.; Lehtonen, M.; Fotuhi-Firuzabad, M. Optimal Distribution Network Automation Considering Earth Fault Events. *IEEE Trans. Smart Grid* **2015**, *6*, 1010–1018. [CrossRef]
44. Wang, N.; Zhao, F. An Assessment of the Condition of Distribution Network Equipment Based on Large Data Fuzzy Decision-Making. *Energies* **2020**, *13*, 197. [CrossRef]
45. Ding, D.; Han, Q.-L.; Ge, X.; Wang, J. Secure State Estimation and Control of Cyber-Physical Systems: A Survey. *IEEE Trans. Syst. Man Cybern.-Syst.* **2021**, *51*, 176–190. [CrossRef]
46. Ledva, G.S.; Mathieu, J.L. Separating Feeder Demand Into Components Using Substation, Feeder, and Smart Meter Measurements. *IEEE Trans. Smart Grid* **2020**, *11*, 3280–3290. [CrossRef]
47. Tao, X.L.; Liu, Y.; Zhao, F.; Yang, C.S.; Wang, Y. Graph database-based network security situation awareness data storage method. *Eurasip J. Wirel. Commun. Netw.* **2018**, *2018*, 294. [CrossRef]
48. Wei, Y.P.; Wu, D.Z.; Terpenny, J. Robust Incipient Fault Detection of Complex Systems Using Data Fusion. *IEEE Trans. Instrum. Meas.* **2020**, *69*, 9526–9534. [CrossRef]
49. Chen, K.; Mahfoud, R.J.; Sun, Y.; Nan, D.; Wang, K.; Alhelou, H.H.; Siano, P. Defect Texts Mining of Secondary Device in Smart Substation with GloVe and Attention-Based Bidirectional LSTM. *Energies* **2020**, *13*, 4522. [CrossRef]
50. Xu, Y.; Wang, M.; Fan, W. Defect Data Association Analysis of the Secondary System Based on AFWA-H-Mine. *Energies* **2021**, *14*, 4228. [CrossRef]
51. Tao, J.S.; Umair, M.; Ali, M.; Zhou, J. The Impact of Internet of Things Supported by Emerging 5G in Power Systems: A Review. *CSEE J. Power Energy Syst.* **2020**, *6*, 344–352.
52. Reka, S.S.; Dragicevic, T.; Siano, P.; Prabaharan, S.R.S. Future Generation 5G Wireless Networks for Smart Grid: A Comprehensive Review. *Energies* **2019**, *12*, 2140.
53. Basnet, M.; Ali, M.H. Exploring cybersecurity issues in 5G enabled electric vehicle charging station with deep learning. *IET Gener. Transm. Distrib.* **2021**, *15*, 3435–3449. [CrossRef]
54. Ahmad, T.; Zhang, D.D. Using the internet of things in smart energy systems and networks. *Sustain. Cities Soc.* **2021**, *68*, 102783. [CrossRef]
55. Strielkowski, W.; Dvorak, M.; Rovny, P.; Tarkhanova, E.; Baburina, N. 5G Wireless Networks in the Future Renewable Energy Systems. *Front. Energy Res.* **2021**, *9*, 654. [CrossRef]
56. Maksimovic, M.; Forcan, M. 5G New Radio channel coding for messaging in Smart Grid. *Sustain. Energy Grids Netw.* **2021**, *27*, 100495. [CrossRef]
57. Premsankar, G.; Piao, G.; Nicholson, P.K.; Di Francesco, M.; Lugones, D. Data-Driven Energy Conservation in Cellular Networks: A Systems Approach. *IEEE Trans. Netw. Serv. Manag.* **2021**, *18*, 3567–3582. [CrossRef]
58. Ge, L.; Qin, Y.; Liu, J.; Bai, X. Virtual acquisition method of distributed photovoltaic data based on similarity day and BA-WNN. *Electr. Power Autom. Equip.* **2021**, *41*, 8–14.
59. Yin, L.F.; Gao, Q.; Zhao, L.L.; Zhang, B.; Wang, T.; Li, S.Y.; Liu, H. A review of machine learning for new generation smart dispatch in power systems. *Eng. Appl. Artif. Intell.* **2020**, *88*, 103372. [CrossRef]
60. Zhang, L.; Zhang, M.; Ji, W.; Fang, L.; Qin, Y.; Ge, L. Virtual Acquisition Method for Operation Data of Distributed PV Applying the Mixture of Grey Relational Theory and BP Neural Work. *Electr. Power Constr.* **2021**, *42*, 125–131.
61. Rashed Mohassel, R.; Fung, A.; Mohammadi, F.; Raahemifar, K. A survey on Advanced Metering Infrastructure. *Int. J. Electr. Power Energy Syst.* **2014**, *63*, 473–484. [CrossRef]
62. Dua, G.S.; Tyagi, B.; Kumar, V. A Novel Approach for Configuration Identification of Distribution Network Utilizing mu PMU Data. *IEEE Trans. Ind. Appl.* **2021**, *57*, 857–868. [CrossRef]
63. Akrami, A.; Doostizadeh, M.; Aminifar, F. Optimal Reconfiguration of Distribution Network Using mu PMU Measurements: A Data-Driven Stochastic Robust Optimization. *IEEE Trans. Smart Grid* **2020**, *11*, 420–428. [CrossRef]

64. Liu, J.; Zhang, M.; Ge, L.; Ji, W.; Wang, B.; Fang, L.; Zhang, W. Optimal Configuration Method of Photovoltaic Intelligent Edge Terminal Based on Improved Coyote Optimization Algorithm. *Trans. China Electrotech. Soc.* **2021**, *36*, 1368–1379.

65. Yang, X.; Gu, C.; Yan, X.; Li, F. Reliability-Based Probabilistic Network Pricing With Demand Uncertainty. *IEEE Trans. Power Syst.* **2020**, *35*, 3342–3352. [CrossRef]

66. Chen, S.; Li, P.; Ji, H.; Yu, H.; Yan, J.; Wu, J.; Wang, C. Operational flexibility of active distribution networks with the potential from data centers. *Appl. Energy* **2021**, *293*, 116935. [CrossRef]

67. Huang, W.; Zheng, W.; Hill, D.J. Distribution Network Reconfiguration for Short-Term Voltage Stability Enhancement: An Efficient Deep Learning Approach. *IEEE Trans. Smart Grid* **2021**, *12*, 5385–5395. [CrossRef]

68. DE Vasconcelos, F.M.; Santos Rocha, C.H.; Meschini Almeida, C.F.; Pereira, D.d.S.; Leite Rosa, L.H.; Kagan, N. Methodology for Inspection Scheduling in Power Distribution Networks Based on Power Quality Indexes. *IEEE Trans. Power Deliv.* **2021**, *36*, 1211–1221. [CrossRef]

69. Ge, L.; Li, Y.; Zhu, X.; Zhou, Y.; Wang, T.; Yan, J. An Evaluation System for HVDC Protection Systems by a Novel Indicator Framework and a Self-Learning Combination Method. *IEEE Access* **2020**, *8*, 152053–152070. [CrossRef]

70. Torbaghan, S.S.; Suryanarayana, G.; Hoschle, H.; D'Hulst, R.; Geth, F.; Caerts, C.; Van Hertem, D. Optimal Flexibility Dispatch Problem Using Second-Order Cone Relaxation of AC Power Flows. *IEEE Trans. Power Syst.* **2020**, *35*, 98–108. [CrossRef]

71. Mohseni, M.; Joorabian, M.; Ara, A.L. Distribution system reconfiguration in presence of Internet of things. *IET Gener. Transm. Distrib.* **2021**, *15*, 1290–1303. [CrossRef]

72. Gong, C.; Lin, F.; Gong, X.; Lu, Y. Intelligent Cooperative Edge Computing in Internet of Things. *IEEE Internet Things J.* **2020**, *7*, 9372–9382. [CrossRef]

73. Liu, B.; Huang, Q.; Zhao, J.; Hu, W. A Computational Attractive Interval Power Flow Approach With Correlated Uncertain Power Injections. *IEEE Trans. Power Syst.* **2020**, *35*, 825–828. [CrossRef]

74. Aghili, S.J.; Saghafi, H.; Hajian-Hoseinabadi, H. Uncertainty Analysis Using Fuzzy Transformation Method: An Application in Power-Flow Studies. *IEEE Trans. Power Syst.* **2020**, *35*, 42–52. [CrossRef]

75. Wang, Z.W.; Shen, C.; Liu, F.; Gao, F. Analytical Expressions for Joint Distributions in Probabilistic Load Flow. *IEEE Trans. Power Syst.* **2017**, *32*, 2473–2474. [CrossRef]

76. Liu, Q.; Wang, S.X.; Zhao, Q.Y.; Wang, K. Interval power flow calculation algorithm for multi-terminal dc distribution networks considering distributed generation output uncertainties. *IET Gener. Transm. Distrib.* **2021**, *15*, 986–996. [CrossRef]

77. Yang, G.K.; Dong, P.; Liu, M.B.; Wu, H.Y. Research on random fuzzy power flow calculation of AC/DC hybrid distribution network based on unified iterative method. *IET Renew. Power Gener.* **2021**, *15*, 731–745. [CrossRef]

78. Liu, H.; Tang, C.; Han, J.; Li, T.; Li, J.; Zhang, K. Probabilistic load flow analysis of active distribution network adopting improved sequence operation methodology. *IET Gener. Transm. Distrib.* **2017**, *11*, 2147–2153. [CrossRef]

79. Dobakhshari, A.S.; Abdolmaleki, M.; Terzija, V.; Azizi, S. Robust Hybrid Linear State Estimator Utilizing SCADA and PMU Measurements. *IEEE Trans. Power Syst.* **2021**, *36*, 1264–1273. [CrossRef]

80. Ruiz, S.; Ahmadi, H.; Gardašević, G.; Haddad, Y.; Katzis, K.; Grazioso, P.; Petrini, V.; Reichman, A.; Ozdemir, M.K.; Velez, F.; et al. Chapter 6—5G and beyond networks. In *Inclusive Radio Communications for 5G and Beyond*; Oestges, C., Quitin, F., Eds.; Academic Press: Salt Lake City, UT, USA, 2021; pp. 141–186.

81. Ozsoy, B.; Gol, M. A Hybrid State Estimation Strategy with Optimal Use of Pseudo-Measurements. In Proceedings of the IEEE PES Innovative Smart Grid Technologies Conference Europe (ISGT-Europe), Sarajevo, Bosnia and Herzegovina, 21–25 October 2018.

82. Kabiri, M.; Amjady, N. A New Hybrid State Estimation Considering Different Accuracy Levels of PMU and SCADA Measurements. *IEEE Trans. Instrum. Meas.* **2019**, *68*, 3078–3089. [CrossRef]

83. Kong, X.Y.; Chen, Y.; Xu, T.; Wang, C.S.; Yong, C.S.; Li, P.; Yu, L. A Hybrid State Estimator Based on SCADA and PMU Measurements for Medium Voltage Distribution System. *Appl. Sci.* **2018**, *8*, 1527. [CrossRef]

84. Xiao, F.; Xia, Y.J.; Zhou, Y.B.; Zhou, K.P.; Zhang, Z.; Yin, X.G. Comprehensive reliability assessment of smart distribution networks considering centralized distribution protection system. *IEEJ Trans. Electr. Electron. Eng.* **2020**, *15*, 40–50. [CrossRef]

85. Alves, G.; Marques, D.; Silva, I.; Guedes, L.A.; da Silva, M.D. A Methodology for Dependability Evaluation of Smart Grids. *Energies* **2019**, *12*, 1817. [CrossRef]

86. Honarmand, M.E.; Ghazizadeh, M.S.; Hosseinnezhad, V.; Siano, P. Reliability modeling of process-oriented smart monitoring in the distribution systems. *Int. J. Electr. Power Energy Syst.* **2019**, *109*, 20–28. [CrossRef]

87. Xue, P.; Xiang, Y.; Gou, J.; Xu, W.T.; Sun, W.; Jiang, Z.Z.; Jawad, S.; Zhao, H.J.; Liu, J.Y. Impact of Large-Scale Mobile Electric Vehicle Charging in Smart Grids: A Reliability Perspective. *Front. Energy Res.* **2021**, *9*, 241. [CrossRef]

88. Khavari, S.; Dashti, R.; Shaker, H.R.; Santos, A. High Impedance Fault Detection and Location in Combined Overhead Line and Underground Cable Distribution Networks Equipped with Data Loggers. *Energies* **2020**, *13*, 2331. [CrossRef]

89. Gilanifar, M.; Cordova, J.; Wang, H.; Stifter, M.; Ozguven, E.E.; Strasser, T.I.; Arghandeh, R. Multi-Task Logistic Low-Ranked Dirty Model for Fault Detection in Power Distribution System. *IEEE Trans. Smart Grid* **2020**, *11*, 786–796. [CrossRef]

90. Mohanty, R.; Balaji, U.S.M.; Pradhan, A.K. An Accurate Noniterative Fault-Location Technique for Low-Voltage DC Microgrid. *IEEE Trans. Power Deliv.* **2016**, *31*, 475–481. [CrossRef]

91. Jia, K.; Li, M.; Bi, T.; Yang, Q. A voltage resonance-based single-ended online fault location algorithm for DC distribution networks. *Sci. China-Technol. Sci.* **2016**, *59*, 721–729. [CrossRef]

92. Wang, D.; Psaras, V.; Emhemed, A.A.S.; Burt, G.M. A Novel Fault Let-Through Energy Based Fault Location for LVDC Distribution Networks. *IEEE Trans. Power Deliv.* **2021**, *36*, 966–974. [CrossRef]
93. Sadeghi, S.E.; Foroud, A.A. A new approach for static voltage stability assessment in distribution networks. *Int. Trans. Electr. Energy Syst.* **2020**, *30*, e12203. [CrossRef]
94. Hu, S.; Xiang, Y.; Zhang, X.; Liu, J.Y.; Wang, R.; Hong, B.W. Reactive power operability of distributed energy resources for voltage stability of distribution networks. *J. Mod. Power Syst. Clean Energy* **2019**, *7*, 851–861. [CrossRef]
95. Abbasi, S.M.; Karbalaei, F.; Badri, A. The Effect of Suitable Network Modeling in Voltage Stability Assessment. *IEEE Trans. Power Syst.* **2019**, *34*, 1650–1652. [CrossRef]
96. Song, Y.; Hill, D.J.; Liu, T. Static Voltage Stability Analysis of Distribution Systems Based on Network-Load Admittance Ratio. *IEEE Trans. Power Syst.* **2019**, *34*, 2270–2280. [CrossRef]
97. Fonteijn, R.; van Amstel, M.; Nguyen, P.; Morren, J.; Bonnema, G.M.; Slootweg, H. Evaluating flexibility values for congestion management in distribution networks within Dutch pilots. *J. Eng. -Joe* **2019**, *2019*, 5158–5162. [CrossRef]
98. Ge, S.Y.; Xu, Z.Y.; Liu, H.; Gu, C.H.; Li, F.R. Flexibility evaluation of active distribution networks considering probabilistic characteristics of uncertain variables. *IET Gener. Transm. Distrib.* **2019**, *13*, 3148–3157. [CrossRef]
99. Chen, Y.W.; Sun, J.J.; Zha, X.M.; Yang, Y.H.; Xu, F. A Novel Node Flexibility Evaluation Method of Active Distribution Network for SNOP Integration. *IEEE J. Emerg. Sel. Top. Circuits Syst.* **2021**, *11*, 188–198 [CrossRef]
100. Liu, X.O. Research on Flexibility Evaluation Method of Distribution System Based on Renewable Energy and Electric Vehicles. *IEEE Access* **2020**, *8*, 109249–109265. [CrossRef]
101. Guo, X.H.; Li, Y.; Wang, S.Y.; Cao, Y.J.; Zhang, M.M.; Zhou, Y.C.; Yosuke, N. A Comprehensive Weight-Based Severity Evaluation Method of Voltage Sag in Distribution Networks. *Energies* **2021**, *14*, 6434. [CrossRef]
102. Amini, M.; Jalilian, A.; Behbahani, M.R.P. A new method for evaluation of harmonic distortion in reconfiguration of distribution network. *Int. Trans. Electr. Energy Syst.* **2020**, *30*, e12370. [CrossRef]
103. Lamedica, R.; Ruvio, A.; Ribeiro, P.F.; Regoli, M. A Simulink model to assess harmonic distortion in MV/LV distribution networks with time-varying non linear loads. *Simul. Model. Pract. Theory* **2019**, *90*, 64–80. [CrossRef]
104. Bajaj, M.; Singh, A.K. An analytic hierarchy process-based novel approach for benchmarking the power quality performance of grid-integrated renewable energy systems. *Electr. Eng.* **2020**, *102*, 1153–1173. [CrossRef]
105. Du, S.; Li, T.; Yang, Y.; Horng, S.-J. Deep Air Quality Forecasting Using Hybrid Deep Learning Framework. *IEEE Trans. Knowl. Data Eng.* **2021**, *33*, 2412–2424. [CrossRef]
106. Chen, L.; Li, M.; Su, W.; Wu, M.; Hirota, K.; Pedrycz, W. Adaptive Feature Selection-Based AdaBoost-KNN With Direct Optimization for Dynamic Emotion Recognition in HumanRobot Interaction. *IEEE Trans. Emerg. Top. Comput. Intell.* **2021**, *5*, 205–213. [CrossRef]
107. Azizivahed, A.; Arefi, A.; Ghavidel, S.; Shafie-khah, M.; Li, L.; Zhang, J.; Catalao, J.P.S. Energy Management Strategy in Dynamic Distribution Network Reconfiguration Considering Renewable Energy Resources and Storage. *IEEE Trans. Sustain. Energy* **2020**, *11*, 662–673. [CrossRef]
108. Viet Thang, T.; Islam, M.R.; Muttaqi, K.M.; Sutanto, D. An Efficient Energy Management Approach for a Solar-Powered EV Battery Charging Facility to Support Distribution Grids. *IEEE Trans. Ind. Appl.* **2019**, *55*, 6517–6526.
109. Liang, J.; Jing, T.; Niu, H.; Wang, J. Two-Terminal Fault Location Method of Distribution Network Based on Adaptive Convolution Neural Network. *IEEE Access* **2020**, *8*, 54035–54043. [CrossRef]
110. Kong, W.; Ma, K.; Wu, Q. Three-Phase Power Imbalance Decomposition Into Systematic Imbalance and Random Imbalance. *IEEE Trans. Power Syst.* **2018**, *33*, 3001–3012. [CrossRef]
111. Sun, M.; Demirtas, S.; Sahinoglu, Z. Joint Voltage and Phase Unbalance Detector for Three Phase Power Systems. *IEEE Signal Processing Lett.* **2013**, *20*, 11–14. [CrossRef]
112. Li, H.; Shi, C.; Liu, X.; Wulamu, A.; Yang, A. Three-Phase Unbalance Prediction of Electric Power Based on Hierarchical Temporal Memory. *Cmc-Comput. Mater. Contin.* **2020**, *64*, 987–1004. [CrossRef]
113. Han, P.; Pan, W.; Zhang, N.; Wu, H.; Qiu, R.; Zhang, Z. Optimization Method for Artificial Phase Sequence Based on Load Forecasting and Non -dominated Sorting Genetic Algorithm. *Autom. Electr. Power Syst.* **2020**, *44*, 71–80.
114. Zhou, W.; Ardakanian, O.; Zhang, H.T.; Yuan, Y. Bayesian Learning-Based Harmonic State Estimation in Distribution Systems With Smart Meter and DPMU Data. *IEEE Trans. Smart Grid* **2020**, *11*, 832–845. [CrossRef]
115. Wang, K.; Qi, X.; Liu, H. A comparison of day-ahead photovoltaic power forecasting models based on deep learning neural network. *Appl. Energy* **2019**, *251*, 113315. [CrossRef]
116. Cui, C.; Zou, Y.; Wei, L.; Wang, Y. Evaluating combination models of solar irradiance on inclined surfaces and forecasting photovoltaic power generation. *IET Smart Grid* **2019**, *2*, 123–130. [CrossRef]
117. Ogliari, E.; Dolara, A.; Manzolini, G.; Leva, S. Physical and hybrid methods comparison for the day ahead PV output power forecast. *Renew. Energy* **2017**, *113*, 11–21. [CrossRef]
118. Ge, L.; Xian, Y.; Yan, J.; Wang, B.; Wang, Z. A Hybrid Model for Short-term PV Output Forecasting Based on PCA-GWO-GRNN. *J. Mod. Power Syst. Clean Energy* **2020**, *8*, 1268–1275. [CrossRef]
119. Wang, X.; Li, P.; Yang, J. Short-term wind power forecasting based on two-stage attention mechanism. *IET Renew. Power Gener.* **2020**, *14*, 297–304.

120. Chen, Y.; Li, T.; Zhao, C.; Wei, W. Decentralized Provision of Renewable Predictions Within a Virtual Power Plant. *IEEE Trans. Power Syst.* **2021**, *36*, 2652–2662. [CrossRef]
121. Peng, X.; Xu, Q.; Wang, H.; Lang, J.; Li, W.; Cai, T.; Duan, S.; Xie, Y.; Li, C. A Novel Efficient DLUBE Model Constructed by Error Interval Coefficients for Clustered Wind Power Prediction. *IEEE Access* **2021**, *9*, 61739–61751. [CrossRef]
122. Yang, M.; Lin, Y.; Han, X. Probabilistic Wind Generation Forecast Based on Sparse Bayesian Classification and Dempster-Shafer Theory. *IEEE Trans. Ind. Appl.* **2016**, *52*, 1998–2005. [CrossRef]
123. Zhao, Y.; Liu, P.; Wang, Z.; Zhang, L.; Hong, J. Fault and defect diagnosis of battery for electric vehicles based on big data analysis methods. *Appl. Energy* **2017**, *207*, 354–362. [CrossRef]
124. Dong, G.; Chen, Z.; Wei, J.; Zhang, C.; Wang, P. An online model-based method for state of energy estimation of lithium-ion batteries using dual filters. *J. Power Sources* **2016**, *301*, 277–286. [CrossRef]
125. Zhang, Y.; Xiong, R.; He, H.; Shen, W. Lithium-Ion Battery Pack State of Charge and State of Energy Estimation Algorithms Using a Hardware-in-the-Loop Validation. *IEEE Trans. Power Electron.* **2017**, *32*, 4421–4431. [CrossRef]
126. Mao, T.; Yao, J.; Xin, J.; Kang, T.; Deng, D.; Zhao, J. Intelligent Overhaul and Safety Check Management System of 110 kV Distribution Network. *Autom. Electr. Power Syst.* **2013**, *37*, 125–129.
127. Fu, J.; Nunez, A.; De Schutter, B. A Short-Term Preventive Maintenance Scheduling Method for Distribution Networks With Distributed Generators and Batteries. *IEEE Trans. Power Syst.* **2021**, *36*, 2516–2531. [CrossRef]
128. Yongmei, L.I.U.; Wanxing, S. Optimization Model for Distribution Equipment Maintenance Scheduling Based on Network Topology and Genetic Algorithm. *Power Syst. Technol.* **2007**, *31*, 11–15.
129. Dashti, R.; Daisy, M.; Mirshekali, H.; Shaker, H.R.; Aliabadi, M.H. A survey of fault prediction and location methods in electrical energy distribution networks. *Measurement* **2021**, *184*, 109947. [CrossRef]
130. Stephen, B.; Jiang, X.; McArthur, S.D.J. Extracting Distribution Network Fault Semantic Labels From Free Text Incident Tickets. *IEEE Trans. Power Deliv.* **2020**, *35*, 1610–1613. [CrossRef]
131. Tsioumpri, E.; Stephen, B.; McArthur, S.D.J. Weather Related Fault Prediction in Minimally Monitored Distribution Networks. *Energies* **2021**, *14*, 2053. [CrossRef]
132. Shakila, B.; Tuithung, T. Security Enhancement in Smart Distribution Grid with Light-Weight Dynamic Key Encryption. *J. Sci. Ind. Res.* **2019**, *78*, 847–851.
133. Chen, L.; Dong, X.; Wu, Z.; Liu, Z.; Chen, B. Security Analysis and Access Protection of Power Distribution Wireless Private TD-LTE Network. In Proceedings of the China International Conference on Electricity Distribution (CICED), Xi'an, China, 10–13 August 2016.
134. Yajing, L.; Fengjie, S.; Shengjin, L.; Fang, L. Research on security isolation method for wireless public network oriented to smart power distribution service. In Proceedings of the 4th IEEE International Conference on Computer and Communications (ICCC), Chengdu, China, 7–10 December 2018; Institute of Electrical and Electronics Engineers Inc.: Chengdu, China, 2018; pp. 1160–1165.
135. Tao, J.; Michailidis, G. A Statistical Framework for Detecting Electricity Theft Activities in Smart Grid Distribution Networks. *IEEE J. Sel. Areas Commun.* **2020**, *38*, 205–216. [CrossRef]

Article

Distributionally Robust Joint Chance-Constrained Dispatch for Electricity–Gas–Heat Integrated Energy System Considering Wind Uncertainty

Hui Liu [1,2], Zhenggang Fan [1,2], Haimin Xie [1,2,*] and Ni Wang [1,2]

1   School of Electrical Engineering, Guangxi University, Nanning 530004, China; hughlh@gxu.edu.cn (H.L.);
    zhenggfan@163.com (Z.F.); w_n604@163.com (N.W.)
2   Guangxi Key Laboratory of Power System Optimization and Energy Technology, Guangxi University,
    Nanning 530004, China
*   Correspondence: xhm2230@163.com; Tel.: +86-18777122236

Abstract: With the increasing penetration of wind power, the uncertainty associated with it brings more challenges to the operation of the integrated energy system (IES), especially the power subsystem. However, the typical strategies to deal with wind power uncertainty have poor performance in balancing economy and robustness. Therefore, this paper proposes a distributionally robust joint chance-constrained dispatch (DR-JCCD) model to coordinate the economy and robustness of the IES with uncertain wind power. The optimization dispatch model is formulated as a two-stage problem to minimize both the day-ahead and the real-time operation costs. Moreover, the ambiguity set is generated using Wasserstein distance, and the joint chance constraints are used to ensure that the safety constraints (e.g., ramping limit and transmission limit) can be satisfied jointly under the worst-case probability distribution of wind power. The model is remodeled as a mixed-integer tractable programming issue, which can be solved efficiently by ready-made solvers using linear decision rules and linearization methods. Case studies on an electricity–gas–heat regional integrated system, which includes a modified IEEE 24-bus system, 20 natural gas-nodes, and 6 heat-node system, are investigated for verification. Numerical simulation results demonstrate that the proposed DR-JCCD approach effectively coordinates the economy and robustness of IES and can offer operators a reasonable energy management scheme with an acceptable risk level.

Keywords: distributionally robust optimization (DRO); integrated energy system (IES); joint chance constraints; linear decision rules (LDRs); Wasserstein distance

Citation: Liu, H.; Fan, Z.; Xie, H.; Wang, N. Distributionally Robust Joint Chance-Constrained Dispatch for Electricity–Gas–Heat Integrated Energy System Considering Wind Uncertainty. *Energies* **2022**, *15*, 1796. https://doi.org/10.3390/en15051796

Academic Editor: Luigi Fortuna

Received: 23 January 2022
Accepted: 10 February 2022
Published: 28 February 2022

Publisher's Note: MDPI stays neutral with regard to jurisdictional claims in published maps and institutional affiliations.

1. Introduction

In order to achieve the 1.5 °C temperature control target set by the Paris Climate Agreement [1], the proportion of global power generation via renewable sources will continue to rise. By 2050, renewable energy is expected to account for 86% of the power generation source. Wind power, in particular, will meet more than 35% of power demand and become the main source of power generation at that time [2]. However, as the penetration of renewable energy sources (RESs) increases, the power network will be exposed to greater risks due to the uncertainty of RESs. Therefore, there is an urgent need to improve the flexibility of power systems or mitigate the variability. Constructing regional integrated energy systems (IESs) has been proved as an effective way to provide more flexibility to accommodate renewable sources and reduce the impact of uncertainty on the power system [3].

Many researches have concentrated on the optimal dispatch of IESs to cope with the uncertainties associated with renewable energy. Two common strategies include stochastic programming (SP) [4–9] and robust optimization (RO) [10,11]. To handle uncertainties of load demand and renewable energy, Yong et al. [4] propose a low-carbon optimal stochastic operation model using power-to-gas technology. In a building energy system, a multistage-based scenario-driven approach is proposed to deal with solar power uncertainty and nonschedulable load uncertainty [5]. However, stochastic programming either relies on scene samples to approximate deterministic distributions [6] or assumes a predefined probability distribution that random variables follow [7]. As a result, it imposes a substantial computational burden on optimization [8] and adds difficulty to scenario selection [9]. Compared to stochastic programming, the robust optimization approach does not require any assumptions about wind power probability distribution, because it can ensure system-operational robustness by using uncertainty sets to make optimal decisions under the worst renewable fluctuation cases [10]. Robust optimization can provide a more reliable scheme while considering wind power uncertainty [11], but it compromises system cost-effectiveness and may result in over-conservative solutions.

Distributionally robust optimization (DRO), an effective strategy to overcome weaknesses of stochastic programming and robust optimization, loads all possible wind power probability distributions information into an ambiguous set to incorporate uncertain wind power distributions. In addition, DRO can ensure that all the possible wind power probability distributions in the ambiguous set are met by making the best decision under the worst-case probability distribution [12]. There are several studies on distributionally robust energy models based on moment-based ambiguity sets such as mean vector, covariance matrix, and higher moment information [13–15]. Reference [13] develops a two-stage voltage and natural gas pipe pressure management model for photovoltaic power in IES, where the photovoltaic power uncertainty is modeled by an ambiguity set containing the first-order moment and second-order moment information. Using the same moment information, a distributionally robust optimal power flow problem is formulated to solve renewable energy and load uncertainties [14]. Due to the distributions of renewable forecast errors practically containing higher moment information, the first two moment-based ambiguous sets may cause unnecessary conservatism. Therefore, Reference [15] proposes a DRO model for an energy hub system with an energy storage function, where the ambiguity set contains the first two moments and multimodal information of photovoltaic power forecasting errors. Despite all this, there may be the same moment information among different distributions, which makes it difficult to determine the worst-case probability distribution.

The other approach to characterizing ambiguity sets is on account of the statistical distance between the true probability distribution and possible probability distributions. One type of discrepancy-based ambiguity set is established by Kullback–Leibler divergence. To guarantee the safe operation of the natural gas system under hydrogen injection when utilizing power-to-gas technology, Ref. [16] develops a natural gas security DRO programming for IES using a Kullback–Leibler divergence-based ambiguity set to capture wind uncertainty. However, only in the circumstances that potential distributions are supported on a set of limited values, the Kullback–Leibler divergence-based ambiguity set can be observed through historical data [17]. In contrast, the ambiguity sets based on Wasserstein distance, which include all possible probability distributions that have a narrow gap with the discrete empirical distribution, are introduced and have been increasingly used. With a Wasserstein-distance-based ambiguity set to deal with renewable energy uncertainties, a power-flow DRO problem with multi-stage feedback policies is formulated in [18]. In [19], considering dynamic line rating and operational risk, a power flow DRO approach is established, which constructs the ambiguity set via combining the moment information and Wasserstein distance. To avoid the calculation issue arising from a large number of historical data sets, Ref. [20] proposes a distance-based aggregation method, and the Wasserstein-distance ambiguity set is introduced to a distributionally robust unit commitment problem. Regarding wind power uncertainty, the Wasserstein-distance-based ambiguity set has shown good performance in both finite-sample guarantees and confidence sets.

Although the DRO method with chance-constrained problems has been extensively addressed in optimal power flow [21–23], its research on energy optimization and management is still in the early stage. A distributionally robust individual chance-constrained energy dispatch model is put forward for an islanded heat and electricity system in [24] while considering the uncertain renewable generation. However, due to the confidence levels considered separately, the individual chance constraints will result in high-risk costs and may even result in confidence levels as low as 0% for any individual constraint [25]. Therefore, a joint chance-constrained DRO model is proposed for the combined electricity and natural gas system to address renewable energy uncertainty while using the ambiguity set with the confidence bands of the true density function [26]. Joint chance constraints can improve the simultaneous satisfaction of multiple safety conditions with a high probability, but the ambiguity set only includes marginal distribution information, which will result in a conservative solution.

Therefore, a two-stage distributionally robust joint chance-constrained dispatch (DR-JCCD) model is proposed for the electricity–gas–heat IES with the Wasserstein distance-based ambiguity set, considering the wind power uncertainty. The main contributions of this paper are as follows:

1. For the electricity–gas–heat IES, a distributionally robust joint chance-constrained dispatch model is proposed to boost the system flexibility while considering wind power uncertainty.
2. A two-stage scheme is adopted to deal with wind power uncertainty. In the day-ahead stage, energy outputs and reserve capacity of multiple energy devices are scheduled considering the probability distributions of uncertain wind power forecasting errors from historical data, and then the power outputs are adjusted accordingly in the real-time stage. As a result, a cost-effective IES operation is achieved.
3. A Wasserstein distance-based ambiguity set focused on the empirical distribution of wind forecasting errors is established to provide strict finite samples and approximate the behavior of wind uncertainty.
4. The proposed model is transformed into a mixed-integer tractable programming problem by linear decision rules and the linearization approach, which can be solved efficiently by ready-made solvers.

The remainder of this paper is organized as follows. Section 2 presents the detailed mathematical formulation for both the day-ahead and the real-time operation, where the nomenclature could be found in Appendix A. Section 3 develops effective approximate and re-modeled schemes for the distributionally robust joint chance-constrained model under the Wasserstein ambiguity set. Numerical results are shown in Section 4. Finally, conclusions and future work are given in Section 5.

2. DR-JCCD Modeling of IES

2.1. Framework for Electricity–Gas–Heat IES

As illustrated in Figure 1, a typical framework of the electricity, gas, and heat IES uses a constrained transmission infrastructure to coordinate power generation and natural gas resources. The configuration of multiple energy carriers employed in the energy hub includes Combined Heat and Power (CHP), gas-fired generations, and an electric boiler. CHP transforms natural gas into electricity and heat concurrently. Natural gas-fired units transform natural gas into electricity, allowing them to respond swiftly to power fluctuations. An electric boiler is introduced to supply enough heat power flexibly, and three different energy sources are used to meet local electricity, gas, and heat demands.

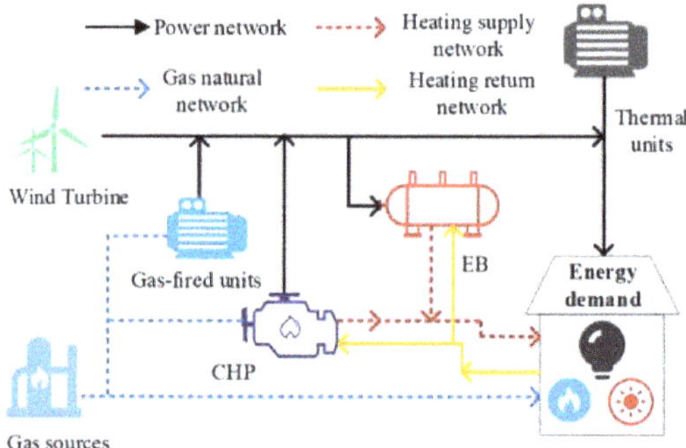

Figure 1. Multi-energy system framework for the IES dispatch.

2.2. Formulation of Day-Ahead Operation

The day-ahead operation schedules the output and reserve capacity of thermal generations, natural gas-fired generations, and CHP. In addition, it determines the natural gas output of gas sources. The objective of the day-ahead operation is to minimize the expected operation cost, including generation cost, reserve capacity costs for traditional units, natural gas-fired units, and CHP, as well as the cost of consuming natural gas from the source, as shown in Equation (1)

$$\min_{\substack{P^{DA},P^{RT}(\zeta),\\ r^+,r^-}} \sum_{\substack{i_e \in I_e, i_g \in I_g,\\ t \in T}} \lambda_{\{\cdot\}} P^{DA}_{\{\cdot\},t} + \lambda_{i_g} G_{gas,t} + \lambda^+_{\{\cdot\}} r^+_{\{\cdot\},t} + \lambda^-_{\{\cdot\}} r^-_{\{\cdot\},t} + E^P[\gamma_c P^w_{\{\cdot\},t}(\zeta)] \tag{1}$$

The detail constraints for day-ahead operation include constraints of the power system, natural-gas system, heating system, and multiple energy converters, as listed in Equations (2)–(29).

2.2.1. Constraints for the Power System

The reserve capacity from thermal generations, natural gas-fired generations, and CHP is shown in Equations (2) and (3), followed by their power output limits in Equations (4) and (5). The adjustments of multiple energy devices for the uncertain wind power forecasting errors are limited by Equation (6), which must be within the reserve capacity range. The ramping up/down constraints are restricted by Equations (7) and (8) [17]. Energy balance is presented in Equation (9). Constrain Equation (10) is used to ensure that the power flows are well within the capacity limits of the transmission lines. Note that $\{\cdot\}$ is the index and set of thermal units, gas-fired generations, and CHP.

$$0 \leq r^+_{\{\cdot\},t} \leq R^+_{\{\cdot\}}, \quad \{\cdot\} = i_e, gg, chp \tag{2}$$

$$0 \leq r^-_{\{\cdot\},t} \leq R^-_{\{\cdot\}}, \quad \{\cdot\} = i_e, gg, chp \tag{3}$$

$$P^{DA}_{\{\cdot\},t} + r^+_{\{\cdot\},t} \leq P^{max}_{\{\cdot\}}, \quad \{\cdot\} = i_e, gg, chp \tag{4}$$

$$P^{min}_{\{\cdot\}} \leq P^{DA}_{\{\cdot\},t} - r^-_{\{\cdot\},t}, \quad \{\cdot\} = i_e, gg, chp \tag{5}$$

$$-r^-_{\{\cdot\},t} \leq P^w_{\{\cdot\},t}(\zeta) \leq r^+_{\{\cdot\},t}, \quad \{\cdot\} = i_e, gg, chp \tag{6}$$

$$(P^{DA}_{\{\cdot\},t} + r^+_{\{\cdot\},t}) - (P^{DA}_{\{\cdot\},t-1} - r^-_{\{\cdot\},t-1}) \leq P^{RU}_{\{\cdot\}}, \quad \{\cdot\} = i_e, gg, chp \tag{7}$$

$$(P^{DA}_{\{\cdot\},t-1} + r^+_{\{\cdot\},t-1}) - (P^{DA}_{\{\cdot\},t} - r^-_{\{\cdot\},t}) \leq P^{RD}_{\{\cdot\}}, \quad \{\cdot\} = i_e, gg, chp \tag{8}$$

$$\sum_{i_e\in I_e}\left(P_{i_e,t}^{DA}+P_{i_e,t}^{RT}(\zeta)\right)+\sum_{gg}\left(P_{gg,t}^{DA}+P_{gg,t}^{RT}(\zeta)\right)+\left(P_{gt,t}^{DA}+P_{gt,t}^{RT}(\zeta)\right)$$
$$+\sum_{j\in J}\left(\omega_{j,t}^{DA}+\zeta_{j,t}\right)+\sum_{l_e\in L_e}f_{l_e,t}^{DA,ini}=\sum_{k_e\in K_e}P_{k_e,t}^{d}+\sum_{b_e\in B_e}P_{b_e,t}^{EB}+\sum_{l_e\in L_e}f_{l_e,t}^{DA,inj} \tag{9}$$

$$\begin{cases} -f_{l_e}^{max}\leq Q^g\left(P_{i_e,t}^{DA}+P_{i_e,t}^{w}(\zeta)\right)+Q^w\left(\omega_{j,t}^{DA}+\zeta_{j,t}\right)-Q^d\left(P_{k_e,t}^{d}+P_{b_e,t}^{EB}\right) \\ Q^g\left(P_{i_e,t}^{DA}+P_{i_e,t}^{w}(\zeta)\right)+Q^w\left(\omega_{j,t}^{DA}+\zeta_{j,t}\right)-Q^d\left(P_{k_e,t}^{d}+P_{b_e,t}^{EB}\right)\leq f_{l_e}^{max}\end{cases} \tag{10}$$

However, constraints Equations (6) and (10) may not always be satisfied [27] due to the uncertain wind power forecasting errors, or the strict restrictions may result in high operational costs. Additionally, individual chance constraints may not be satisfied simultaneously with a certain confidence level. To deal with this problem, the two constraints are converted into joint chance constraints in Equations (11) and (12).

$$P\{-r_{\{\cdot\},t}^{-}\leq P_{\{\cdot\},t}^{RT}(\zeta)\leq r_{\{\cdot\},t}^{+}\}\geq 1-\varepsilon^{gen},\{\cdot\}=i_e,gg,chp \ \ \forall i_e,gg,chp \tag{11}$$

$$P\{-f_{l_e}^{max}\leq Q^g\left(P_{i_e,t}^{DA}+P_{i_e,t}^{RT}(\zeta)\right)+Q^w\left(\omega_{j,t}^{DA}+\zeta_{j,t}\right)-Q^d\left(P_{k_e,t}^{d}+P_{b_e,t}^{EB}\right)\leq f_{l_e}^{max}\}\geq 1-\varepsilon^{grid}\ \forall l_e \tag{12}$$

2.2.2. Constraints for Natural Gas System

Equation (13) explains the relation between the natural gas pressure of gas compressors' headend nodes and terminals. The nodal natural gas balance is given in Equation (14), whose total demand includes gas load, gas consumed by CHP, and gas-fired units. Constraint Equation (15) implies that gas flows only in the positive/negative direction, which cannot exist simultaneously in the gas pipelines. Equation (16) is the Weymouth gas flow equation [28], which describes the relationship between natural gas pressure and natural gas flow in a steady-state condition. The constraints of natural gas pressure at each node and the limits of the natural gas flow in each gas pipeline are represented by Equations (17) and (18), respectively.

$$\psi_{j_g,t}=\rho_c\psi_{i_g,t} \tag{13}$$

$$w_{j_g,t}^{source}+\sum_{i_gj_g\in Z(j_g)}w_{i_gj_g,t}-w_{j_g,t}^{chp}-w_{j_g,t}^{gg}-w_{j_g,t}^{load}=\sum_{j_gk_g\in Z(k)}w_{j_gk_g,t} \tag{14}$$

$$w_{i_gj_g,t}+w_{j_gi_g,t}=0 \tag{15}$$

$$w_{i_gj_g,t}=C_{i_gj_g}\sqrt{\left|\psi_{i_g,t}^2-\psi_{j_g,t}^2\right|} \tag{16}$$

$$\psi_{min}\leq\psi_{i_g,t}\leq\psi_{max} \tag{17}$$

$$w_{i_gj_g,min}\leq w_{i_gj_g,t}\leq w_{i_gj_g,max} \tag{18}$$

2.2.3. Constraints for Heating System

A closed-cycle heating system is employed, which comprises supply and return pipelines. Hot water is selected as the heat medium for transmission, and quality adjustment is adopted by adjusting the temperature of the heat medium. When the water from different pipelines flows into the same node, there will be a mixture of water described by Equations (19) and (20) [29]. Then, Equations (21) and (22) show that the temperature of the output at each node is equal to the mixed water temperature. Equation (23) presents the relationship between heat demand and the heat flow. Furthermore, Equation (24) takes the loss of heat transmission into consideration. The nodal heat balance is given in Equation (25) [30].

$$\sum_{b\in S_i^{-}}\left(T_{b,t}^{ps,outb}\cdot m_{s_b}\right)=T_{i_h,t}^{ms}\cdot\sum_{b\in S_b^{-}}m_{s_b} \tag{19}$$

$$\sum_{b \in S_i^+} (T_{b,t}^{\text{pr,outb}} \cdot m_{r_b}) = T_{i_h,t}^{\text{mr}} \cdot \sum_{b \in S_b^+} m_{r_b} \tag{20}$$

$$T_{b,t}^{\text{ps,inb}} = T_{i_h,t}^{\text{ms}}, b \in S_{i_h}^+ \tag{21}$$

$$T_{b,t}^{\text{pr,inb}} = T_{i_h}^{\text{mr}}, b \in S_{i_h}^- \tag{22}$$

$$d_{i_h,t}^{\text{heat}} = C_p m_{s_b} (T_{i_h,t}^{\text{ms}} - T_{i_h,t}^{\text{mr}}) \tag{23}$$

$$T_{b,t}^{\text{outb}} = (T_{b,t}^{\text{inb}} - T^{\text{a}}) \exp(-\frac{\varsigma^b L^b}{C_p m^b}) + T^{\text{a}} \tag{24}$$

$$H_{chp,t}^{\text{DA}} + H_{EB,t}^{\text{DA}} = \sum_{i_h \in I_h} d_{i_h,t}^{\text{heat}} + \sum_{i_h j_h \in J(j_h)} H_{i_h j_h,t}^{\text{loss}} \tag{25}$$

2.2.4. Constraints for Multiple Energy Converters

Various energy converters are applied to make use of natural gas and electricity resources conjointly to realize the policy to adopt a balanced energy mix. The constraints Equations (26)–(29) detail the energy input–output relationship energy converters, i.e., CHP, gas-fired generations, and electric boiler.

$$H_{chp,t}^{\text{DA}} = \eta_{\text{he}}^{chp} P_{chp,t}^{\text{DA}} \tag{26}$$

$$P_{chp,t}^{\text{DA}} = \eta_{\text{ge}}^{chp} w_{i_g,t}^{chp,\text{DA}} \tag{27}$$

$$P_{gg,t}^{\text{DA}} = \eta_{\text{ge}}^{gg} w_{i_g,t}^{gg,\text{DA}} \tag{28}$$

$$H_{EB,t}^{\text{DA}} = \eta_{\text{he}}^{EB} P_{b_e,t}^{EB,\text{DA}} \tag{29}$$

2.3. Formulation of Real-Time Operation

The real-time stage considers the re-dispatch and adjustive actions to address wind power uncertainty. The second-stage objective function contains two parts: (I) fines for the overrated or underrated schedule in the day-ahead stage and (II) fines for wind power curtailment or load shedding, as shown in Equation (30).

$$\min_{\substack{P^{DA}, P^{RT}(\zeta), \\ r^+, r^-}} \sum_{\substack{i_e \in I_e, i_g \in I_g, \\ t \in T}} \lambda_{\{\cdot\}} |P_{\{\cdot\},t}^{DA} - P_{\{\cdot\},t}^{RT}| + \lambda_{i_g} G_{gas,t} + \lambda_{ke}^{shed} \sum_{k_e \in K_e} l_{ke,t}^{shed} + \lambda_{ie}^{spil} |\sum_{j \in J} \omega_{j,t}^{RT} - (\omega_{j,t}^{DA} + \zeta_{j,t})| \tag{30}$$

Constraints for the real-time operation are the same as the constraints in the day-ahead stage, Equations (7), (8), (10) and (13)–(29), and at the same time include Equations (31)–(34). The constraint Equation (31) requires the load shedding quantity to be no more than actual energy demands. Wind power curtailment quantity is restricted by Equation (32). Constraint Equation (33) limits the adjusted power outputs of traditional generations, gas-fired generations, and CHPs. Equation (34) is the real-time power balance constraint considering wind spills and load shedding.

$$0 \leq \sum_{k_e \in K_e} l_{ke,t}^{shed} \leq \sum_{k_e \in K_e} P_{k_e,t}^{\text{d}} + \sum_{b_e \in B_e} P_{b_e,t}^{EB} \tag{31}$$

$$0 \leq P_{\text{spi},t}^{\text{RT}} \leq \sum_{j \in J} (\omega_{j,t}^{DA} + \zeta_{j,t}) \tag{32}$$

$$P_{\{\cdot\},t}^{DA} - r_{\{\cdot\},t}^- \leq P_{\{\cdot\},t}^{RT} \leq P_{\{\cdot\},t}^{DA} + r_{\{\cdot\},t}^+, \{\cdot\} = i_e, gg, chp \tag{33}$$

$$\sum_{i_e \in I_e} (P_{i_e,t}^{DA} + P_{i_e,t}^{RT}) + \sum_{gg} (P_{gg,t}^{DA} + P_{gg,t}^{RT}) + (P_{chp,t}^{DA} + P_{chp,t}^{RT}) + \sum_{j \in J} \omega_{j,t}^{RT} +$$
$$\sum_{l_e \in L_e} f_{l_e,t}^{DA,ini} + \sum_{k_e \in K_e} l_{k_e,t}^{shed} = \sum_{k_e \in K_e} P_{k_e,t}^{d} + \sum_{b_e \in B_e} P_{b_e,t}^{EB,RT} + \sum_{l_e \in L_e} f_{l_e,t}^{DA,inj} + P_{spi,t}^{RT} \tag{34}$$

3. Proposed Solution Method

The proposed two-stage DR-JCCD problem cannot be solved easily due to the wind power uncertainty and the chance constraints. To find a solution to this problem, firstly, a Wasserstein distance-based ambiguity set is used to collect the uncertain wind power distributions information. Then, based on this ambiguity set, the day-ahead objective function is reformulated through linear decision rules by considering the decision variables' ambiguity in the worst-case expectation. Moreover, the joint chance constraints are transformed into tractable constraints through the Bonferroni approximation and the Worst-Conditional Value-at-Risk approximation. Furthermore, the Weymouth gas flow equation of the proposed model will increase computational burden due to its nonlinear and nonconvex nature. Therefore, the linear programming technique is utilized to solve the gas flow equation.

3.1. Basic Formulation

The proposed model is described as

$$\min_{x \in X} c'x + \sup_{P \in D_\zeta} E^P[Q(x,\zeta)] \tag{35}$$

$$s.t. \; Ax' < b \tag{36}$$

$$P\{g(\bar{f})\} \geq 1 - \varepsilon^{grid} \tag{37}$$

$$P\{h(P_{\{\cdot\},t}^{RT}(\zeta))\} \geq 1 - \varepsilon^{gen} \tag{38}$$

$$\min_{y} f'y \tag{39}$$

$$s.t. \; Ex + Fy + G\zeta \leq h \tag{40}$$

The objective function Equation (35) is to minimize the day-ahead operation cost and the expected cost caused by the energy adjustments Equation (1) combined. x is the decision including the energy output, reserve capacity, and adjustment of multiple energy devices. D_ζ indicates the ambiguity set containing possible probability distribution P of wind data. Equations (36)–(38) present the day-ahead constraints. The real-time model is shown in Equations (39) and (40), where f represents the coefficient of decision variables in Equation (39).

3.2. Wasserstein Distance-Based Ambiguity Set

In order to estimate the probability distributions of wind power uncertainty, it is important to build an effective ambiguity set. Although potential probability distribution is uncertain, an enormous number of recorded historical data are accessible. Therefore, an empirical distribution $P_N = \frac{1}{N}\sum_{k=1}^{N} \delta_{\zeta^k}$ can be considered as the approximate substitution for the true distribution P, where δ_{ζ^k} presents the Dirac measure on the wind power forecasting error sample ζ^k [31]. To estimate the distance between the true distribution P and an empirical distribution P_N, the Wasserstein distance is defined as follows.

Definition of the Wasserstein distance [32]: The Wasserstein distance $d_w(P_1,P_2) : R^W \times R^W \to R$ is defined via

$$d_w(P_1,P_2) = \inf \left\{ \int_{R^w \times R^w} \|\zeta_1 - \zeta_2\| \prod(d\zeta_1, d\zeta_2) \right\} \tag{41}$$

where $\|\zeta_1 - \zeta_2\|$ is the distance between random variables ζ_1 and ζ_2. Additionally, 1-norm is applied in this paper due to its superior numerical tractability. $M(\Xi)$ denotes all probability measures of wind power uncertainty supported on the polyhedron $\Xi = \{\zeta \in R^W : H\zeta \leq h\}$. The Wasserstein distance serves to establish an array of ambiguity sets. Every ambiguity set differs from the empirical distribution within the preset distance [33]:

$$D_\zeta \triangleq \{P \in M(\Xi) : W(P, P_N) \leq \rho\} \tag{42}$$

3.3. Reformulation of Objective Function

It is complex and time-consuming to directly find the exact solution to DRO problems when the decision variables are coupled with random variables. Therefore, the use of LDRs [34], which is a typical approximate method that can deal with the coupling relationship between decision variables and uncertain parameters, is used to approximate the model [35]. In this context, the objective function Equation (35) can be reformulated as a conic program [32].

$$
\begin{aligned}
&\max_{P \in D_\zeta} E^P[\gamma_c^T(Y_0 + Y\zeta)] \\
&= \begin{cases}
\max\limits_{P \in D_\zeta} \int_\Xi \gamma_c^T(Y_0 + Y\zeta)P(\zeta)d\zeta \\
s.t.\ \frac{1}{N}\sum\limits_{i=1}^{N}\int_\Xi \|\zeta - \hat{\zeta}_i\|P_i(d\zeta) \leq \rho \quad \forall i \leq N
\end{cases} \\
&= \begin{cases}
\min\limits_{\lambda^o,s^o,\gamma^o} \lambda^o\rho + \frac{1}{N}\Sigma_{i=1}^N s_i^o \\
s.t.\ \gamma_c^T(Y_0 + Y\hat{\zeta}_i) + \gamma_i^{oT}(h - H\hat{\zeta}_i) \leq s_i^o \quad \forall i \leq N \\
\quad\quad \|H^T\gamma_i^o - (Y_0 + Y)^T c\|_* \leq \lambda^o \quad \forall i \leq N
\end{cases}
\end{aligned}
\tag{43}
$$

where γ_i^o, λ^o, and s^o are auxiliary variables.

3.4. Approximation of Joint Chance Constraints

Joint chance constraints Equations (11) and (12) include a series of constraints of energy output adjustment and transmission lines separately. We consider the two joint chance constraints in a general form.

$$P[A_l x + (B_l Y + C_l)\zeta \leq b_l \ \forall l \leq L] \geq 1 - \varepsilon \tag{44}$$

where l represents the index of energy devices or transmission lines and is the total quantity of energy devices or transmission lines. ε is a predefined confidence level.

Then, the joint chance constraints can be divided into L individual chance constraints whose confidence level is $\varepsilon_l = \varepsilon/L$ by the Bonferroni conservative approximation [36]:

$$
\begin{aligned}
&\min_{P \in D_\zeta} P[A_l x + (B_l Y + C_l)\zeta \leq b_l] \geq 1 - \varepsilon_l \\
&= \min_{P \in D_\zeta} P[A_l x + (B_l Y + C_l)\zeta - b_l \leq 0] \geq 1 - \varepsilon_l
\end{aligned}
\tag{45}
$$

The worst-case Conditional Value-at-Risk approximation can be used to transform Equation (45) [24] into

$$
\begin{aligned}
&\max_{P \in D_\zeta} P - CVaR_{\varepsilon l}[A_l x(B_l Y + C_l)\zeta - b_l] \\
&= \max_{P \in D_\zeta} \inf_{\tau_l} \left\{ \tau_l + \frac{L}{\varepsilon}E^P[(A_l x(B_l Y + C_l)\zeta - b_l - \tau_l)^+]\right\} \\
&= \max_{P \in D_\zeta} \inf_{\tau_l} \left\{ E^P[\max\{\tau_l, \frac{L}{\varepsilon}(A_l x(B_l Y + C_l)\zeta - b_l)\}]\right\}
\end{aligned}
\tag{46}
$$

which can be rewritten as

$$\inf_{\tau_l,\lambda_l,s_l,\gamma_l} \lambda_l\rho + N^{-1}\sum_{i=1}^{N} s_{il}$$
$$\text{s.t. } \tau_l \leq s_{il} \qquad\qquad\qquad \forall i, \forall l$$
$$\tfrac{L}{\varepsilon}(A_l x - b_l) + \tfrac{L}{\varepsilon}(B_l Y + C_l)\hat{\zeta_i} + (1 - \tfrac{L}{\varepsilon})\tau_l$$
$$+\gamma_{il}^T(h - H\hat{\zeta_i}) \leq s_{il} \qquad\qquad \forall i, \forall l \tag{47}$$
$$\|H^T\gamma_{il} - \tfrac{L}{\varepsilon}(B_l Y + C_l)^T\|_* \leq \lambda_k \qquad \forall i, \forall l$$
$$\gamma_{il} \geq 0 \qquad\qquad\qquad\qquad \forall i, \forall l$$

3.5. Reformulation of the Weymouth Gas Flow Equation

The Weymouth gas flow Equation (16), applied to characterize the natural gas flow, is nonlinear and nonconvex. These properties make the optimization of natural gas system operation an NP-hard problem. The problem can be solved by mixed-integer linear programming techniques [37]. Among linear programming techniques, piecewise linear functions describing nonlinearities and binary variables can avoid local optima due to nonconvexities.

Assuming that natural gas flows from pipe node i_g to pipe node j_g, the variable $\psi_{i_g,t}^2$ is replaced by the second-order conic $\psi'_{i_g,t}$, representing the pressure of the pipe node i_g. Then, Equation (13) can be transformed into the constraint as follows:

$$w_{i_g j_g,t}^2 = C_{ij}^2(\psi'_{i_g,t} - \psi'_{j_g,t}) \tag{48}$$

Next, the square term of each pipeline on the left hand of the equation can be piecewise approximated into m linear segments shown in Figure 2.

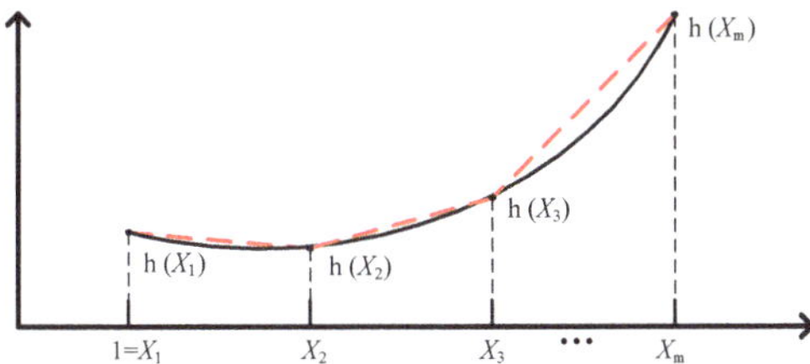

Figure 2. Approximation of a nonlinear function.

When the direction of the gas flow is from node i_g to node j_g, the range of the gas flow $w_{i_g j_g,t}$ is from 0 to $w_{i_g j_g,\max}$, which is the maximum flow in the pipelines. The transformed model of $w_{i_g j_g,t}^2$ is shown as:

$$0 \leq A_{i_g j_g,t}^l \leq \delta_{i_g j_g,t}^l d_{i_g j_g}^l \quad (l=1)$$
$$\delta_{i_g j_g,t}^l d_{i_g j_g}^{l-1} \leq A_{i_g j_g,t}^l \leq \delta_{i_g j_g,t}^l d_{i_g j_g}^l \quad (l \geq 2)$$
$$\sum_{i=1}^{m} \delta_{i_g j_g,t}^l = 1 \;\; \delta_{i_g j_g,t}^l = 0,1; \;\; A_{i_g j_g,t}^l = \omega_{i_g j_g,t}$$
$$f_{i_g j_g}^l = 0 \;\; (l=1); \;\; f_{i_g j_g}^l = (d_{i_g j_g}^{l-1})^2 \;\; (l \geq 2) \tag{49}$$
$$k_{i_g j_g}^l = \frac{d_{i_g j_g}^l}{f_{i_g j_g}^l - f_{i_g j_g}^{l-1}} \;\; l \geq 2$$
$$d_{i_g j_g}^l = \frac{\omega_{i_g j_g,\max} l}{m}$$
$$\psi_{\min}^2 \leq \psi'_{i_g,t} \leq \psi_{\max}^2$$

Finally, the Weymouth gas flow equation is transformed into the form as:

$$C^2_{i_g j_g}(\psi'_{i_g,t} - \psi'_{j_g,t}) = \sum_{i=1}^{m} [(A^l_{i_g j_g,t} - d^{l-1}_{i_g j_g})k^l_{i_g j_g} + f^l_{i_g j_g}\delta^l_{i_g j_g,t}] \tag{50}$$

Therefore, the two-stage distributionally robust joint chance-constrained dispatch model considering wind power uncertainty can be converted into a mixed-integer conic reformulation with Equations (1) and (30) as the objective function and Equations (2)–(5), (7)–(9), (11)–(15), (17)–(29), (31)–(34), (43), (47), and (50) as the constraints, which can be solved directly by calling Gurobi solver under Matlab.

4. Case Study

As shown in Figure 3, a regionally integrated energy system for electricity, gas, and heating comprising a modified IEEE 24-bus system, 20 natural gas nodes, and 6 heat nodes, was used to test the validity of the proposed DR-JCCD model. The detailed information for the generators and natural gas sources are given in Appendix B. The power and the gas subsystems have three coupled points: buses 13 and 23 in the power subsystem are severally connected with natural gas nodes 3 and 19 via two gas-fired generators, and a CHP at bus 22 is linked with gas node 6. The heat demand is satisfied by the gas turbine at power bus 22, and an electric boiler supplied by power bus 13.

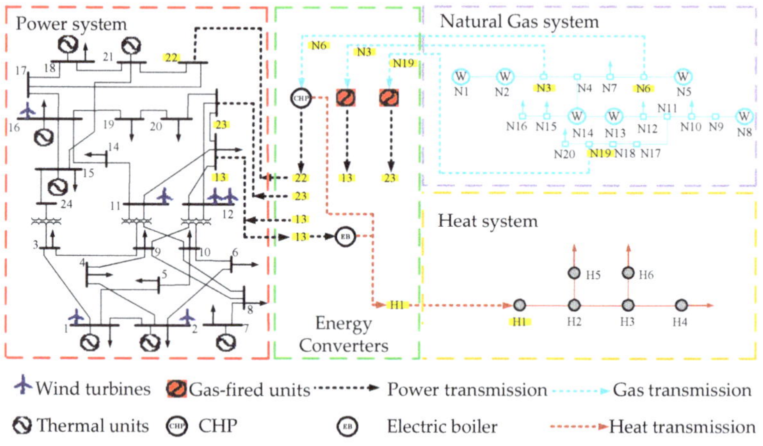

Figure 3. Modified 24-electricity bus, 20-natural gas node, and 6-heat node system for regional IES.

4.1. Robust Performance with Different Sample Sizes

In general, by using another dataset diverse from the experimental one, out-of-sample performance is a helpful tool for assessing the robustness of the optimal schedules [37]. Additionally, sampling errors that comes from limited historical data may cause poor out-of-sample performance in operation. In this condition, the empirical evidence based on out-of-sample forecasting errors is used to measure the robustness of the model in this paper.

As illustrated in Figure 4, an unwise decision that ignores ambiguity (by setting $\rho = 0$) has a large out-of-sample size, which is the maximum cost (5.37×10^7). Therefore, this unwise decision is costlier than a more advanced decision that takes the ambiguity of uncertain wind power into account by setting an appropriate distance ρ. The largest difference in cost is up to 5.1×10^5. In short, the distance ρ precisely regulates the conservativeness of the optimal decision. A considerable distance will make optimal decisions more independent of the characteristics of the historical data and offer stronger robustness to energy adjustments and reserve policies.

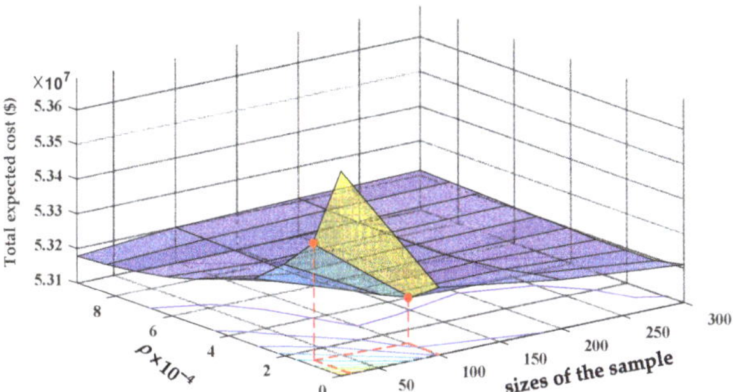

Figure 4. The acquisition cost of out-of-sample size influenced by training sample sizes and Wasserstein distance.

Furthermore, a model considering uncertain variable ambiguity is more competitive for larger sample sizes N. For example, as shown in Figure 4, the acquisition cost of out-of-sample sizes is higher when using an N = 20 training sample ($5.35 $\times$ 10^7) versus an N = 100 training sample ($5.32 $\times$ 10^7). This is due to the fact that a smaller sample size results in a poorer robustness choice in the first stage, necessitating a higher cost for adjustment at the second stage. It is now obvious that the out-of-sample acquisition cost decreases gradually as the training sample size increases. It further implies that, with adequate data support, the proposed solution is fairly robust to wind uncertainty.

4.2. The Influence of Different Confidence Levels

Because risk levels of renewable energy uncertainty can be estimated by confidence level or the parameter ε, we investigate the influence of a series of confidence levels on the total cost. The whole system cost with different parameters $1 - \varepsilon$, including 99%, 95%, 90%, 75%, 65%, 55%, and 45%, are summarized in Figure 5.

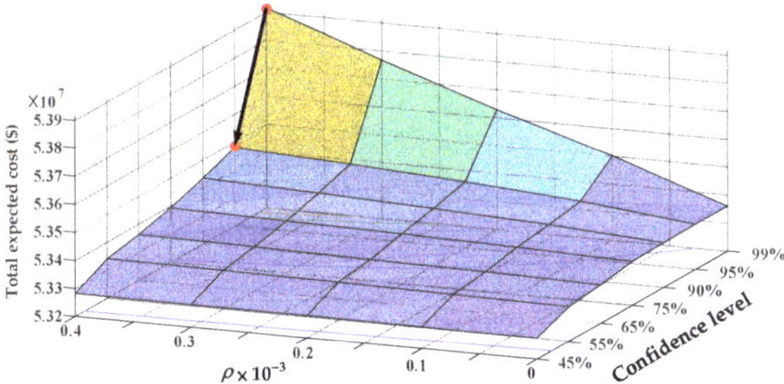

Figure 5. The impact of different confidence levels on the operational cost.

With the same ρ, when the ε increases from 0.01 to 0.55, the marginal cost gradually decreases from $5.39 $\times$ 10^7 to $5.32 $\times$ 10^7, as shown in Figure 5. The largest marginal cost difference is $7 $\times$ 10^5 among the listed confidence levels. This is because the lower the decision-risk maker's tolerance, the greater the necessity for the reserve to balance the unpredictability of wind turbine output, and the greater the amount of natural gas consumed. Meanwhile, a small ε value represents a low tolerance level of risk-taking. It means that as the level of confidence diminishes, the marginal cost decreases. In particular, when the confidence level change from 99% to 95%, the marginal cost dramatically decreases by $5 $\times$ 10^5, which accounts for 71% of the largest marginal cost difference. Therefore, it will cost more to achieve more a reliable operation of the system. It acts as a reminder to decision-makers that in practice, they should choose an adequate confidence level to avoid the significant costs associated with high-reliability standards.

4.3. Comparisons among DR-JCCD, RO, and SP

To assess the effectiveness of the proposed DR-JCCD in balancing robustness and economy, this subsection compares the operating costs among the DR-JCCD approach, RO, and SP methods.

As shown in Figure 6, the total expected costs of the DR-JCCD model increase as the confidence level increases. The expected costs of the DR-JCCD model are always between the RO and SP, regardless of how the expenses change. Meanwhile, the RO approach has a higher total expected cost than its counterparts by at least 1×10^4 and at most 6.4×10^5 due to overly conservative decisions on energy reserve and dispatch. In particular, when the Wasserstein distance is zero, the total expected cost of DR-JCCD is equal to the anticipated expenses of SP. According to the definition of the Wasserstein distance, the ambiguity set contains only empirical probability distributions of wind power on this condition. When the Wasserstein radius approaches infinity and the confidence level reaches 100%, DR-JCCD almost degrades into RO, where the cost of DR-JCCD is only 0.02% lower than the cost of RO. This is because the former has the tendency to contain all probability distributions.

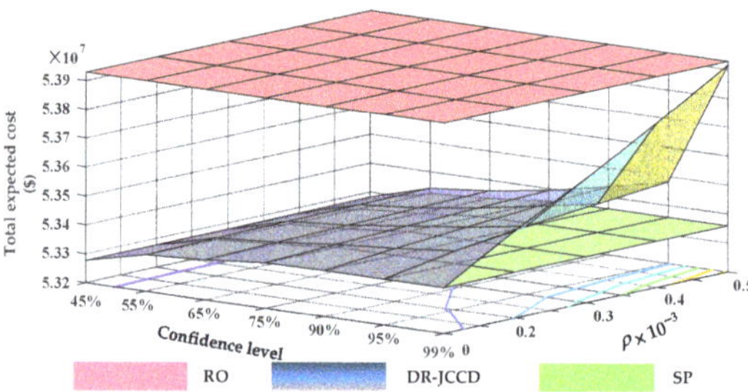

Figure 6. Comparison of total expected costs among DR-JCCD with different ε, RO, and SP.

Moreover, the acquisition cost of out-of-sample size of the DR-JCCD model decreases with the increase in training sample size. However, this cost is always between RO and SP, as illustrated in Figure 7.

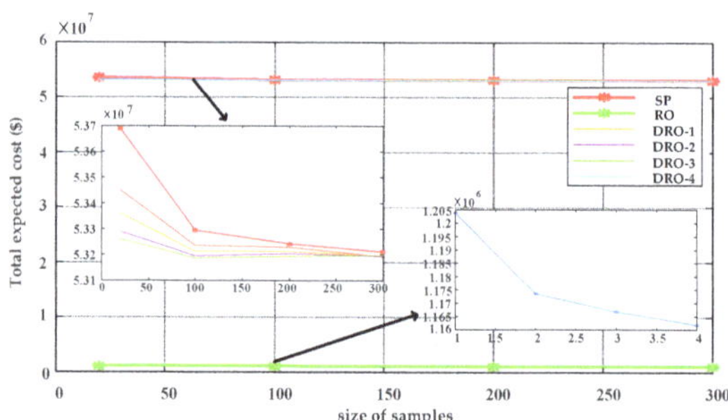

Figure 7. Comparison acquisition cost of out-of-sample size among DR-JCCD with different sizes of training samples, RO, and SP.

Moreover, the acquisition cost of out-of-sample size of the DR-JCCD model decreases with the increasing training sample size. However, this cost is always in-between RO and SP, as illustrated in Figure 7.

From a different perspective, RO, which dispatches and reserves more energy in the worst case of wind power generation, has the most robustness of the three techniques. However, taking accurate dispatch and reserve into account, the proposed approach has lower operational costs than RO. This is due to the fact that it bases its decisions on the worst-case probability distribution of wind power generation, which means more information on uncertain wind power considered. It also has a higher energy reserve than the SP to cope with the wind power uncertainty.

To sum up, by employing partial distributional information, the DR-JCCD approach realizes that the robustness of the model is well-balanced with its economy.

4.4. Analysis of Energy Conversions in Energy Balance

Different confidence levels will result in various solutions in multi-energy management. Here, $1 - \varepsilon = 95\%$ is chose to show the energy mutual assistance effect of gas-fired generations and CHP in the gas system. Note that positive values in Figure 8 represent the result of day-ahead operation, while negative values represent the result of real-time dispatch.

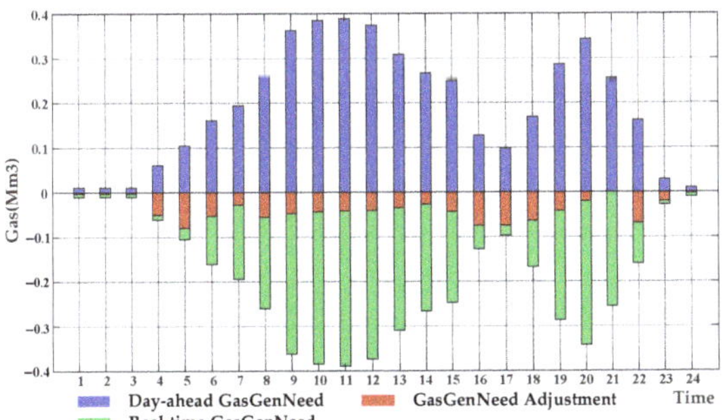

Figure 8. Natural gas consumed in day-head and real-time operation.

A small amount of natural gas is transformed into power at 1:00–3:00 and 24:00 owing to the large output of wind power, as illustrated in Figure 8. At this time, the results of the day-ahead operation will not be adjusted in real-time. Because the influence of the expectation of wind power real-time deviation at the first stage is considered, the adjustment may always take a cut action to revise the results over day-ahead dispatch between 4:00 and 23:00, where the maximum adjustment amount accounts for 19.83% of the day-ahead dispatch. It can reduce unnecessary costs and verify the robustness of the decisions made at the first stage. Meanwhile, by CHP and gas-fired units, the natural gas system effectively supports the power grid under wind power uncertainty.

The output of the heating system is shown in Figure 9. Additionally, the difference between the heating source and heating demand is precisely the transmission loss, as shown in Figure 9a,b. According to Figure 9c, the CHP and electric boiler convert the electricity in order to meet the heating demand, where CHP is the dominant heating source with 83% of the heating capacity and the electric boiler assists with 17% of the heating capacity. When the wind power is abundant at 1:00–5:00 and 23:00–24:00, the electric boiler transforms more wind power to heat, enhancing the IES capacity of wind power utilization in IES.

4.5. Effect of Gas-Fired Generations on Electric Peak Shaving

Peak shaving in a multi-energy system proactively adjusts actions of energy utilization to broaden energy sources or reduce short-term multi-energy demand at peak periods. As a device of multi-energy cooperation, gas-fired generations have a positive effect on Electric Peak shaving, as illustrated in Figure 10. At the peak of power consumption, gas-fired generation systems quickly adjust their output to meet the power demand, thereby reducing the regulatory burden of the power system. Especially at 11:00, the supply of gas-fired generators reaches its maximum, accounting for 22.26% of the power demand. This demonstrates the advantages of multi-energy cooperation in the proposed model. This is particularly the case when the regulation resource is limited or the regulation cost is too high in a subsystem.

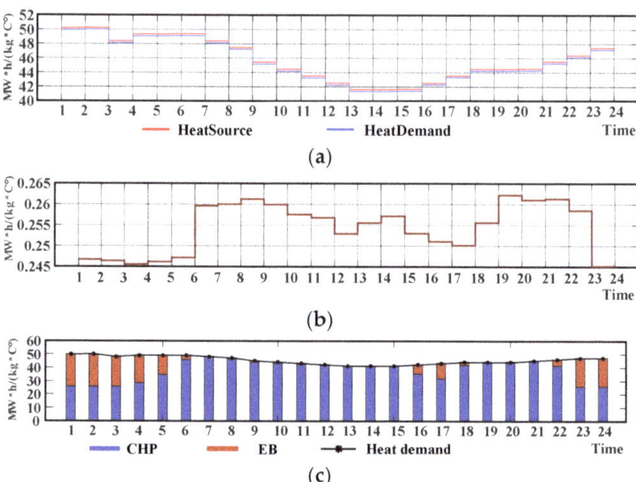

Figure 9. The output of the heating system. (**a**) The heat balance of the heating system. (**b**) The transmission loss of heating system. (**c**) The output of heat sources.

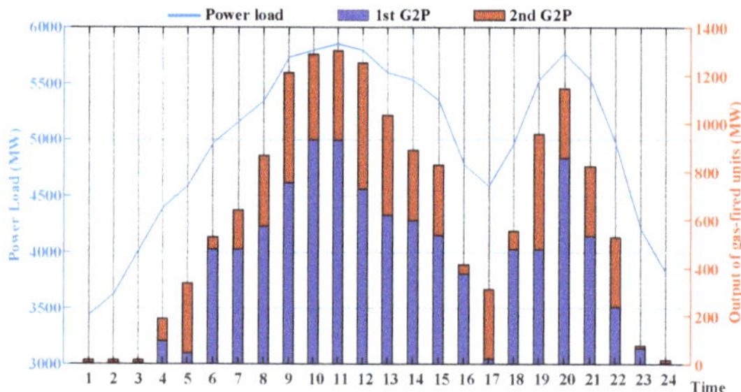

Figure 10. Actions of gas-fired generations in Electric Peak-shaving.

5. Conclusions

A two-stage distributionally robust joint chance-constrained dispatch model for electricity–gas–heat IES with wind power uncertainty is investigated in this paper. The wind power generation uncertainty is captured in the model by employing the worst-case probability distributional information in an ambiguity set based on Wasserstein distance. In light of the operational risk caused by wind power uncertainty, the joint chance constraints ensure that multiple safety conditions are met simultaneously with a high confidence level. Next, by the linear decision rules and linear incremental method, the problem is reformulated as a mixed-integer tractable optimization issue. The effectiveness of the proposed model is corroborated on an electricity–gas–heating regional integrated energy system with a modified IEEE 24-bus system, with 20 natural gas nodes and 6 heat nodes. Notably, the proposed DR-JCCD method can pay 1.3% less than the RO method and achieve a more robust out-of-sample performance than the SP approach at risk, which is a shaving of 22.3% on the acquisition cost of out-of-sample size. Hence, the proposed model achieves a good balance between economy and robustness. Furthermore, the higher the risk preference of the decision-maker, the cheaper the operating cost of the optimization solutions will be, but this will also lessen the robustness of this scheme. To put it differently, the DR-JCCD method can provide decision-makers with information on cost and risk.

With the increase in the number of wind power data, the statistical wind power characteristics are closer to the true wind power distribution. However, a large number of data sets will bring about a calculation issue. Therefore, it is necessary to consider effective scene reduction technology in future work. For more flexible energy management, the IES can take more functional interdependent coupling devices into account and integrate a multi-energy demand response. In addition, multiple uncertainties can be considered in future work, such as various uncertain energy supply and energy demand.

Author Contributions: Conceptualization, H.L.; methodology, H.L. and Z.F.; software, Z.F.; validation, H.L., H.X., N.W.; formal analysis, H.X.; investigation, Z.F.; resources, Z.F. and H.X.; data curation, Z.F.; writing—original draft preparation, Z.F. and H.X.; writing—review and editing, H.X.; visualization, H.L. and N.W.; supervision, H.L. and N.W. All authors have read and agreed to the published version of the manuscript.

Funding: This research was funded in part by the National Key Research and Development Program of China (Grant No. 2019YFE0118000), in part by the National Natural Science Foundation of China (Grant No. 61963003), and in part by the Natural Science Foundation for Distinguished Young Scholars of Guangxi (Grant No. 2018GXNSFFA281006).

Institutional Review Board Statement: Not applicable.

Informed Consent Statement: Written informed consent has been obtained from the patient(s) to publish this paper.

Data Availability Statement: Not applicable.

Conflicts of Interest: The authors declare no conflict of interest.

Appendix A

Table A1. Nomenclature.

Nomenclature			
Indices and sets		η_{ge}^{gg}	Efficiency coefficient of gas-fired units.
t, T	Index and set of time periods.	η_{he}^{EB}	Efficiency coefficient of electric boiler.
l_e, L_e	Index and set of transmission lines.	ε	Confidence levels of chance constraints.
i_e, L_e	Index and set of traditional units.	$\lambda_{i_e}, \lambda_{gg}, \lambda_{chp}$	Cost coefficients for traditional units, gas-fired units, and CHPs.
gg, GG	Index and set of gas-fired units.	$\lambda_{i_e}^{+}, \lambda_{i_e}^{-}, \lambda_{gg}^{+}$	Cost coefficients up and down reserve of traditional units, gas-fired units, and CHPs.
Chp, CHP	Index and set of CHPs.	$\lambda_{gg}^{-}, \lambda_{chp}^{+}, \lambda_{chp}^{-}$	
j, J	Index and set of wind farms.	λ_{i_g}	Cost coefficient for natural gas source.
k_e, K_e	Index and set of power demand.	$\lambda_{k_e}^{shed}, \lambda_{j}^{spil}$	Cost coefficient for load shedding and wind spilling.
b_e, B_e	Index and set of electric boilers.	**Variables**	
i_g, j_g, k_g, Z	Index and set of gas nodes.	$r_{i_e,t}^{+}, r_{i_e,t}^{-}, r_{gg,t}^{+}$	Up and down reserve capacity of traditional units, gas-fired units, and CHPs.
b	Index and set of heat pipes.	$r_{gg,t}^{-}, r_{chp,t}^{+}, r_{chp,t}^{-}$	
i_h, I_h	Index and set of heat nodes.	$P_{i_e,t}^{DA}, P_{gg,t}^{DA}, P_{chp,t}^{DA}$	Active power output of traditional units, gas-fired units, and CHPs day head.
S_{ih}^{-}, S_{ih}^{+}	Index and set of heat pipes at the end/head of node i_h.		
Parameters		$P_{i_e,t}^{w}(\zeta), P_{gg,t}^{w}(\zeta),$ $P_{chp,t}^{w}(\zeta)$	The adjustment of traditional units, gas-fired units and CHPs responding to uncertain wind forecasting errors.
$R_{ie}^{+}, R_{ie,}^{-} R_{gg}^{+}$ $R_{gg}^{-}, R_{chp,}^{+} R_{chp}^{-}$	Maximum up and maximum down reserve capacity of traditional units, gas-fired units, and CHPs.	$P_{b_e,t}^{EB}$	Active power consumed by electric boiler.

Table A1. *Cont.*

Nomenclature			
$P_{i_e}^{max}$, $P_{i_e}^{min}$,	Maximum and minimum limits of active power output of traditional units, gas-fired units, and CHPs.	$l_{k_e,t}^{shed}$	Load shedding at bus k_e in period t.
P_{gg}^{max}, P_{gg}^{min},		$P_{j,t}^{RT,spi}$	Wind spilling of wind farm j in period t.
P_{chp}^{max}, P_{chp}^{min} S_t^{up}		$P_{i_e,t}^{RT}$, $P_{gg,t}^{RT}$	The real-time adjustments of traditional units, gas-fired units and CHPs.
$P_{i_e}^{RU}$, $P_{i_e}^{RD}$		$P_{chp,t}^{RT}$	
P_{gg}^{RU}, P_{gg}^{RD}	Maximum ramp-up and ramp-down rate of traditional units, gas-fired units, and CHPs.	$f_{l_e,t}^{ini}$, $f_{l_e,t}^{inj}$	Injected active power flow at bus i_e and j_e. terminal services
P_{chp}^{RU}, P_{chp}^{RD}			
$f_{l_e}^{max}$	Maximum active power flow of line l_e.	$\psi_{i_g,t}$	Pressure of gas node i_g.
Q^g, Q^w, Q^d	Matrices of power transfer distribution factors.	$w_{i_g,t}^{source}$	Output of natural gas sources.
$P_{k_e,t}^d$, $W_{i_g}^{load}$	Electricity, gas load in period t.	$w_{i_g,j_g,\ t}$	Gas flow of pipeline i_g,j_g.
$\omega_{j,t}^{DA}$	Forecasting output of wind farm j in period t.	$w_{l_g,\ t}^{chp}$, $w_{l_g,\ t}^{gg}$	Injected gas flow of gas turbine and gas-fired units.
$\zeta_{j,t}$	Uncertain wind forecasting errors.	$T_{b,t}^{ps,inb}$, $T_{b,t}^{ps,outb}$	Inlet/Outlet temperature of feed piping b in period t.
ρ_c	Gas compressor coefficient.		
$C_{i_g j_g}$	The coefficient for Weymouth equation.	$T_{b,t}^{pr,inb}$, $T_{b,t}^{pr,outb}$	Inlet/Outlet temperature of return piping b in period t.
m_{s_b}, m_{r_b}	Heating water mass of feed/return piping.		
$d_{i_h,t}^{heat}$	Heating demand at heat piping node i_h in period t.	$T_{i_h,\ t}^{ms}$, $T_{i_h,\ t}^{mr}$	Mixed temperature at node i_h of feed/return piping in period t.
C_p	Specific heat capacity of water.		
T^a	Ambient temperature.	$H_{chp,t}^{DA}$, $H_{EB,t}^{DA}$	Heat output of CHPs and electric boilers in period t day ahead.
ζ^b	Heat transfer coefficient of heat piping b.		
L^b	Length of heat piping b.	$H_{i_g,j_h,\ t}^{loss}$	Heat loss of heat piping in period t.
η_{ge}^{chp}, η_{he}^{chp}	Efficiency coefficient of CHPs.		

Appendix B

Table A2. Parameters of traditional units, gas-fired units and CHPs.

No.	$P_{(\cdot)}^{max}$ (MW)	$P_{(\cdot)}^{min}$ (MW)	$\lambda_{(\cdot)}$ (k\$/MWh)	Bus	Ramp up (MW)	Ramp down (MW)	Type
1	304	40	17.5	1	150	150	0
2	304	40	20	2	150	150	0
3	600	70	15	7	300	300	0
4	1182	60	22.5	13	590	590	1
5	120	30	30	15	60	60	0
6	310	30	22.5	15	105	105	0
7	310	30	25	16	105	105	0
8	800	50	5	18	400	400	0
9	800	50	7.5	21	400	400	0
10	652	50	22.5	22	325	325	2
11	620	60	15	23	310	310	1
12	700	40	22.5	23	350	350	0

Where type 0\1\2 represents traditional units\gas-fired units\CHPs.

Table A3. Parameters of natural gas sources.

No.	Gas Node	$w_{i_g j_g,\ min}(\mathbf{Mm}^3)$	$w_{i_g j_g,\ max}(\mathbf{Mm}^3)$	λ_{i_g} ($/\mathbf{Mm}^3$)
1	1	0.9	1.7391	85,000
2	2	0	1.26	85,000
3	5	0	0.72	85,000
4	8	1.0	2.3018	62,000
5	13	0	0.27	62,000
6	14	0	1.44	62,000

References

1. UNFCCC. Adoption of the Paris agreement (1/CP.21). In *United Nations Framework Convention on Climate Change*; Paris, France, 2015. Available online: https://unfccc.int/resource/docs/2015/cop21/eng/10a01.pdf (accessed on 28 October 2021).
2. Orths, A.; Anderson, C.L.; Brown, T.; Mulhern, J.; Pudjianto, D.; Ernst, B.; O'Malley, M.; McCalley, J.; Strbac, G. Flexibility From Energy Systems Integration: Supporting Synergies Among Sectors. *IEEE Power Energy Mag.* **2019**, *17*, 67–78. [CrossRef]
3. Zhao, P.; Gu, C.; Cao, Z.; Ai, Q.; Xiang, Y.; Ding, T.; Lu, X.; Chen, X. Water-Energy Nexus Management for Power Systems. *IEEE Trans. Power Syst.* **2021**, *36*, 2542–2554. [CrossRef]
4. Li, Y.; Zou, Y.; Tan, Y.; Cao, Y.; Liu, X.; Shahidehpour, M.; Tian, S.; Bu, F. Optimal Stochastic Operation of Integrated Low-Carbon Electric Power, Natural Gas, and Heat Delivery System. *IEEE Trans. Sustain. Energy* **2018**, *9*, 273–283. [CrossRef]
5. Liu, P.; Fang, Y.; Amin, K.M. Multi-stage Stochastic Optimal Operation of Energy-efficient Building with Combined Heat and Power System. *Electr. Mach. Power Syst.* **2014**, *42*, 327–338. [CrossRef]
6. Wang, J.; Botterud, A.; Bessa, R.; Keko, H.; Carvalho, L.; Issicaba, D.; Sumaili, J.; Miranda, V. Wind power forecasting uncertainty and unit commitment. *Appl. Energy* **2011**, *88*, 4014–4023. [CrossRef]
7. Zhao, C.; Wang, Q.; Wang, J.; Guan, Y. Expected Value and Chance Constrained Stochastic Unit Commitment Ensuring Wind Power Utilization. *IEEE Trans. Power Syst.* **2014**, *29*, 2696–2705. [CrossRef]
8. Zheng, Q.P.; Wang, J.; Liu, A.L. Stochastic Optimization for Unit Commitment—A Review. *IEEE Trans. Power Syst.* **2015**, *30*, 1913–1924. [CrossRef]
9. Xiong, P.; Jirutitijaroen, P. Stochastic unit commitment using multi-cut decomposition algorithm with partial aggregation. In Proceedings of the 2011 IEEE Power and Energy Society General Meeting, Detroit, MI, USA, 24–28 July 2011; pp. 1–8.
10. Li, P.; Wu, Q.; Yang, M.; Li, Z.; Hatziargyriou, N.D. Distributed Distributionally Robust Dispatch for Integrated Transmission-Distribution Systems. *IEEE Trans. Power Syst.* **2021**, *36*, 1193–1205. [CrossRef]
11. Yan, M.; Zhang, N.; Ai, X.; Shahidehpour, M.; Kang, C.; Wen, J. Robust Two-Stage Regional-District Scheduling of Multi-carrier Energy Systems With a Large Penetration of Wind Power. *IEEE Trans. Sustain. Energy* **2018**, *10*, 1227–1239. [CrossRef]
12. Delage, E.; Ye, Y. Distributionally Robust Optimization Under Moment Uncertainty with Application to Data-Driven Problems. *Oper. Res.* **2010**, *58*, 595–612. [CrossRef]
13. Zhao, P.; Lu, X.; Cao, Z.; Gu, C.; Ai, Q.; Liu, H.; Bian, Y.; Li, S. Volt-VAR-Pressure Optimization of Integrated Energy Systems With Hydrogen Injection. *IEEE Trans. Power Syst.* **2021**, *36*, 2403–2415. [CrossRef]
14. Zhang, Y.; Shen, S.; Mathieu, J.L. Distributionally Robust Chance-Constrained Optimal Power Flow with Uncertain Renewables and Uncertain Reserves Provided by Loads. *IEEE Trans. Power Syst.* **2017**, *32*, 1378–1388. [CrossRef]
15. Zhao, P.; Gu, C.; Huo, D.; Shen, Y.; Hernando-Gil, I. Two-Stage Distributionally Robust Optimization for Energy Hub Systems. *IEEE Trans. Ind. Inform.* **2020**, *16*, 3460–3469. [CrossRef]
16. Zhao, P.; Gu, C.; Hu, Z.; Xie, D.; Hernando-Gil, I.; Shen, Y. Distributionally Robust Hydrogen Optimization with Ensured Security and Multi-Energy Couplings. *IEEE Trans. Power Syst.* **2021**, *36*, 504–513. [CrossRef]
17. Zhu, R.; Wei, H.; Bai, X. Wasserstein Metric Based Distributionally Robust Approximate Framework for Unit Commitment. *IEEE Trans. Power Syst.* **2019**, *34*, 2991–3001. [CrossRef]
18. Guo, Y.; Baker, K.; Dall'Anese, E.; Hu, Z.; Summers, T.H. Data-based Distributionally Robust Stochastic Optimal Power Flow, Part I: Methodologies. *IEEE Trans. Power Syst.* **2019**, *34*, 1483–1492. [CrossRef]
19. Wang, C.; Gao, R.; Qiu, F.; Wang, J.; Xin, L. Risk-Based Distributionally Robust Optimal Power Flow with Dynamic Line Rating. *IEEE Trans. Power Syst.* **2018**, *33*, 6074–6086. [CrossRef]
20. Zheng, X.; Chen, H. Data-Driven Distributionally Robust Unit Commitment with Wasserstein Metric: Tractable Formulation and Efficient Solution Method. *IEEE Trans. Power Syst.* **2020**, *35*, 4940–4943. [CrossRef]
21. Xie, W.; Ahmed, S. Distributionally Robust Chance Constrained Optimal Power Flow with Renewables: A Conic Reformulation. *IEEE Trans. Power Syst.* **2018**, *33*, 1860–1867. [CrossRef]
22. Fang, X.; Cui, H.; Yuan, H.; Tan, J.; Jiang, T. Distributionally-robust chance constrained and interval optimization for integrated electricity and natural gas systems optimal power flow with wind uncertainties. *Appl. Energy* **2019**, *252*, 113420. [CrossRef]
23. Wang, J.; Wang, C.; Liang, Y.; Bi, T.; Shafie-khah, M.; Catalao, J.P.S. Data-Driven Chance-Constrained Optimal Gas-Power Flow Calculation: A Bayesian Nonparametric Approach. *IEEE Trans. Power Syst.* **2021**, *36*, 4683–4698. [CrossRef]

24. Shi, Z.; Liang, H.; Huang, S.; Dinavahi, V. Distributionally Robust Chance-Constrained Energy Management for Islanded Microgrids. *IEEE Trans. Smart Grid* **2019**, *10*, 2234–2244. [CrossRef]
25. Baker, K.; Toomey, B. Efficient relaxations for joint chance constrained AC optimal power flow. *Electr. Power Syst. Res.* **2017**, *148*, 230–236. [CrossRef]
26. Zhang, Y.; Zheng, F.; Shu, S.; Le, J.; Zhu, S. Distributionally robust optimization scheduling of electricity and natural gas integrated energy system considering confidence bands for probability density functions. *Int. J. Electr. Power Energy Syst.* **2020**, *123*, 106321. [CrossRef]
27. Duan, C.; Jiang, L.; Fang, W.; Liu, J.; Liu, S. Data-Driven Distributionally Robust Energy-Reserve-Storage Dispatch. *IEEE Trans. Ind. Inform.* **2018**, *14*, 2826–2836. [CrossRef]
28. Liu, F.; Bie, Z.; Wang, X. Day-Ahead Dispatch of Integrated Electricity and Natural Gas System Considering Reserve Scheduling and Renewable Uncertainties. *IEEE Trans. Sustain. Energy* **2019**, *2*, 1. [CrossRef]
29. Li, Z.; Wu, W.; Shahidehpour, M.; Wang, J.; Zhang, B. Combined Heat and Power Dispatch Considering Pipeline Energy Storage of District Heating Network. *IEEE Trans. Sustain. Energy* **2016**, *7*, 12–22. [CrossRef]
30. Dai, Y.; Chen, L.; Min, Y.; Chen, Q.; Hao, J.; Hu, K.; Xu, F. Dispatch Model for CHP With Pipeline and Building Thermal Energy Storage Considering Heat Transfer Process. *IEEE Trans. Sustain. Energy* **2019**, *10*, 192–203. [CrossRef]
31. Wang, C.; Gao, R.; Wei, W.; Shafie-khah, M.; Bi, T.; Catalão, J.P.S. Risk-Based Distributionally Robust Optimal Gas-Power Flow with Wasserstein Distance. *IEEE Trans. Power Syst.* **2019**, *34*, 2190–2204. [CrossRef]
32. Esfahani, P.M.; Kuhn, D. Data-driven Distributionally Robust Optimization Using the Wasserstein Metric: Performance Guarantees and Tractable Reformulations. *Math. Program.* **2018**, *171*, 115–166. [CrossRef]
33. Duan, C.; Fang, W.; Jiang, L.; Yao, L.; Liu, J. Distributionally Robust Chance-Constrained Approximate AC-OPF With Wasserstein Metric. *IEEE Trans. Power Syst.* **2018**, *33*, 4924–4936. [CrossRef]
34. Shi, Z.; Liang, H.; Dinavahi, V. Distributionally robust multi-period energy management for CCHP-based microgrids. *IET Gener. Transm. Distrib.* **2020**, *14*, 4097–4107. [CrossRef]
35. Chen, X.; Sim, M.; Sun, P.; Zhang, J. A Linear Decision-Based Approximation Approach to Stochastic Programming. *Oper. Res.* **2008**, *56*, 344–357. [CrossRef]
36. Gicquel, C.; Cheng, J. A joint chance-constrained programming approach for the single-item capacitated lot-sizing problem with stochastic demand. *Ann. Oper. Res.* **2018**, *264*, 123–155. [CrossRef]
37. Correa-Posada, C.; Sanchez-Martion, P. Gas Network Optimization: A Comparison of Piecewise Linear Models. 2014. Available online: http://www.optimization-online.org/DB_FILE/2014/10/4580.pdf (accessed on 12 June 2021).

Article

A Novel Denoising Auto-Encoder-Based Approach for Non-Intrusive Residential Load Monitoring

Xi He [1,*], Heng Dong [1], Wanli Yang [2] and Jun Hong [1]

[1] Department of Electrical and Information Engineering, Hunan Institute of Technology, Hengyang 421002, China; 2014001882@hnit.edu.cn (H.D.); 2013001763@hnit.edu.cn (J.H.)
[2] Department of Electrical and Information Engineering, Hunan University, Changsha 410006, China; yangwanli542@hnu.edu.cn
* Correspondence: 2019001008@hnit.edu.cn

Abstract: Mounting concerns pertaining to energy efficiency have led to the research of load monitoring. By Non-Intrusive Load Monitoring (NILM), detailed information regarding the electric energy consumed by each appliance per day or per hour can be formed. The accuracy of the previous residential load monitoring approach relies heavily on the data acquisition frequency of the energy meters. It brings high overall cost issues, and furthermore, the differentiating algorithm becomes much more complicated. Based on this, we proposed a novel non-Intrusive residential load disaggregation method that only depends on the regular data acquisition speed of active power measurements. Additionally, this approach brings some novelties to the traditionally used denoising Auto-Encoder (dAE), i.e., the reconfiguration of the overlapping parts of the sliding windows. The median filter is used for the data processing of the overlapping window. Two datasets, i.e., the Reference Energy Disaggregation Dataset (REDD) and TraceBase, are used for test and validation. By numerical testing of the real residential data, it proves that the proposed method is superior to the traditional Factorial Hidden Markov Model (FHMM)-based approach. Furthermore, the proposed method can be used for energy data, disaggregation disregarding the brand and model of each appliance.

Keywords: load disaggregation; denoising auto-encoder; REDD dataset; TraceBase dataset; machine learning

Citation: He, X.; Dong, H.; Yang, W.; Hong, J. A Novel Denoising Auto-Encoder-Based Approach for Non-Intrusive Residential Load Monitoring. *Energies* **2022**, *15*, 2290. https://doi.org/10.3390/en15062290

Academic Editor: Djaffar Ould-Abdeslam

Received: 14 February 2022
Accepted: 18 March 2022
Published: 21 March 2022

Publisher's Note: MDPI stays neutral with regard to jurisdictional claims in published maps and institutional affiliations.

1. Introduction

At present, the household electric meter can only measure total electricity consumption, and not the individual electric consumption of various loads. Energy disaggregation is the computational process of distinguishing individual power consumptions of an electrical appliance from the mixed measurement. The application of NILM can help households reduce their cost of energy consumption. According to related studies, with the energy consumption information of each appliance, users can realize energy conservation of more than 12% [1]. In addition, with the increasing installation of renewable energy, the distribution network needs faster and more accurate demand-side response capability. The realization of this capability depends on load disaggregation [2,3].

The load disaggregation of residential electrical equipment is an important direction of smart grid research. The user's electrical equipment has the characteristics of wide variety, large scale, and large differences in the load characteristics [4]. At present, with the pilot and promotion of load disaggregation for residential users, many local load monitoring devices have been deployed. In actual use, it is found that load monitoring devices generally undergo sample data training or learning process in advance. The difficulties in field use are threefold: firstly, due to the low efficiency of the algorithm, the real-time performance of load disaggregation is difficult to guarantee; secondly, due to the wide variety of electrical equipment and complex working conditions, it is difficult to find an algorithm to accurately identify each electrical equipment, and thirdly, when users deploy new devices, they often cannot be identified correctly, which brings great limitations to field usage. Therefore, it is necessary to consider adopting a method to solve the problem of online disaggregation and synchronization of local load disaggregation equipment [5,6].

Load disaggregation can be divided into intrusive methods based on hardware devices and non-intrusive methods based on software algorithms (Nonintrusive Load Monitoring—NILM) [7–9]. In 1992, Hart addressed the energy data disaggregation problem for the first time using Finite State Machine (FSM), which led to the new approaches based on Hidden Markov Models (HMM), and Factorial Hidden Markov Models (FHMM) [10–13]. The essence of these methods is to model the specific electrical signatures or features of each device, either manually or automatically. Ref. [14] proposed an intrusive load disaggregation method based on distributed power Measurement and Actuation Units (MAUs). MAUs are connected between the device plug and the power outlet. The MAU device can measure the power consumption of a single device and control the power failure of the device for demand-side response. Because the invasive method requires additional installation of equipment, the user's responsiveness is relatively low. More research on load disaggregation focus on non-invasive methods. For example, [15] separates the high frequency collected load current data to build a load feature library to realize non-intrusive automatic load monitoring of adaptive users and [16] proposes a non-intrusive load disaggregation method based on generalized regression neural network. This method needs to obtain data such as power, harmonics, switching time, and so on. Ref. [17] proposes separating the superimposed loads based on the transient reactive power characteristics of the load at opening moment, and the coded Particle Swarm Algorithm (E-PSO) is deployed for disaggregation. The above studies all have high load disaggregation accuracy; however, all of them have high requirements for data measurement. Whether it is the high-frequency load current data or the transient waveform when the load is turned on, the ordinary electric meter needs to be transformed before these data can be obtained, adding additional cost to the customers.

In recent years, some scholars proposed to only use low-frequency single measurement for load disaggregation [18–21]. Ref. [18] uses the effective value of current to identify the load, and Ref. [19] only uses the steady-state time domain active and reactive power to identify the turn-on or turn-off status of electrical equipment. A common defect of these methods is that the disaggregation accuracy is poor when multiple loads with similar steady-state waveforms are turned on at the same time.

In terms of disaggregation algorithms, load disaggregation based on machine learning methods is known as a research hotspot [22–28]. Various mature machine learning algorithms are applied to load disaggregation, such as Factorial Hidden Markov Model (FHMM), Artificial Neural Network (ANN), decision tree, etc. In these studies, Deep Neural Networks (DNNs) seem to have certain advantages in both the accuracy and handiness. Ref. [27] proposed a Fully Convolutional Noise Reduction Encoder Algorithm (FCN-dAE) for load disaggregation of non-residential large buildings. This algorithm can train the weight coefficients more effectively in the process of time series modeling. It has a more stable gradient, which simplifies and speeds up the training process. Three difference neural network architectures have been investigated and compared by Kelly and Knottenbelt in [10].

This paper proposes a non-intrusive load disaggregation method that only relies on a single active power measurement at a conventional data acquisition rate. This method requires less measurement and does not require additional installation of hardware and equipment or modification of existing electric energy meters. In terms of the algorithm, this paper is based on the improved Denoising Auto-Encoder algorithm, which can better distinguish loads with similar steady-state power waveforms. Compared with the literature [28], this paper obtains the adjacent maximum value through the maximum pooling operation in the encoding stage, so that the activation function in the analysis window is more independent, and the length of the feature map and the elements of the fully connected layer can also be reduced. Two datasets, i.e., the Reference Energy Disaggregation Dataset (REDD) and TraceBase, are used for test and validation. By numerical test of the real residential data, it proves that the proposed method is superior to the traditional Factorial Hidden Markov Model (FHMM)-based approach. Besides, the proposed method can be used for energy data disaggregation, disregarding the brand and model of each appliance.

This study is organized as follows: Section 2 briefly reviews the four mainstream datasets for NILM, i.e., the REDD, TraceBase, UK-DALE, and Dataport. In Section 3, the proposed disaggregation algorithm is introduced. It elaborates the improvements of the dAE and the two-step procedure of implementing the modified algorithm. Section 4 discusses the test, results, and performance of the proposed method. The proposed DAE network is trained on REDD and TraceBase datasets, and the test results are compared with an FHMM-based approach. Section 5 presents the research conclusions.

2. Dataset Review and Comparison

There are many open-source datasets for non-invasive load disaggregation research worldwide. The commonly used ones are as follows:

(1) **REDD dataset** [29]. Its full name is the Reference Energy Disaggregation Dataset, developed by J. Kolter and M. Johnson of MIT, and is the first dataset for NILM research. The REDD dataset provides high-frequency data sampled at 15 kHz and low-frequency data sampled at 0.5 Hz and 1 Hz. A total of 10 households, 119 days, 268 devices, 1 T electricity consumption data were recorded. Figure 1 is an example of the REDD dataset, showing the electricity usage of various devices in a household over the course of a day. The REDD dataset can be processed with Excel, which is easy to operate. The data download website is: http://redd.csail.mit.edu (accessed on 25 November 2021).

(2) **TraceBase dataset** [14]. The TraceBase dataset was developed by A. Reinhardt of Darmstadt University in Germany. It monitors and records more than ten homes and offices, 31 different types of equipment, 122 devices, and 1270 pieces of load electricity data. Figure 2 shows the electricity consumption of a dishwasher over a period. The entry on the left is time, and the two numbers on the far right represent the average active power consumption within 1 s and 8 s, respectively. The TraceBase dataset is also stored in the form of an Excel table. The format of the data entry is shown in Figure 2. The data download website is: http://www.TraceBase.org (accessed on 25 November 2021).

(3) **UK-DALE dataset** [30]. Developed by J. Kelly and W. Knottenbelt of Imperial College London, the UK-DALE dataset provides 16 kHz energy consumption data for the whole house and 1/6 Hz energy consumption data for a single device. It is the first dataset for load disaggregation in the UK. This dataset recorded the electricity consumption data of five households, one of which was monitored for up to 655 days. The monitoring equipment recorded the active power of a single device as well as the apparent power of the entire house every 6 s, with the voltage and current of three households sampled at 44.1 kHz but reduced to 16 kHz when stored. In addition, the active power, apparent power, and voltage RMS were calculated according to the measured voltage and current, and the calculation frequency was 1 Hz. This dataset is a file in HDF5 (Hierarchical Data Format) format, which needs to be read and analyzed with NILMTK, a non-intrusive load monitoring tool. However, the NILMTK package needs to be loaded and configured with Anaconda software, which is relatively complicated to use.

(4) **Dataport dataset** [31]. The Dataport dataset was developed by Pecan Street company and is the most comprehensive dataset for NILM research. In total, it contains up to 722 households' power consumption data and individual device's power consumption data. Its data sampling rate is low, sampling once a minute. The Dataport dataset is free for member universities, but a paid download is required for commercial use. Like the UK-DALE dataset, this dataset also requires the use of the NILMTK tool for data analysis and statistics.

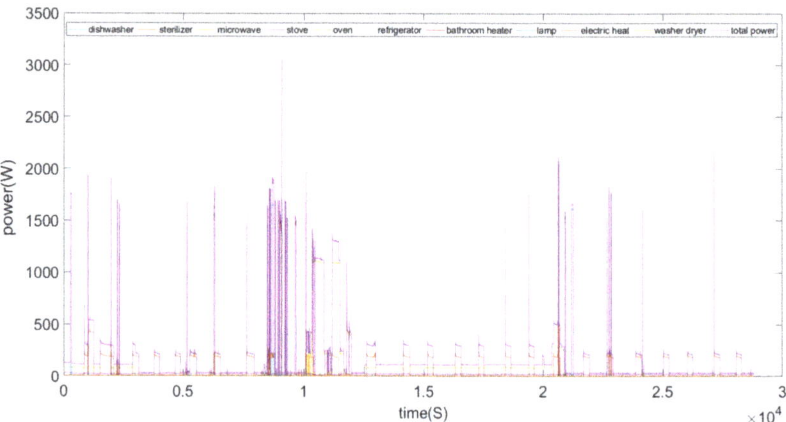

Figure 1. The power waveform of each load in a household in one day.

07/02/2012 10:02:35;9;9
07/02/2012 10:02:36;109;23
07/02/2012 10:02:38;104;34
07/02/2012 10:02:39;100;55
07/02/2012 10:02:41;100;79
07/02/2012 10:02:42;98;90
07/02/2012 10:02:44;102;98
07/02/2012 10:02:45;102;98
07/02/2012 10:02:47;107;100
07/02/2012 10:02:48;104;102

Figure 2. The data format of the TraceBase dataset.

This paper only uses low-frequency active power data. Considering that the REDD dataset and the TraceBase dataset are relatively simple to use, and the data volume is sufficient for machine learning training, the REDD dataset and the TraceBase dataset are used for sample training and method verification.

3. The Proposed Load Disaggregation Algorithm

Usually, a household has multiple electrical devices turned on at the same time, so its total active power is composed of the sub-power of each electrical device. What we need to do is to extract the power characteristics of each electrical device and use it to separate the individual power consumption from the total power mixture. This separation process can be regarded as noise reduction in image processing or speech recognition. Typical noise reduction treatments include removing noise from old photos, or removing noise from a piece of sound, or even filling in the unclear parts of an image. The essence of load disaggregation is load decomposition. The total mixed power can be regarded as the picture or recording that needs to be processed, and the power generated by other unconcerned equipment can be regarded as "noise".

3.1. Improved Denoising Auto-Encoder Algorithm

The Auto-Encoder algorithm (AE) belongs to unsupervised learning and does not require labeling of training samples. AE consists of a three-layer network. First, the input layer is encoded and compressed, stored in the intermediate layer (or called the encoding layer), and then the intermediate layer is decoded, and a reconstructed new vector is output in the output layer. So, in essence, AE consists of two processes: encoding and decoding. In the encoding process, the deterministic mapping f_θ maps the input vector $\mathbf{x}$ to a hidden agent $\mathbf{y}$, and f_θ is the encoder. A typical encoder adopts the nonlinear affine mapping model shown in Equation (1).

$$f_\theta(\mathbf{x}) = s(\mathbf{Wx} + \mathbf{b}) \tag{1}$$

where $\theta = \{\mathbf{W}, \mathbf{b}\}$ represents the parameter set, $\mathbf{W}$ is the weight matrix of $d' \times d$, and $\mathbf{b}$ is the offset vector of d'. In the decoding process, the previously obtained hidden agent $\mathbf{y}$ is mapped back to reconstruct a d-dimensional vector $\mathbf{z}$ in the input space, $\mathbf{z} = g_{\theta'}(\mathbf{y})$. $g_{\theta'}$ is the decoder. A typical decoder adopts the squeezed nonlinear radial mapping model shown in Equation (2).

$$f_\theta(\mathbf{x}) = s(\mathbf{Wx} + \mathbf{b}) \tag{2}$$

where $\theta' = \{\mathbf{W'}, \mathbf{b'}\}$. The meanings of $\mathbf{W'}$ and $\mathbf{b'}$ are similar to those of $\mathbf{W}$ and $\mathbf{b}$ in Formula (1). It should be noted that the d-dimensional vector $\mathbf{z}$ obtained after decoding is not a reconstruction of the input vector $\mathbf{x}$ in the full sense, but a reconstruction of probability theory, because the probability distribution parameters of $p(X \mid Z = z)$ (especially its mean) may increase the probability of $\mathbf{x}$. One of the simplest compression methods is to reduce the dimensionality of the input vector, so linear AE with only a single hidden layer can be regarded as a special principal component analysis method (PCA). But unlike PCA, AE can contain multiple layers and the network function can be nonlinear.

Denoising Auto-encoder (dAE) is a special autoencoder whose purpose is to separate a "clean" target signal from a noisy input, proposed by P. Vencent et al. in 2008 [32]. The dAE algorithm first artificially adds a random "noise" signal $\tilde{x}(\tilde{x} \sim q_D(\tilde{\mathbf{x}}|\mathbf{x}))$ to the input vector $\mathbf{x}$. Similar to an auto-encoder, dAE maps the noisy input signal $\tilde{x}$ to a hidden agent $y = f_\theta(\tilde{x}) = s(W\tilde{x} + b)$, which constructs a decoded output vector $\mathbf{z} = g_{\theta'}(\mathbf{y})$. The structure of the denoising autoencoder is shown in Figure 3. The parameters θ and θ' are trained to minimize the average reconstruction error during training, i.e., to make the output $\mathbf{z}$ as close as possible to the original uncontaminated input vector $\mathbf{x}$, so that $\mathbf{z}$ is now a deterministic function of $\tilde{\mathbf{x}}$. It is worth noting that although dAE is still to minimize the reconstruction loss between the original input $\mathbf{x}$ and the reconstructed agent $\mathbf{y}$, it still needs to maximize the lower bound of mutual information between the original input $\mathbf{x}$ and the reconstructed agent $\mathbf{y}$. However, at this time $\mathbf{y}$ is obtained by using deterministic mapping for "polluted" input, so its feature extraction and learning ability is stronger than traditional autoencoders.

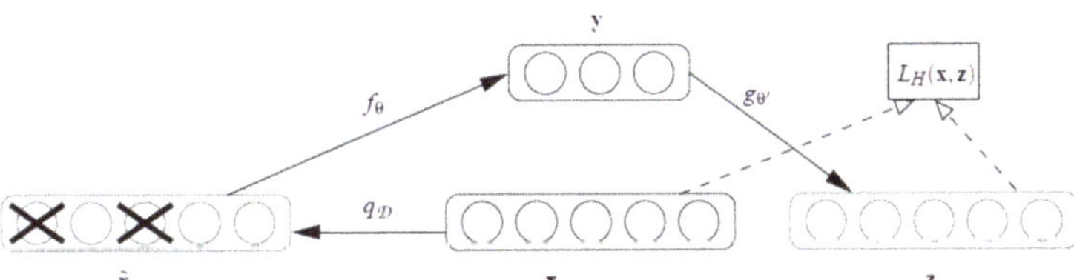

Figure 3. The structure of Denoising Autoencoder (the signal obtained by adding random noise to the original input $\mathbf{x}$, f_θ is the encoder, $\mathbf{y}$ is the intermediate proxy after encoding and mapping, $g_{\theta'}$ is the decoder, $\mathbf{z}$ is the reconstruction input, and $L_H(\mathbf{x, z})$ is the reconstruction loss, which is used to measure the reconstruction error).

In the load separation stage, the load identification method based on dAE generally uses a sliding window to analyze the input mixed power signal $y(t)$, and the length of the sliding window is determined by the use time of the corresponding electrical equipment. Therefore, for a mixed power obtained by turning on multiple devices at the same time, the sliding windows will be overlapped. Traditional denoising autoencoder-based load decomposition methods use the average value of the overlapping parts to reconstruct this overlapping window [10]. A problem with this approach is that when a device's on-time is only included in this overlapping window for a small fraction of time, the load identification results can be significantly higher than the actual power usage. As the window slides, the identified error will further increase. Here we use the median filter to process the overlapping part, that is, the output signal of the overlapping part is the result of $y(t)$ after median filtering. Specifically, because the power change of the overlapping window is relatively small, the output value of the overlapping window can be replaced by the statistical median of all values in a neighborhood of a certain size. This neighborhood is called a window. The wider the window, the smoother the output will be, but it may also wipe out useful signal features. Therefore, the size of the window should be determined according to the actual hybrid power characteristics.

3.2. Decomposition Steps Based on Improved DAE

The problem of non-intrusive load identification can be expressed by Equation (3).

$$y(t) = \sum_{i=1}^{N} y_i(t) + e(t) \tag{3}$$

where $y_i(t)$ represents the electrical quantity of a single electrical device, and this electrical quantity may be power, voltage, or current. Without loss of generality, we consider it the active power value. $y(t)$ indicates the total electricity consumption of this household. $e(t)$ represents the total measurement error, where we consider the measurement error to be 0. N represents the number of electrical appliances in this household. Therefore, according to Formula (3), the NILM problem is to use the algorithm to obtain the power consumption value of a single electrical device when only the total load power is known. We transform the load decomposition into a noise reduction problem, as shown in Equation (4).

$$y(t) = y_k(t) + c_k(t), \quad k = 1, 2, \ldots, N \tag{4}$$

$$c_k(t) = \sum_{i=1, i \neq k}^{N} y_i(t) \tag{5}$$

where $c_k(t)$ represents the sum of the power of all other devices except device k, and $y_k(t)$ represents the load k that needs to be separated. Therefore, to obtain the value of the active power consumed by the load k of interest, one only needs to separate $c_k(t)$ from the total load $y_k(t)$.

The separation steps based on the improved dAE algorithm are as follows:

Stage 1: Encoding the network:

1. One or more one-dimensional convolutional layers process the original total input power value to generate a set of feature maps;
2. Each convolutional layer sequentially goes through a linear activation function, a maximum pooling layer, an additional convolutional layer, and a pooling layer, and finally forms a fully connected multilayer perceptron;
3. The fully connected layer is processed by the modified linear unit (ReLU) activation function to end the entire encoding process.

Stage 2: Decoding the network:

4. Upsampling the fully connected multilayer perceptron through deconvolution;
5. Up-pooling the results in 4 (the inverse process of max-pooling);
6. Continue to upsample the results in 5 through deconvolution;
7. Obtain the decoded and reconstructed noise reduction signal.

In stage 1 and step 2, the adjacent maxima are obtained through the maximum pooling operation, so that the activation function positions in the analysis window are more independent, and the length of the feature map and the number of fully connected layer elements can also be reduced. The modified linear unit (ReLU) activation function compares the magnitude of the input with zero and outputs a larger value, thereby avoiding negative values of the load power after decomposition. The goal of this modified dAE training network is to minimize the mean squared error (MSE) between the output and the activation function of the device to be separated, using a stochastic gradient descent (SGD) method for training parameter optimization. Unlike traditional dAE, which requires artificially adding noise data to the input data, in NILM research, only the power of non-research objects is used as noise. It can be seen that the noise reduction automatic coding for NILM research is not equivalent to the traditional image or sound noise reduction but uses noise reduction as a training standard to better learn how to extract useful features, so as to better construct high-level acting.

4. Performance Evaluation

In this section, the proposed improved dAE network is trained on the measured data of REDD and TraceBase, and the test results are compared with the factorial Hidden Markov Model (FHMM) algorithm [28]. All codes are in Python language, and NILMTK and Pandas tools are used to analyze the data. The neural network training environment is Win10 Home Edition, Intel i5-10210U processor, 8 G memory, and NVIDIA GeForce MX110 graphics card.

4.1. Performance Metrics

The evaluation of the NILM algorithm can be divided into two aspects: the accuracy of energy decomposition and the correctness of equipment state detection. In terms of energy decomposition, the evaluation indicators are authenticity, accuracy, and F_1 index, which are represented by $R_i^{(E)}$, $P_i^{(E)}$, and $F_1^{(E)}$, respectively. The specific calculation formulas of the first two indicators are shown in Formulas (6) and (7).

$$R_i^{(E)} = \frac{\sum_{t=1}^{T} \min(\hat{y}_i(t), y_i(t))}{\sum_{t=1}^{T} y_i(t)} \tag{6}$$

$$P_i^{(E)} = \frac{\sum_{t=1}^{T} \min(\hat{y}_i(t), y_i(t))}{\sum_{t=1}^{T} \hat{y}_i(t)} \tag{7}$$

where $\hat{y}_i(t)$ represents the separated energy signal, $y_i(t)$ represents the real energy consumption of the device, and T represents the total number of samples. In order to analyze the overall performance of the load disaggregation algorithm, we analyze the average authenticity and accuracy of all equipment, and calculate as follows:

$$R^{(E)} = \frac{1}{N} \sum_{i=1}^{N} R_i^{(E)} \tag{8}$$

$$P^{(E)} = \frac{1}{N} \sum_{i=1}^{N} P_i^{(E)} \tag{9}$$

where $R^{(E)}$ and $P^{(E)}$ represent the average value obtained by considering the authenticity and accuracy of all equipment load resolution, respectively, reflecting the overall performance of the NILM algorithm. The metric $F_1^{(E)}$ is the geometric mean of authenticity and accuracy, calculated as follows:

$$F_1^{(E)} = 2\frac{R^{(E)}P^{(E)}}{R^{(E)} + P^{(E)}} \tag{10}$$

In addition, we also define the standard error NEP of load identification, which is used to represent the sum of the deviation between the equipment energy consumption obtained after decomposition and the standard energy consumption. This deviation sum is normalized by the total real equipment energy consumption, and its calculation formula is:

$$\text{NEP}_i = \frac{\sum_{t=1}^{T} |y_i(t) - \hat{y}_i(t)|}{\sum_{t=1}^{T} y_i(t)} \tag{11}$$

The detection of equipment status refers to the detection of the on/off status of the equipment, which can be decomposed into four indicators, true positive (TP), false positive (FP), false negative (FN), and true negative (TN). The specific definitions of the four indicators are as follows:

$$\text{TP}_i = \sum_{t=1}^{T} (s_i(t) = on, \quad \hat{s}_i(t) = on) \tag{12}$$

$$\text{FP}_i = \sum_{t=1}^{T} (s_i(t) = off, \quad \hat{s}_i(t) = on) \tag{13}$$

$$\text{FN}_i = \sum_{t=1}^{T} (s_i(t) = on, \quad \hat{s}_i(t) = off) \tag{14}$$

$$\text{TN}_i = \sum_{t=1}^{T} (s_i(t) = off, \quad \hat{s}_i(t) = off) \tag{15}$$

In Equations (12)–(15), $s_i(t)$ and $\hat{s}_i(t)$ represent the real state and identification state of the device i at time t, respectively, and on and off represent the two states of the device. The authenticity and accuracy of identification based on device status are defined as:

$$R_i^{(S)} = \frac{\text{TP}_i}{\text{TP}_i + \text{FN}_i}, \quad P_i^{(S)} = \frac{\text{TP}_i}{\text{TP}_i + \text{FP}_i} \tag{16}$$

Similarly, considering the authenticity and accuracy of all equipment status detection and identification, the indicators are obtained:

$$R^{(S)} = \frac{1}{N} \sum_{i=1}^{N} R_i^{(S)}, \quad P^{(S)} = \frac{1}{N} \sum_{i=1}^{N} P_i^{(S)} \tag{17}$$

Thus, the index $F_1^{(S)}$ based on the device state is obtained:

$$F_1^{(S)} = \frac{2R^{(S)} P^{(S)}}{R^{(S)} + P^{(S)}} \tag{18}$$

In addition, we also use the Matthews Correlation Coefficient (MCC) as the identification accuracy index, which is defined as:

$$\text{MCC}_i = \frac{\text{TP}_i \text{TN}_i - \text{FP}_i \text{FN}_i}{\sqrt{(\text{TP}_i + \text{FP}_i)(\text{TP}_i + \text{FN}_i)(\text{TN}_i + \text{FP}_i)(\text{TN}_i + \text{FN}_i)}} \tag{19}$$

The overall Matthews Correlation Coefficient is

$$\text{MCC} = \frac{1}{N} \sum_{i=1}^{N} \text{MCC}_i \tag{20}$$

The value of MCC is in the range of $[-1, 1]$. The larger the value is, the more accurate the identification is, and the value of 0 is a random prediction.

4.2. Test Result

4.2.1. Performance Test Using REDD Dataset

In this REDD dataset, Household 1 and Household 2 data were selected as test subjects. The data is updated every 3 s, so it contains a total of 28,800 pieces of data in one day. In order to verify the effectiveness of the proposed dAE-based algorithm, we tested and compared the load decomposition effects of 10 kinds of electrical equipment in Household 1 and 8 kinds of electrical equipment in Household 2, respectively. Among them, the 10 kinds of electrical equipment in family 1 are oven, refrigerator, dishwasher, sterilizer, lamp, dryer, microwave oven, bathroom heater, electric heater, stove. The 8 kinds of electrical equipment in Household 2 are kitchen appliance 1, kitchen appliances 2, lamp, stove, microwave, dryer, refrigerator, dishwasher.

In the process of data training, considering that the device may show different power waveforms in different time periods, for each device, 10 days of data are selected for training, and the other 10 days of data are used for testing and verification. Therefore, a total of 576,000 pieces of data are used. In the REDD dataset, the power consumption data of all 10 electrical devices exceeds 600,000.

To keep it concise, only the power decomposition results of three electrical appliances in Household 1 are presented, namely dishwasher, refrigerator, and lamp (shown in Figure 4). The abscissa in the figure is the time, and the unit is seconds. Because we hope to better observe the load disaggregation effect of the improved dAE algorithm and the FHMM algorithm, only the power waveform during the time when the device is turned on is selected, so the abscissa time only lasts for 6000 s, that is 2000 data points. In Figure 4, the waveform of line 1 represents the actual power curve of the load, the waveform of line 2 represents the load identification result based on the improved dAE algorithm, the waveform of line 3 represents the load identification result based on the standard DAE algorithm, and the waveform of line 4 represents the load identification result based on the FHMM algorithm.

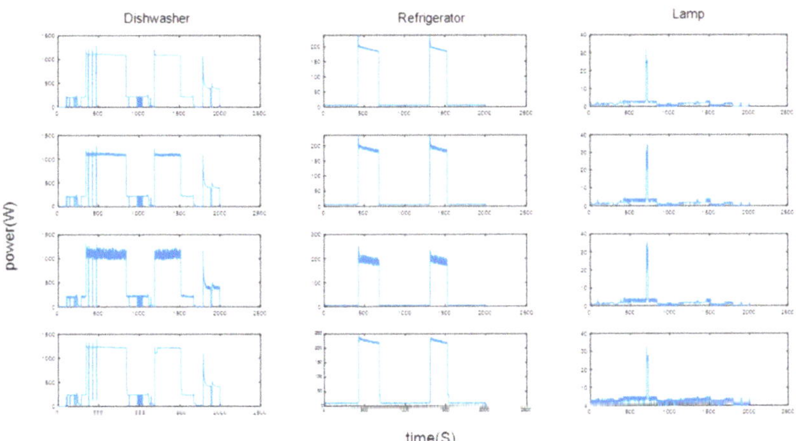

Figure 4. Identification results of three devices in home 1. (Line 1: the actual power curve of the load; Line 2: the load identification result based on the improved dAE algorithm; Line 3: the load identification result based on the standard dAE algorithm; Line 4: the load identification result based on the FHMM algorithm).

It should be noted that commonly used household electrical equipment can be divided into three categories from the operating state: single state class, continuous change class, and multi-state class:

Single state class: This means that there is only one stable state after the device is turned on, and the power generally remains unchanged, such as lamps, kettles, microwave ovens, etc.

Continuous change type: This means that the power of the device will have a continuous increase/decrease process during the process of turning on/off, such as TV (power change 50 W–75 W), computer (80 W–100 W), etc.

Multi-state class: Refers to the device having multiple power states during operation, such as refrigerators, washing machines, dishwashers, dryers, etc.

Among these three types of electrical equipment, the identification of single-state and continuous-change types is relatively simple, while the multi-state type is easily confused with other equipment due to its great difference in power in different state stages.

As can be seen from Figure 4, for lamps belonging to the single-state category, the identification effects of the three algorithms are good, which can well reflect the on and off states of the device, and the calculation of the power consumption value is also relatively accurate. For the dishwashers and refrigerators belonging to the multi-state category, the load identification effect based on the improved dAE algorithm is better, which is reflected in two aspects: (1) It decomposes the real power consumption value of the equipment more accurately; (2) It detects the different state stages of the equipment more accurately, thereby reducing the probability of misjudgment.

Figure 5 shows the usage of the dishwasher in Household 1 on a certain day, and its usage time is in the interval of 10,000–12,000 s. This interval is enlarged and the identification results of the two algorithms are compared, as shown in Figure 6.

It can be clearly seen from the figure that the load identification algorithm based on the improved dAE only has a little jitter in the high-power operation state; the jitter error does not exceed 5%, and can well fit the switching process between the states. Overall, the identification method based on FHMM has a higher power decomposition result; the amplitude is close to 20% and cannot accurately represent the load switching process. The result from standard dAE is also included for comparison, from which we can see that it has much more fluctuation. Especially at the time 1100 s, there is a big spike.

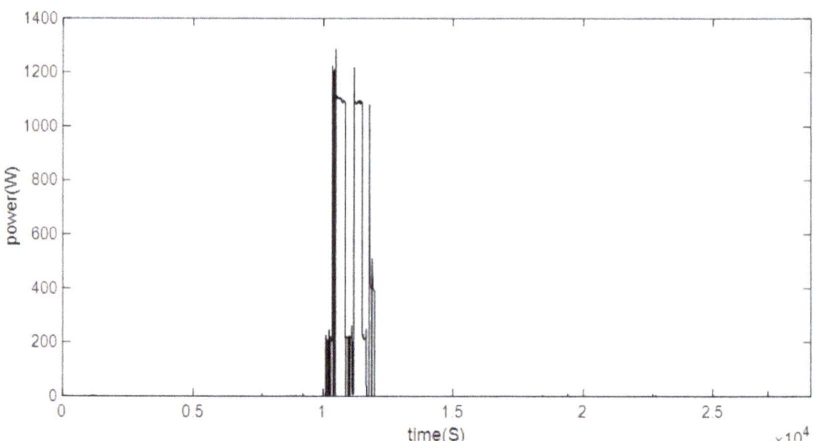

Figure 5. The actual daily energy consumption of Household 1's dishwasher.

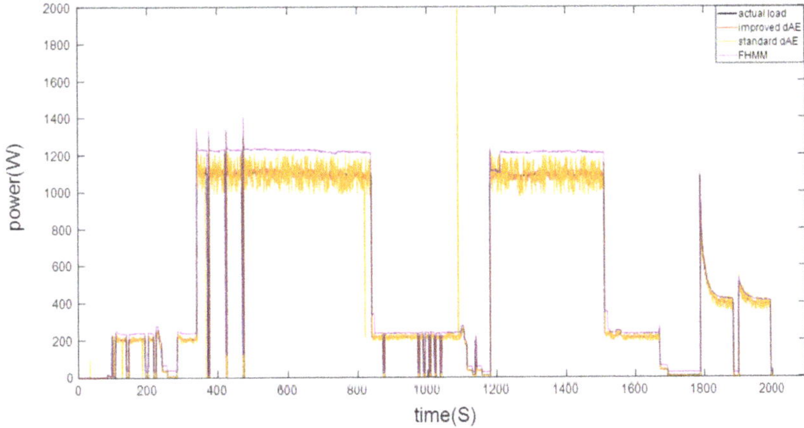

Figure 6. Actual energy consumption of dishwasher in Household 1 in one day.

Table 1 compares the four indicators of the three algorithms. These four indicators are defined and explained in Section 4.1. They represent the accuracy of energy consumption disaggregation (the bigger the better), the accuracy of the device status detection (the bigger the better), the NEP, which represents the deviation of the power disaggregation result from the actual value (the smaller the better), and the Matthews Correlation Coefficient (MCC), which represents the accuracy of the state detection (the closer to 1 the better). Due to space limitations, the table only lists the comparison of 5 kinds of equipment. It can be seen from the table that all indicators obtained by the improved dAE algorithm are better than the FHMM algorithm. The percentage of improvement regarding improved dAE and standard dAE is listed on the far-right side of the table, and the bold font indicates better performance of the proposed algorithm.

Table 1. Comparison of identification indexes of several equipment using REDD dataset.

Algorithm	Index	Oven	Refrigerator	Dish Washer	Lamp	Washer Dryer	Overall Performance	Improvement *
FHMM	$F_1^{(E)}\%$	33.2	22.7	50.0	45.3	80.3	46.30	
	$F_1^{(S)}\%$	78.6	42.6	21.5	36.3	52.3	46.26	
	NEP	2.652	0.709	3.222	1.562	0.441	1.7172	
	MCC	0.223	0.420	0.478	0.423	0.652	0.4392	
Standard dAE	$F_1^{(E)}\%$	42.6	45.6	70.5	59.6	85.4	60.74	
	$F_1^{(S)}\%$	82.6	58.1	44.9	55.0	66.2	61.36	
	NEP	1.852	0.652	1.256	1.006	0.333	1.020	
	MCC	0.455	0.558	0.658	0.455	0.742	0.574	
Improved dAE	$F_1^{(E)}\%$	78.5	66.84	88.8	69.0	99.3	80.50	**32.5%**
	$F_1^{(S)}\%$	92.3	65.96	65.3	67.5	74.5	73.11	**19.1%**
	NEP	0.389	0.520	0.226	0.265	0.225	0.325	**68.1%**
	MCC	0.674	0.685	0.783	0.885	0.898	0.785	**36.8%**

* percentage of improvement regarding the improved dAE and standard dAE.

Due to the variety of types of household electrical appliances, there may be differences in the power consumption behavior of different types of equipment. To test the generality of the algorithm, we trained the network using the data of Household 1, Household 3, and Household 4, and the trained network decomposes the ensemble power of Household 2. Figure 7 shows the results of identifying each device in Household 2 after using the data of Household 1 for network training. Due to space limitations, only the comparison results of three devices are shown, namely stove, microwave, and sterilizer. In Figure 7, the waveform of line 1 represents the actual power curve of the load, the waveform of line 2 represents the load identification result based on the improved dAE algorithm, the waveform of line 3 represents the load identification result based on the standard dAE algorithm, and the waveform of line 4 represents the load identification result based on the FHMM algorithm. It can be seen from the figure that, for single-state microwave ovens and sterilizers, all three algorithms can properly identify the equipment, while for stoves with multiple states, the improved dAE algorithm is obviously better than the standard dAE or FHMM algorithm.

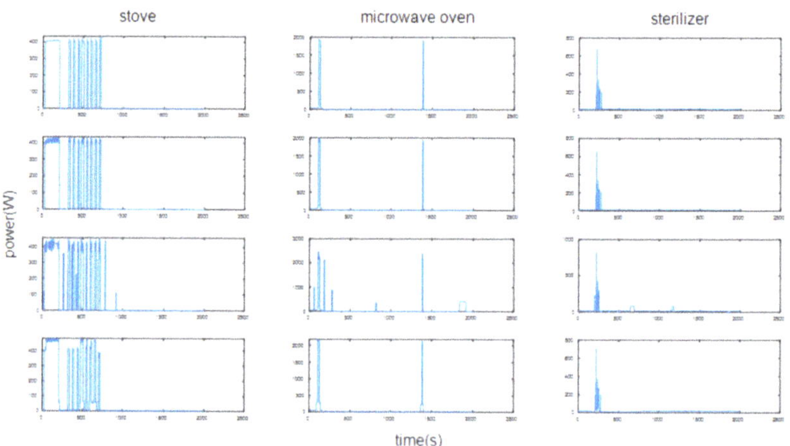

Figure 7. Identification result of Household 2 after using the network trained by dataset of Household 1 (Line 1: the actual power curve of the load; Line 2: the load identification result based on the improved dAE algorithm; Line 3: the load identification result based on the standard dAE algorithm; Line 4: the load identification result based on the FHMM algorithm).

4.2.2. Performance Test Using TraceBase Dataset

The TraceBase dataset contains 31 different types of devices, 122 devices, and 1270 pieces of load power consumption data. The data collection interval is 1–2 s. We used two algorithms to identify 20 of these devices, and selected the identification results of TV sets, desktop computers, and electric irons to display, as shown in Figure 8. In order to better illustrate the pattern of electric iron, the abscissa axis is truncated from time 0 to 800 s because the power assumption is 0 afterwards. As can be seen from the figure, the improved dAE algorithm has obvious advantages in both identifying the power consumption of the real equipment and detecting the different stages of the equipment.

Figure 9 compares the recognition performance of the three algorithms on a desktop computer from 15,000 s to 25,000 s. It can be seen from the figure that the jitter error of the load identification algorithm based on the improved dAE does not exceed 4%, and it can well fit the switching process between states, while the decomposition method based on standard dAE and FHMM are not accurate at the time of load start and stop. Additionally, the overall decomposed load power is too high. Table 2 compares the four indexes of the three algorithms. It can be seen from the table that all indicators obtained by the improved dAE algorithm are better than the standard dAE or FHMM algorithm, and the overall performance value is listed on the second far-right side of the table.

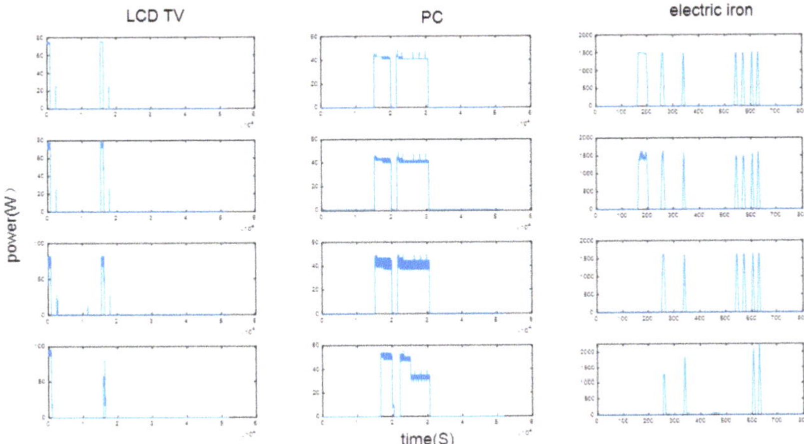

Figure 8. Identification Results of the three devices in TraceBase. (Line 1: the actual power curve of the load; Line 2: the load identification result based on the improved dAE algorithm; Line 3: the load identification result based on the standard dAE algorithm; Line 4: the load identification result based on the FHMM algorithm).

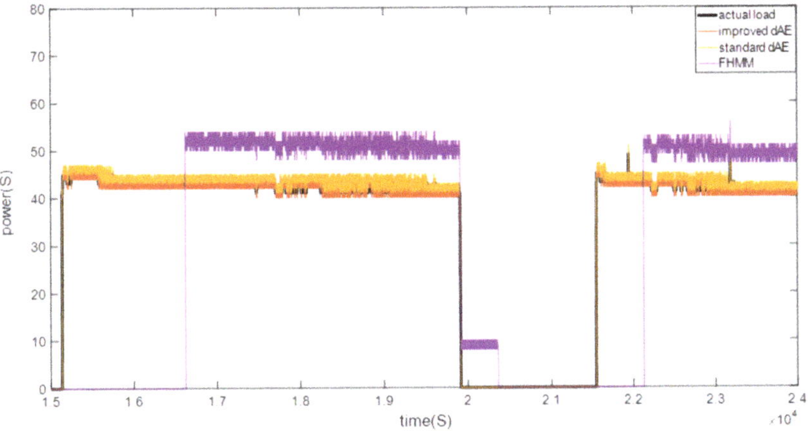

Figure 9. PC load identification results over a period of 15,000–25,000 s.

Table 2. Comparison of identification indexes of several equipment using TraceBase dataset.

Algorithm	Index	Coffee Machine	LCD-TV	Desktop	Wash Machine	Electric Iron	Overall Performance	Improvement *
FHMM	$F_1^{(E)}\%$	65.6	45.3	33.3	35.9	39.6	43.94	
	$F_1^{(S)}\%$	60.4	56.3	35.6	52.52	54.1	51.784	
	NEP	0.744	0.523	2.250	6.235	9.601	3.8706	
	MCC	0.732	0.729	0.420	0.452	0.333	0.5332	
Standard dAE	$F_1^{(E)}\%$	74.2	66.6	44.3	55.2	52.3	58.52	
	$F_1^{(S)}\%$	71.0	74.2	45.3	62.3	65.4	63.64	
	NEP	0.653	0.387	2.023	5.236	6.200	2.9000	
	MCC	0.741	0.774	0.625	0.650	0.661	0.6902	

Table 2. *Cont.*

Algorithm	Index	Coffee Machine	LCD-TV	Desktop	Wash Machine	Electric Iron	Overall Performance	Improvement *
Improved dAE	$F_1^{(E)}$%	87.3	77.6	55.6	65.2	87.6	**74.66**	27.60%
	$F_1^{(S)}$%	88.9	85.3	65.4	72.3	74.1	**77.20**	21.30%
	NEP	0.520	0.125	1.985	1.690	3.652	**1.5944**	45.00%
	MCC	0.812	0.874	0.898	0.870	0.704	**0.8316**	20.50%

* percentage of improvement regarding the improved dAE and standard dAE.

5. Conclusions

This paper proposes a non-intrusive load identification method that only relies on single active power measurements at a conventional sampling rate. This method is based on the Denoising Auto-Encoder (dAE) algorithm, which regards the total mixing power as a picture or a recording that needs to be processed, and the power generated by other unconcerned devices as "noise". The load power of the individual equipment is disaggregated from the total mixed power.

In the performance evaluation test, the REDD and TraceBase datasets are used to compare the effectiveness between the proposed method and the Factorial Hidden Markov Model (FHMM) algorithm, and four specific metrics for power disaggregation and state detection performance are introduced. The test results show that the proposed method has obvious advantages in both identifying the actual power consumption of the device and detecting the state of the device. In addition, the proposed algorithm has good generality and can effectively identify the same equipment of different models or brands.

Author Contributions: Conceptualization, methodology, X.H.; software, validation, X.H. and H.D.; data curation, manuscript review and revision, W.Y. and J.H. All authors have read and agreed to the published version of the manuscript.

Funding: This research was funded by Provincial Science and Technology Joint Fund Project, 2021–2023, grant number 2021JJ50082.

Institutional Review Board Statement: Not applicable.

Informed Consent Statement: Not applicable.

Data Availability Statement: Not applicable.

Conflicts of Interest: The authors declare no conflict of interest.

References

1. Ruano, A.; Hernandez, A.; Ureña, J.; Ruano, M.; Garcia, J. NILM techniques for intelligent home energy management and ambient assisted living: A review. *Energies* **2019**, *12*, 2203. [CrossRef]
2. Sadeghianpourhamami, N.; Ruyssinck, J.; Deschrijver, D.; Dhaene, T.; Develder, C. Comprehensive feature selection for appliance classification in NILM. *Energy Build.* **2017**, *151*, 98–106. [CrossRef]
3. Figueiredo, M.B.; Almeida, A.; Ribeiro, B. An Experimental Study on Electrical Signature Identification of Non-Intrusive Load Monitoring (Nilm) Systems. In Proceedings of the International Conference on Adaptive and Natural Computing Algorithms, Ljubljana, Slovenia, 14–16 April 2011; Springer: Berlin/Heidelberg, Germany, 2011; pp. 31–40.
4. Bouhouras, A.S.; Gkaidatzis, P.A.; Panagiotou, E.; Poulakis, N.; Christoforidis, G. CA NILM algorithm with enhanced disaggregation scheme under harmonic current vectors. *Energy Build.* **2019**, *183*, 392–407. [CrossRef]
5. Tu, C.; He, X.; Shuai, Z.; Jiang, F. Big data issues in smart grid—A review. *Renew. Sustain. Energy Rev.* **2017**, *79*, 1099–1107. [CrossRef]
6. Biansoongnern, S.; Plangklang, B. Nonintrusive load monitoring (NILM) using an Artificial Neural Network in embedded system with low sampling rate. In Proceedings of the 2016 13th International Conference on Electrical Engineering/Electronics, Computer, Telecommunications and Information Technology (ECTI-CON), Chiang Mai, Thailand, 28 June–1 July 2016; IEEE: Piscataway, NJ, USA, 2016; pp. 1–4.
7. Huber, P.; Calatroni, A.; Rumsch, A.; Paice, A. Review on deep neural networks applied to low-frequency nilm. *Energies* **2021**, *14*, 2390. [CrossRef]

8.   Revuelta Herrero, J.; Lozano Murciego, Á.; López Barriuso, A.; Hernández de la Iglesia, D.; Villarrubia González, G.; Corchado Rodríguez, J.M.; Carreira, R. Non Intrusive Load Monitoring (Nilm): A State of the Art. In Proceedings of the International Conference on Practical Applications of Agents and Multi-Agent Systems, Porto, Portugal, 21–23 June 2017; Springer: Cham, Switzerland, 2017; pp. 125–138.
9.   Machlev, R.; Tolkachov, D.; Levron, Y.; Beck, Y. Dimension reduction for NILM classification based on principle component analysis. *Electr. Power Syst. Res.* **2020**, *187*, 106459. [CrossRef]
10. Kelly, J.; Knottenbelt, W. Neural nilm: Deep neural networks applied to energy disaggregation. In Proceedings of the 2nd ACM International Conference on Embedded Systems for Energy-Efficient Built Environments, Seoul, Korea, 4–5 November 2015; pp. 55–64.
11. Batra, N.; Singh, A.; Whitehouse, K. Neighbourhood nilm: A big-data approach to household energy disaggregation. *arXiv* **2015**, arXiv:1511.02900.
12. Nalmpantis, C.; Vrakas, D. Machine learning approaches for non-intrusive load monitoring: From qualitative to quantitative comparation. *Artif. Intell. Rev.* **2019**, *52*, 217–243. [CrossRef]
13. Yang, F.; Liu, B.; Luan, W.; Zhao, B.; Liu, Z.; Xiao, X.; Zhang, R. FHMM Based Industrial Load Disaggregation. In Proceedings of the 2021 6th Asia Conference on Power and Electrical Engineering (ACPEE), Chongqing, China, 8–11 April 2021; IEEE: Piscataway, NJ, USA, 2021; pp. 330–334.
14. Reinhardt, A.; Baumann, P.; Burgstahler, D.; Hollick, M.; Chonov, H.; Werner, M.; Steinmetz, R. On the accuracy of appliance identification based on distributed load metering data. In Proceedings of the IFIP Conference on Sustainable Internet & Ict for Sustainability, Pisa, Italy, 4–5 October 2012; IEEE Computer Society: Washington, DC, USA, 2012.
15. Bouhouras, A.S.; Milioudis, A.N.; Labridis, D.P. Development of distinct load signatures for higher efficiency of NILM algorithms. *Electr. Power Syst. Res.* **2014**, *117*, 163–171. [CrossRef]
16. Desai, S.; Alhadad, R.; Mahmood, A.; Chilamkurti, N.; Rho, S. Multi-state energy classifier to evaluate the performance of the nilm algorithm. *Sensors* **2019**, *19*, 5236. [CrossRef]
17. Puente, C.; Palacios, R.; González-Arechavala, Y.; Sánchez-Úbeda, E.F. Non-intrusive load monitoring (NILM) for energy disaggregation using soft computing techniques. *Energies* **2020**, *13*, 3117. [CrossRef]
18. Dinesh, C.; Nettasinghe, B.W.; Godaliyadda, R.I.; Ekanayake MP, B.; Ekanayake, J.; Wijayakulasooriya, J.V. Residential appliance identification based on spectral information of low frequency smart meter measurements. *IEEE Trans. Smart Grid* **2016**, *7*, 2781–2792. [CrossRef]
19. Gillis, J.M.; Morsi, W.G. Non-intrusive load monitoring using semi-supervised machine learning and wavelet design. *IEEE Trans. Smart Grid* **2017**, *8*, 2648–2655. [CrossRef]
20. Li, D.; Dick, S. Non-intrusive load monitoring using multi-label classification methods. *Electr. Eng.* **2020**, *103*, 607–619. [CrossRef]
21. Ahmadi, H.; Marti, J.R. Load decomposition at smart meters level using eigenloads approach. *IEEE Trans. Power. Syst.* **2015**, *30*, 3425–3436. [CrossRef]
22. Cominola, A.; Giuliani, M.; Piga, D.; Castelletti, A.; Rizzoli, A.E. A hybrid signature-based iterative disaggregation algorithm for non-intrusive load monitoring. *Appl. Energy.* **2017**, *185*, 331–344. [CrossRef]
23. Makonin, S.; Popowich, F.; Bajić, I.V.; Gill, B.; Bartram, L. Exploiting hmm sparsity to perform online real-time nonintrusive load monitoring. *IEEE Trans. Smart Grid* **2016**, *7*, 2575–2585. [CrossRef]
24. Tsai, M.S.; Lin, Y.H. Modern development of an adaptive non-intrusive appliance load monitoring system in electricity energy conservation. *Appl. Energy* **2012**, *96*, 55–73. [CrossRef]
25. Mauch, L.; Yang, B. A new approach for supervised power disaggregation by using a deep recurrent LSTM network. In Proceedings of the 2015 IEEE Global Conference on Signal and Information Processing (GlobalSIP), Orlando, FL, USA, 14–16 December 2015; pp. 63–67.
26. Figueiredo, M.; De Almeida, A.; Ribeiro, B. Home electrical signal disaggregation for non-intrusive load monitoring (NILM) systems. *Neurocomputing* **2012**, *96*, 66–73. [CrossRef]
27. García-Pérez, D.; Pérez-López, D.; Díaz-Blanco, I.; González-Muñiz, A.; Domínguez-González, M.; Vega AA, C. Fully-convolutional denoising auto-encoders for NILM in large non-residential buildings. *IEEE Trans. Smart Grid* **2020**, *12*, 2722–2731. [CrossRef]
28. Bonfigli, R.; Felicetti, A.; Principi, E.; Fagiani, M.; Squartini, S.; Piazza, F. Denoising autoencoders for non-intrusive load monitoring: Improvements and comparative evaluation. *Energy Build.* **2018**, *158*, 1461–1474. [CrossRef]
29. Kolter, J.Z.; Johnson, M.J. REDD: A public data set for energy disaggregation research. *Artif. Intell.* **2011**, *25*, 59–62.
30. Kelly, J.; Knottenbelt, W. The UK-DALE dataset, domestic appliance-level electricity demand and whole-house demand from five UK homes. *Sci. Data* **2015**, *2*, 150007. [CrossRef] [PubMed]
31. Parson, O.; Fisher, G.; Hersey, A.; Batra, N. Dataport and NILMTK: A building data set designed for non-intrusive load monitoring. In Proceedings of the 3rd IEEE Global Conference on Signal & Information Processing, Orlando, FL, USA, 14–16 December 2015; IEEE: Piscataway, NJ, USA, 2015; pp. 210–214.
32. Vincent, P.; Larochelle, H.; Lajoie, I.; Bengio, Y.; Manzagol, P.A.; Bottou, L. Stacked denoising autoencoders: Learning useful representations in a deep network with a local denoising criterion. *J. Mach. Learn. Res.* **2010**, *11*, 3371–3408.

Article

Short Text Classification for Faults Information of Secondary Equipment Based on Convolutional Neural Networks

Jiufu Liu *, Hongzhong Ma, Xiaolei Xie and Jun Cheng

College of Energy and Electrical Engineering, Hohai University, Nanjing 210098, China;
19960072@hhu.edu.cn (H.M.); 20060042@hhu.edu.cn (X.X.); 20060011@hhu.edu.cn (J.C.)
* Correspondence: 19940063@hhu.edu.cn; Tel.: +86-138-1408-1399

Abstract: As the construction of smart grids is in full swing, the number of secondary equipment is also increasing, resulting in an explosive growth of power big data, which is related to the safe and stable operation of power systems. During the operation of the secondary equipment, a large amount of short text data of faults and defects are accumulated, and they are often manually recorded by transportation inspection personnel to complete the classification of defects. Therefore, an automatic text classification based on convolutional neural networks (CNN) is proposed in this paper. Firstly, the topic model is used to mine the global features. At the same time, the word2vec word vector model is used to mine the contextual semantic features of words. Then, the improved LDA topic word vector and word2vec word vector are combined to absorb their respective advantages and utilizations. Finally, the validity and accuracy of the model is verified using actual operational data from the northwest power grid as case study.

Keywords: secondary equipment; CNN; short text classification

Citation: Liu, J.; Ma, H.; Xie, X.; Cheng, J. Short Text Classification for Faults Information of Secondary Equipment Based on Convolutional Neural Networks. *Energies* **2022**, *15*, 2400. https://doi.org/10.3390/en15072400

Academic Editor: Abu-Siada Ahmed

Received: 11 February 2022
Accepted: 22 March 2022
Published: 24 March 2022

Publisher's Note: MDPI stays neutral with regard to jurisdictional claims in published maps and institutional affiliations.

1. Introduction

With the accelerated development of the power system, the number of secondary equipment is also increasing. Therefore, there is an explosive emergence of power big data, which hides massive information. This is related to the safe and stable operation of the power system [1–4]. However, a small proportion of these data can be used to mine important information, and research on these data has become a current hot topic. Among these data, the first category is the time-series structured data represented by output power, temperature and humidity of the equipment and its environment, and the light intensity of the optical module. This type of data mining work is relatively mature; the other is based on semi-structured and unstructured data represented by text, images, audio, etc., which are difficult to express using relational databases. The low value density of these data restricts the mining of unstructured data [5].

During the operation of the secondary equipment, a lot of short text data of faults and defects have been accumulated. These data are often manually recorded by the transportation inspection personnel and rely on the experience of professionals to complete the classification of defects. However, due to the subjective and empirical constraints of transport inspection personnel, the fault data are difficult to classify accurately. At the same time, the high volume of fault data requires a great deal of human participant involvement, and efficiency is difficult to be guaranteed. Moreover, the text information of secondary equipment faults has a short length and sparse semantic features. The improvement of the classification model for short text data is also the focus and hotspot [6].

The earliest text classification can be traced back to an article published by Maron in 1961 on the method of text classification using the Bayesian formula. In the next 20-odd years, a series of classification rules were manually built on the basis of expert knowledge to construct a classifier. This method often requires the experience and knowledge of a large number of expert engineers in related fields, which is difficult to effectively promote [7]. In addition to the development of disciplines such as artificial intelligence, machine learning, pattern recognition, and statistical theory, text classification technology has entered a more intelligent automatic classification era, and text classification methods based on expert knowledge and experience have gradually withdrawn from the historical stage. Using Bayes' [8,9] neural network [10] and support vector machine [11] and other methods to liberate people from heavy tasks, and with high classification efficiency and accuracy, the machine learning methods have developed rapidly in the field of Chinese text classification. Benefits from the development of the machine learning, neural network method are the most prominent [12]. According to some papers, it can be concluded that the long and short-time neural network models used to mine context features has a significant effect on the classification of long document text data [13,14], and the convolutional neural network model has a significant effect on the classification of short text data [15]. In [16], the CNN model was proposed for brain tumor classification. In [17], a feature fusion method based on an ensemble convolutional neural network and deep neural network was used for bearing fault diagnosis. In [18], an enhanced convolutional neural network was designed and analyzed.

The text classification technology is also widely used in professional fields, such as social science information, biomedicine, and so on [19]. There are also endless categories of patents [20], academic papers, academic news, and even the content of WeChat public accounts. In social media, the classification of user emotion recognition is an important part [21]. In e-commerce, user evaluation of products can help companies understand user satisfaction with products [22]. In biomedicine, intelligent triage can save a lot of medical resources and improve the quality and efficiency of services [23].

Text data mining in the power industry is still in the emerging field, and foreign countries have studied the relationship between the historical fault data and the weather to further predict the fault of the substation. However, these text mining methods are mostly based on traditional machine learning methods, seldom adopt deep learning methods, and lack research on the classification of a specific device type or the fault text data itself. Generally speaking, the text mining technology is still in its infancy in the field of electric power, especially the research on text information of secondary equipment faults; most of the research is only based on traditional machine learning methods, and the classification model lacks pertinence [24]. Moreover, due to the short text length and lack of sufficient context for semantic feature analysis, when mining this type of text data, it is easy to cause high-dimensional information features to be sparse, resulting in a serious lack of semantic relations, and ultimately resulting in poor classification results [25]. Considering that some faulty text data are short and the traditional convolutional neural network is insufficient for feature extraction, in this paper, convolution kernels of various sizes are used to extract features from short text data.

Based on the above discussions, this paper focuses on a mass of short text data produced in the secondary equipment operation production management system and conducts related research on automatic text classification based on convolutional neural networks. In order to solve the problem of poor topic focus and sparse text density in short text data, an improved LDA topic model was proposed based on the Relevance formula for the problem of insignificant characteristics caused by excessive repetition of feature text information [26]. By setting different weighting coefficients to adjust the sampling of words, the problem of repeated vocabulary of different types of defective data was solved. Afterwards, the RLDA model and the word2vec [27] model were combined together, and the document-topic vector was constructed using the RLDA subject word model to obtain the global features. At the same time, the local features attained by using the word2vec word vector technology to mine the latent semantic features were combined. Construct the input matrix of convolutional neural network. Considering the superiority of convolutional neural for feature extraction at the level of short text information word vectors, convolutional neural networks were employed for extracting feature text vectors and classifying text vectors. The traditional convolutional neural network uses a single size convolution kernel to extract features. When faced with different document lengths, the classification results are not ideal. On the basis of the original convolution model, this paper proposes to use deep convolution kernels of multiple sizes to mine text features in depth to enhance their ability to extract locally sensitive information. Finally, the actual operation data of a northwestern power system company were used to conduct a comparative experiment to test the validity of the presented model and the accuracy of the classification algorithm in this paper.

2. Lower-Level Modeling and Optimization

2.1. Data Analysis

This paper randomly selects 1000 defect text data from a power company in a northwestern province from 2015 to 2019, according to the length of the string statistics as shown in Figure 1.

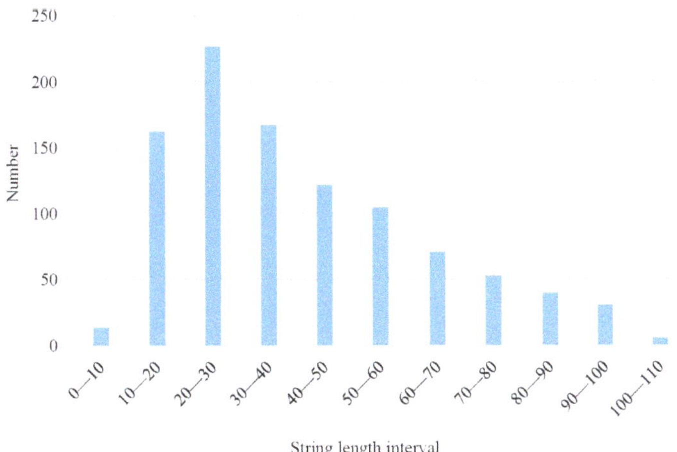

Figure 1. Interval diagram of text information length distribution of secondary equipment faults in a northwestern power system.

The fault data selected were caused by the manufacturing of the same auxiliary device from two different devices recorded by the same person in charge of a northwest power system in January 2017. The two devices belong to the 220 kV plant, and the same batch was delivered by the same company. The secondary equipment protection device of the model PSIU601GC-B-E1 omitted the same data and the content that has less influence on the classification result. The content is shown in Table 1.

Table 1. Example of the text information record of a second equipment failure in a northwestern company in January.

Number	Defects and Treatment Methods	Defect Classification
1	The main set B intelligent terminal sends a GOOSE link interruption general alarm, which does not affect the normal operation of the B set protection and needs to be exited for inspection.	Serious defect
2	General alarm of GOOSE in A set of intelligent terminals of Yanhua Temple Line, data of GOOSE network is interrupted.	Critical defect

Compared with the general Chinese short text, the text of the secondary equipment fault defect of the power system not only contains the unique attributes of the Chinese language family and the Asian-European language family, but also has the following characteristics:

(1) Fault and defect data are deeply involved in the professional field of power systems, including many low-frequency words such as electrical professional vocabulary, equipment names, and equipment models. Because the same vocabulary is in different fields, it brings different common names or abbreviations, such as GIS, which represents the geographic location information system at the large level of the power system, and gas insulated combined electrical appliances at the device level.

(2) Due to the classification of the secondary equipment based on the fault category, the same fault location, such as the problem of the display screen, has different defect level definition results according to the display screen, blue screen, and display failure.

(3) Most of the fault data are based on the data manually recorded by the transport inspection personnel. The details of the text records are slightly different. The text length of each piece of defect data varies greatly. The shortest data are less than ten characters, and the longest data can be up to more than 100 morphemes.

(4) The defect data of different fault categories have high similarity and lack sufficient semantic co-occurrence. Traditional text mining methods have limitations for short text data mining and classification with high similarity.

Through the above feature analysis of short text, it is not ideal to directly apply the topic model to text classification.

2.2. Text Classification Process for Chinese Characters

For text classification for Chinese characters, the machine learning method is always utilized to find the correspondence between text features and its categories, and relevant technology is used for automatic classification of new text because of its laws.

The steps of the aforementioned text classification model for Chinese characters can be summarized as follows. Firstly, the preprocessing of the text is completed, where the unnecessary information is removed, such as clauses, word segmentation, and stop words. This step is implemented according to the text length and text. The specific content is related. Then, the text can be expressed, namely, the text is transmitted into a computer-recognizable and processed form, which is usually expressed by a matrix or a vector. The text representation affects the effect of later text classification because it is related to the extraction of text features. Then, a suitable classifier is selected to classify the text and output the predicted classification result. Finally, the aforementioned two results of the classifier are compared (practical and predicted results). If the prediction results meet a prior standard, the training is completed, where this standard could be the prediction accuracy rate and iterations. Otherwise, the corresponding parameters need to be adjusted by means of the comparison result, and the classification is re-classified until the classification prediction result reaches the standard.

3. Short Text Data Model of Secondary Equipment Faults in Power Systems Based on LDA Topic Model and Convolutional Neural Network

3.1. Improved Text Classification Process

The quality of text representation directly affects the effect of final classification. Transforming Chinese language into a structured language that can be recognized by the computer is the process of feature extraction and semantic abstraction of Chinese text. The traditional LDA model uses the external corpus or the method of merging short texts to improve the semantic information between words, but the word vector captured by the topic model is the word bag model, that is, the two phrases "a before B" and "B before a" are characterized as the same word vector after extracting the features from the topic model. However, most of the original data in this paper were based on the manual records of operation and maintenance personnel, and it is difficult for different people to form a standardized recording method. In the face of short text feature mining with poor context dependence such as fault data, the classification result obtained by directly using LDA model is poor.

In this paper, the RLDA model was used to extract global features to construct the subject word vector, and the word2vec model was used to mine the potential feature vector extracted by local features. The two features were combined to absorb their respective advantages as the input of the convolutional neural network.

3.1.1. Text Preprocessing

Consulting the published work [28], the collected short text data can be labeled as serious, critical, and general defects for the secondary equipment. In a ratio of 7:2:1, the obtained text short messages could be defined as training set, verification set, and test set. The top 30 terms in terms of frequency without text pre-processing are shown in Figure 2.

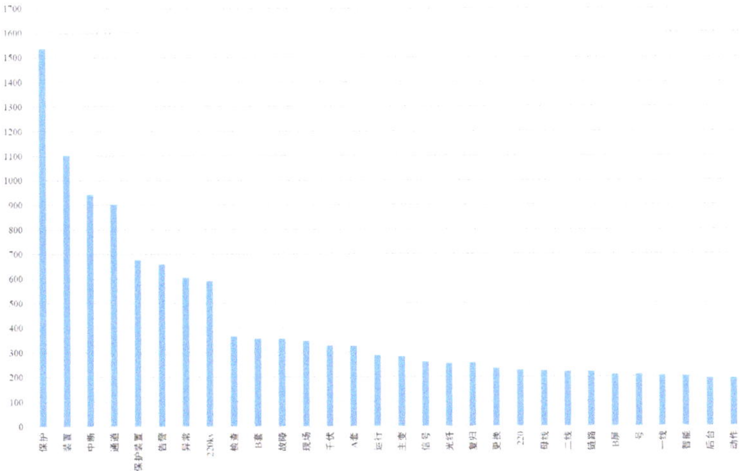

Figure 2. The top 30 words without text pre-processing.

Analyzing and summarizing the natural language characteristics of the defective text data, the secondary equipment defect text data cleaning was based on the following steps:

(1) Remove useless characters. Defective text generally involves a great deal of spaces. Some useless characters should be filtered, such as punctuation and so on, because they are not related to the text content. In Chinese, the words "I" and "do" are used a lot. By utilizing excessive words, the accuracy of the segmentation is increased and the efficiency is decreased. Meanwhile, the words "no" and "yes" are also used a lot in prepositions, conjunctions, and adverbs. These words are usually meaningless.

(2) English characters are uniformly given in the form of lowercase. In the secondary equipment, the recording format of the defect text is not standardized due to the fact that it includes a lot of English characters, such as "10 KV", "10 kv", and "10 Kv" for the description of transformer grades. They all represent the same voltage level, but the recording format is different.

(3) The repeated records and fragmentary text are detected and removed. When the defect records are uploaded, some problems, including data loss and repeated data entry, are produced easily for operation and maintenance personnel due to improper operations. The text classification and information mining are not easy to implement by using such data, where these data should be processed in advance to guarantee the quality of the text.

(4) A professional dictionary should be constructed for secondary equipment. The establishment of a special dictionary corresponding to the professional field is the

basic work of text mining in various professional fields. The quality and quantity of words included in its professional dictionary determine the accuracy of word segmentation and the part of speech tagging in text preprocessing. Due to the large number and miscellaneous types of electrical secondary equipment, the number of words related to the construction of this field is very huge, and there are thousands of words describing the equipment itself, such as the transformer station names, equipment protection proper terms, and so on.

3.1.2. Text Classification Model by Using LDA

The LDA topic model features based on short text data from secondary equipment are explained as follows:

(1) Initializing model parameters α, β and K, where α, β and K are the denoted prior parameters file-theme distribution parameter, theme-word distribution parameter, and the number of themes K, respectively [26].
(2) Traverse and classify short text data, and for each word w_i in terms of the list L_i, build $\theta_i = Dirichlet(\alpha)$ where θ_i and L_i, stand for the document-topic distribution and the adjacent word of w_i respectively.
(3) Suppose that Z satisfies the Dirichlet prior distribution, where Z is the potential word set. Moreover, the computational formula $\phi_Z = Dirichlet(\beta)$ is utilized in this step, in which ϕ_Z stands for the topic-word distribution.
(4) In view of each word in L_i, choose words $Z_j \sim \theta_i$ and $w_j \sim \phi_{Z_j}$ with $Z_j \sim \theta_i$ and $w_j \sim \phi_{Z_j}$ being potential and neighboring words, respectively; attain short texts with the help of the documents. Then, the subject matter is inferred from the secondary device short text data on the basis of the following expression:

$$P(w_i|d) = \frac{f_d(w_i)}{Len(d)} \tag{1}$$

where $f_d(w_i)$ represents the frequency of the words in the document, and $Len(d)$ stands for the length of the short text d.

Inspired by [26], the expectation of the topic distribution for document-generating words can be regarded as the distribution of document-generating topics:

$$P(z|d) = \sum_{w_i \in W_d} P(z|w_i)P(w_i|d) \tag{2}$$

where $P(z|d)$, W_d, and $P(z|w_i)$ are the probability of the text generating words, the short text set, and the probability of the word generating topics, respectively.

The LDA topic generation model was established. Then, we needed to implement the Gibbs sampling estimation based on the corresponding model parameters and give the number of iterations. Finally, the topic distribution matrix of any text in the corpus could be obtained after completing the model training.

3.1.3. Improved LDA Topic Analysis Model Based on Relevance Formula

In this paper, the LDA topic model was improved by introducing a weighting coefficient λ in the topic correction layer to realize the model's potential topic extraction and topic correction function for secondary equipment fault text information. The proposed model is shown in Figure 3. The Relevance formula is as follows:

$$r(w,k|\lambda) = \lambda \cdot \log(\phi_{k,w}) + (1-\lambda) \cdot \frac{\phi_{k,w}}{p_w} \tag{3}$$

where $r(w,k|\lambda)$ represents the degree of relevance of word w and topic k under the set weight coefficient λ. The value range of λ is ($0 \leq \lambda \leq 1$). $\phi_{k,w}$ is the probability distribution matrix of the words w under the topic k, and the marginal probability of the words under the topic-term matrix ϕ.

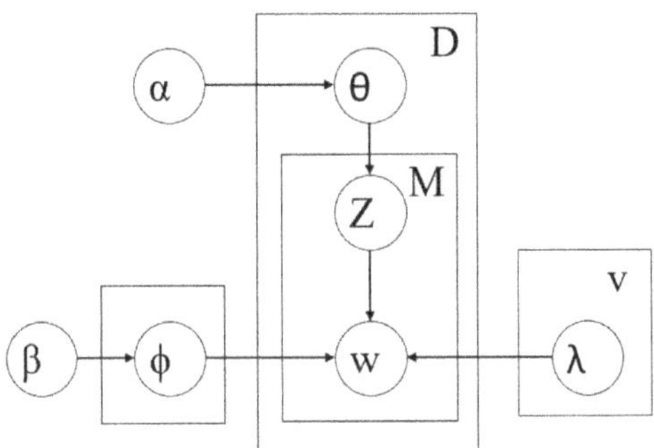

Figure 3. RLDA model structure diagram.

From Equation (3), we can dynamically adjust the relationship between words and subjects by establishing weight coefficients. When the weight coefficient λ is close to 1, the more frequent the words appear in the word frequency, the higher contribution to the document theme, that is, the more frequent words in the default document are more relevant to the topic; when the weight coefficient λ is close to 0, the improved model indicates that the word appears more frequently in the selected topic, but less frequently in other topics; that is, the words and topics generally appear concomitantly.

3.2. Fusion of Word2vec Model and RLDA Model

In order to increase the interpretability of the text feature vector to the text representation, the improved LDA subject word model was proposed based on the Relevance formula to extract the global features to construct the subject word vector, and the latent feature vector extracted using the word2vec algorithm. By combining two features, the following new text feature representation is given by

$$v'_m = [z_m^T, \theta_m^T]^T \tag{4}$$

where z_m is the latent semantic vector representation of the document, θ_m is the latent text-topic vector of the text extracted based on the improved topic model of the Relevance formula, v'_m is the combined semantic feature representation vector, and T is the transpose operation on the matrix.

The topic vector and the latent semantic vector are different in the dimension representation of the word vector. In order to eliminate the influence of the difference in magnitude generated by the fusion of the two vectors on the final classification result, this paper summarizes the two vectors z_m and θ_m. In a one-way combination, the processing method is as follows:

$$v_m = \left[\frac{z_m^T}{\|z_m\|}, \frac{\theta_m^T}{\|\theta_m\|} \right]^T \tag{5}$$

The vectors combined by normalization not only regularize the length and eliminate the gap in magnitude between the two vectors, but also the new vectors generated by the fusion have both topical and potentially topical features.

In the following, the text classification model is constructed based on convolutional neural network. By means of the convolutional neural network, a four-layer model was developed, which is shown in Figure 4.

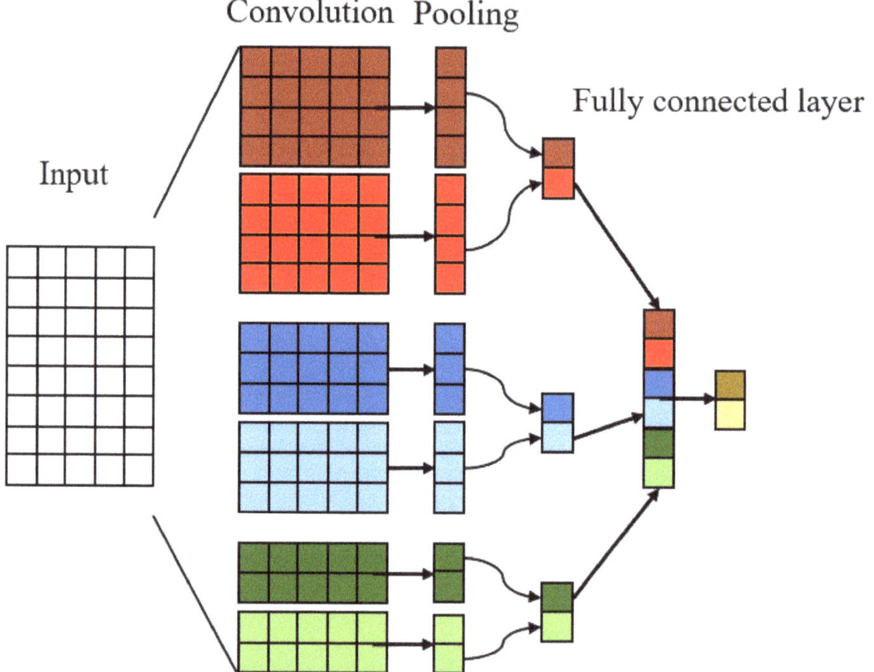

Figure 4. Text convolutional neural network model.

The detailed design are presented as following four parts:

(1) The first layer

The first layer could be defined as the input layer. In this layer, a length of text data was selected, and the vectorization of the text data was implemented with the help of step C. Employing the matrix $I \in \mathbb{R}^{m \times n}$ as the input and defining the number of words as m, m represents the number of rows in the input layer. Similarly, we defined the dimension of the text vector as n, which can represent the columns of the input layer. Then, all word data could be divided into word vectors of equal dimensions, namely, the number of columns is the same in the input layer. Accordingly, matrix $I \in \mathbb{R}^{m \times n}$ was constructed. During the training process, we employed the stochastic gradient descent method to adjust the word vector.

(2) The second layer

The second layer was named as the convolution layer. Each scale includes two convolution kernels that have the scales of $3 \times n, 4 \times n, 5 \times n$ Then, for the input matrix $I \in \mathbb{R}^{m \times n}$ of the input layer, we needed to implement the convolution operation and acquire the matrix features of the input layer. The corresponding result vector could be attained ($c_i (i = 1, 2, 3, 4, 5, 6)$), which was input to the pooling layer for data compression. Meanwhile, the activation function ReLU was used to activate the convolution result. After each convolution operation, one convolution result will be obtained:

$$r_i = W \cdot I_{i:i+h-1} \tag{6}$$

where the size of $i = 1, 2, \cdots, s - h + 1$ and $I_{i:i+h-1}$ are the size of convolution kernel, which represents the number i of $h \times n$ matrix block from top to bottom when the matrix I is operated in sequence; "$\cdot$" means that the elements at the corresponding positions of two matrix blocks are multiplied first and then added. Meanwhile, the activation function ReLU was used to activate the convolution result. Nonlinear processing was carried out for each convolution result r_i, and the result c_i was obtained after each operation. The formula is as follows:

$$c_i = \mathrm{ReLU}(r_i + b) \tag{7}$$

where b is the offset coefficient. Each such operation will produce a nonlinear result c_i. Because $i = 1, 2, \cdots, s - h + 1$, after $s - h + 1$ convolution operations on the input matrix from top to bottom, we should arrange the results in order, and obtain the vector of the convolution layer $c \in R^{s-h+1}$, which is shown as:

$$c = [c_1, c_2, \ldots, c_{s-h+1}] \tag{8}$$

(3) The third layer

We defined the third layer as the poling layer and employed the maximum pooling method for pooling. For the convolution result vector c_i, the largest element was chosen as the feature value, which is defined as $p_j (j = 1, 2, 3, 4, 5, 6)$. Then, the value p_j was injected in succession into the vector $p \in \mathbb{R}^{6 \times 1}$, which was input to the output layer of the next layer. Vector p stands for the global features of the text data, and it can reduce the dimensionality of the features and enhance the efficiency of classification.

(4) The fourth layer

Here, the output layer was utilized to name the fourth layer. We plugged the pooling layer completely into the output layer. In the pooling layer, we selected the vector p as an input, which was classified with the help of a SoftMax classifier. Then, the final classification result was output. The probability was computed using SoftMax classification, which is as follows:

$$L(p_j) = \frac{e^{p_j}}{\sum\limits_{j=1}^{6} e^{p_j}} \tag{9}$$

where the formula (9) refers to the probability that belongs to the secondary device category.

The fault level was output for the secondary equipment. The traditional convolutional neural network used a single size convolution kernel to extract features. When faced with different document lengths, the classification results were not ideal. On the basis of the original convolution model, the deep convolution kernels of multiple sizes were utilized to mine text features in depth to enhance their ability to extract locally sensitive information, so that they can represent more feature information. To make a clear statement, the overall flow chart of the proposed model in this paper is shown in Figure 5.

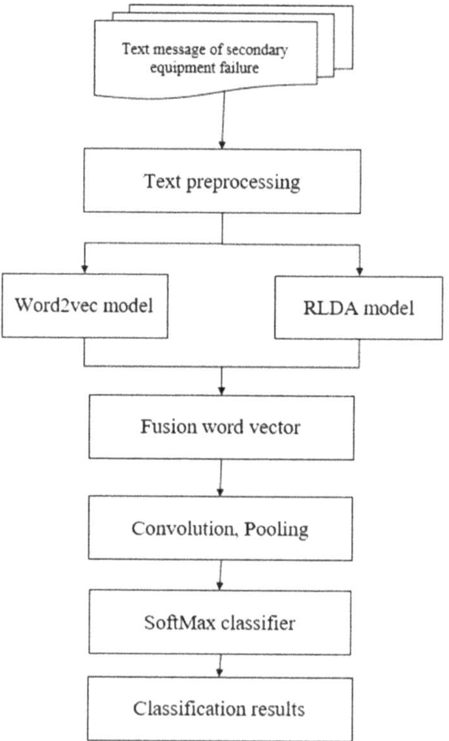

Figure 5. The flow chart of the proposed text classification algorithm.

4. Case Study

4.1. RLDA Model Experiment

In order to compare the advantages and disadvantages of the LDA model and the improvement of the LDA model based on the Relevance formula in terms of prediction ability and generalization ability, this experiment used the theme consistency (coherence score) indicator. Generally, the larger the value, the stronger the predictive ability and generalization ability of the model, indicating that the model was more practical. According to the characteristics of the experimental data set, the main parameter values set by the text are shown in Table 2, and K represents the number of topics contained.

Table 2. Parameter setting of the RLDA model.

Parameter	Value
Hyperparameter α	50/K
Hyperparameter β	0.01
Gibbs sampling iterations	1000
Input word vector	Word2vec
Filter size	(3,4,5)
Number of filters per size	100
Activation function	ReLU
Pooling strategy	1-max pooling
Dropout rate	0.5

In this paper, the comparison experiment was carried out by changing the value of the number of topics K. Under different values of the number of topics, the corresponding coherence score value of the improved LDA model based on the Relevance formula was calculated according to the theme consistency calculation formula. The experimental comparison results are shown in Figures 6–9. As shown in Figure 6, as the number of topics continued to increase, the coherence score had a process of increasing first, then decreasing, and then slowly smoothing out. The score is the highest when the number of topics is about seven to eight.

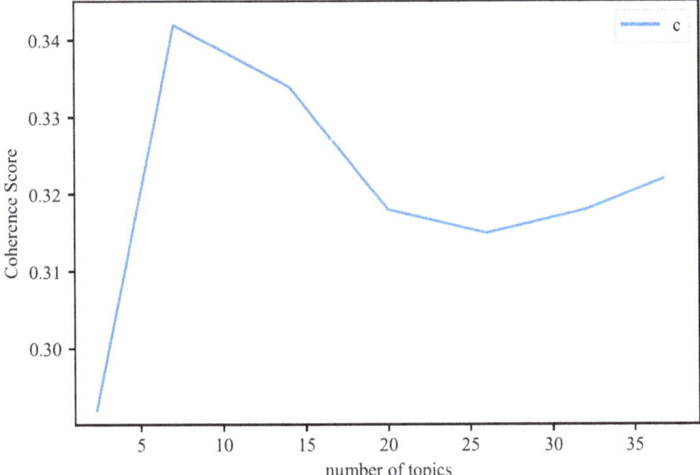

Figure 6. The relationship between the coherence score and number of topics.

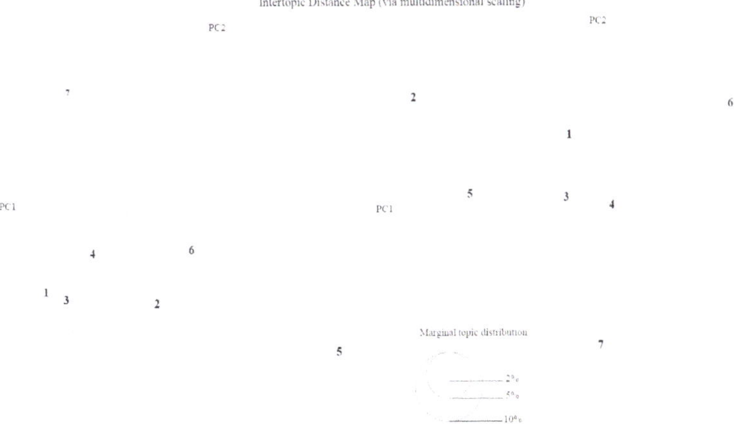

Figure 7. Theme map of the theme model with eight topics (**left**) and seven topics (**right**).

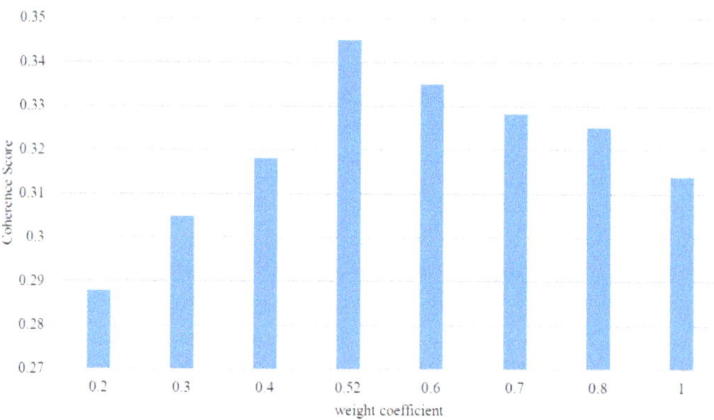

Figure 8. Diagram of the relationship between λ and the consistency score.

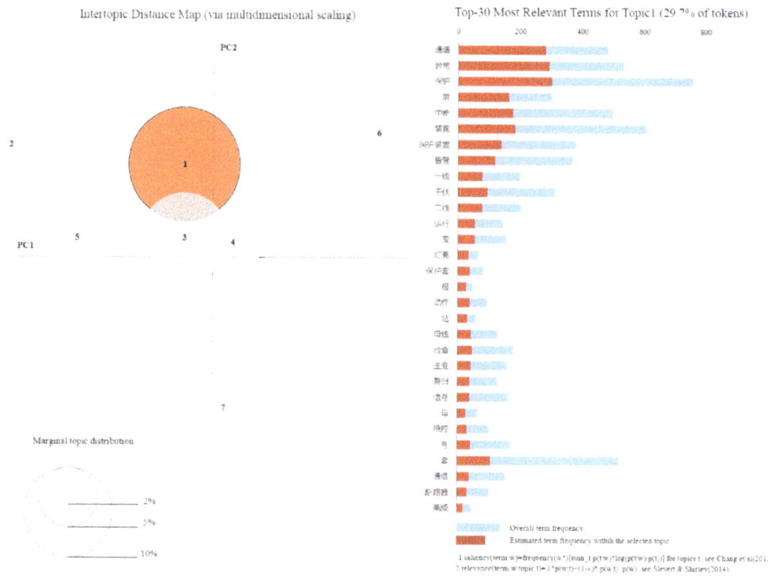

Figure 9. Top 30 words of topic distribution and relativity of topic 1 when λ was 0.52.

With the help of the LDAvis toolkit, the model topics with topic number seven and topic number eight were reduced to a two-dimensional plane for visual display. The results are shown in Figure 7. The left half is the topic model with the number of topics eight, and the right half is the theme model with theme seven. The greater the degree of topic intersection, the greater the difficulty of distinguishing the topic. The degree of intersection between the topics of the model with eight topics was much greater than that of the model with seven topics. Therefore, this article was in the pursuit of model generalization ability.

When the weighting factor λ was close to 1, it indicated a high frequency of occurrence in the word frequency and a high contribution from its document topic. We can conclude that the relevance to the topic was higher in the default document. When the weight coefficient λ was close to 0, the improved model indicated that the word appeared more frequently in the selected topic, but less frequently in other topics; that is, the generality between words and topics appeared. Considering the influence of relevance and concomitates, the consistency score of the model was repeatedly calculated, and the result was found to be the best when λ was 0.52. The relationship between the weight coefficient and the consistency score is shown in Figure 8. When λ was 0.52, the relationship between the theme of topic 1 and the words is shown in Figure 9.

4.2. Results and Analysis of Evaluation Index of Classification Effect

Text classification effect evaluation is an important module of text classification. It usually uses the mixed matrix as the basis, also known as the error matrix. It is usually expressed in two-dimensional tables. The classification results can be visually analyzed through the confusion matrix [29,30]. The confusion matrix is shown in Table 3.

Table 3. Mixed matrix of classification results.

Classification Category	Manually Marked as Belonging to	Manually Marked as Not Belonging to
Classifiers marked as belonging to	*TP*	*TN*
Classifier marked as not belonging to	*FP*	*FN*

For classification results, internationally recognized evaluation indicators were used: accuracy rate P, recall rate R, and *F1* values. The calculation formula is as follows:

$$P = \frac{TP}{TP + FP} \tag{10}$$

$$R = \frac{TP}{TP + FN} \tag{11}$$

$$F1 = \frac{2 \times P \times R}{P + R} \tag{12}$$

where *TP* indicates the number of samples that a certain type of text is correctly identified as a class, *FP* indicates the number of samples that a certain type of text has to be identified as other classes, and *FN* indicates that the text of other types is confirmed as the number of samples of the class. In order to verify the effectiveness of the improved input feature matrix, the CNN text classification method based on word2vec was compared with the experiments in this paper. It compared precision P, recall R, and *F1* values.

In the next study, we will test the superiority of the presented method of this paper, which was compared with traditional machine learning methods such as SVM, LR, KNN, and other models to find the accuracy of each algorithm model on the same data set. The experimental results are shown in Table 4.

Table 4. Comparison of the experimental results with machine learning methods.

Classifier Name	F1 Value (%)
LR	51.20
SVM	54.53
KNN	51.20
CNN	55.36
WORD2VEC + CNN	63.63
LDA + CNN	63.00
WORD2VEC + TEXTCNN	78.54
WORD2VEC + RLDA + TEXTCNN	81.69

Compared with the traditional machine learning methods LR, SVM, and KNN, due to the large corpus short text in this experiment, the *F1* values of the results were basically around 50%, and the accuracy of the highest SVM model classification results was only 54.53%. The accuracy of the typical CNN model classification results is only 55.36%. The effect of machine learning classification was not ideal. The traditional LDA topic model extracts features and lacks contextual semantic information, which makes it difficult to achieve ideal results in short text classification. The $F1$ value of the experiment was only 63.00%. Compared with the traditional CNN, the model of WORD2VEC + TEXTCNN was 14.91% higher than WORD2VEC + CNN. The text was improved on the traditional LDA theme model. The weight coefficient λ was used to adjust the relationship between words and subjects. Finally, the $F1$ of the WORD2VEC + RLDA + TEXTCNN model was the highest, up to 81.69%, whether it was with traditional machines. Compared with the traditional convolutional neural network learning algorithm, the $F1$ results were significantly improved. Therefore, the generalization ability and practicability of the model constructed in this paper have satisfied the possibility of practical application.

5. Discussion

Aiming at the problem of multi-type and complex secondary equipment in power systems and the low accuracy of word segmentation results, in this paper, a stop words dictionary and a professional dictionary in the field of secondary equipment in power system were constructed. An improved LDA topic analysis model based on the Relevance formula was proposed. By setting different weight coefficients, the feature similar words in texts with different defect categories were separated to solve the problem of feature sparseness. An improved algorithm was proposed by integrating the improved LDA topic model with word2vec, where the global features were mined by using the topic model, and the context semantic features were mined by using the latent semantic word vector model, which can better extract the short text features. The multi-scale convolution kernel was used to extract features to enhance its ability to extract local sensitive information, and further to conduct in-depth mining of text semantic information.

There are also some problems, such as a large number of professional dictionaries in the field of secondary equipment are constructed in the preprocessing process, which improves the professionalism of this model to some extent. However, the direct application of this model to other fields is likely to lead to poor generalization ability. All these topics are left for the future and ongoing research topics.

6. Conclusions

In this paper, for the problem of short text information of secondary equipment faults in the power system and the high repetition of words between different defect categories, an LDA topic model based on the Relevance formula was built to dynamically adjust the correlation between topics and words. In addition, considering that the topic model itself has insufficient ability to extract short text features, the word2vec latent semantic feature vectors were fused to compensate for contextual semantic information. Considering that some fault text data were short, the traditional convolutional neural network had insufficient feature extraction, and multiple sizes of convolution kernels were used to extract features from short text data. Finally, using the fault text data generated by the actual operation of a power system company in a northwestern province to verify the method in this paper, the results showed that the algorithm has a certain practicality.

Author Contributions: J.L. created models, developed methodology, wrote the initial draft, and designed computer programs; H.M. supervised and leaded responsibility for the research activity planning and presented the critical review; X.X. and J.C. conducted the research and investigation process and edited the initial draft; J.L. and H.M. reviewed the manuscript and synthesized the study data. All authors have read and agreed to the published version of the manuscript.

Funding: This work was supported by the National Natural Science Foundation of China (51577050).

References

1. Jin, H.; Liu, X.; Liu, W.; Xu, J.; Li, W.; Zhang, H.; Zhou, J.; Zhou, Y. Analysis on Ubiquitous Power Internet of Things Based on Environmental Protection. *IOP Conf. Ser. Earth Environ. Sci.* **2019**, *300*, 42077–42083. [CrossRef]
2. Chen, K.; Mahfoud, R.J.; Sun, Y.; Nan, D.; Wang, K.; Haes Alhelou, H.; Siano, P. Defect Texts Mining of Secondary Device in Smart Substation with GloVe and Attention-Based Bidirectional LSTM. *Energies* **2020**, *13*, 4522. [CrossRef]
3. Bakr, H.M.; Shaaban, M.F.; Osman, A.H.; Sindi, H.F. Optimal Allocation of Distributed Generation Considering Protection. *Energies* **2020**, *13*, 2402. [CrossRef]
4. Liu, G.; Zhao, P.; Qin, Y.; Zhao, M.; Yang, Z.; Chen, H. Electromagnetic Immunity Performance of Intelligent Electronic Equipment in Smart Substation's Electromagnetic Environment. *Energies* **2020**, *13*, 1130. [CrossRef]
5. Yuan, Q. Research on New Technology of Power System Automation Based on Ubiquitous Internet of Things Technology. *IOP Conf. Ser. Earth Environ. Sci.* **2020**, *440*, 32101–32110. [CrossRef]
6. Yu, X.; Xue, Y. Smart Grids: A Cyber-Physical Systems Perspective. *Proc. IEEE* **2016**, *104*, 1058–1070. [CrossRef]
7. Collobert, R.; Weston, J. A Unified Architecture for Natural Language Processing: Deep Neural Networks with Multitask Learning. In Proceedings of the 25th International Conference on Machine Learning, Helsinki, Finland, 5 July 2008; pp. 160–167.
8. El Hindi, K.; Alsalman, H.; Qasem, S. Building an Ensemble of Fine-Tuned Naive Bayesian Classifiers for Text Classification. *Entropy* **2018**, *20*, 857. [CrossRef]
9. Xu, S. Bayesian Naïve Bayes Classifiers to Text Classification. *J. Inform. Sci.* **2018**, *44*, 48–59. [CrossRef]
10. Collobert, R.; Weston, J.; Bottou, L. Natural Language Processing (Almost) from Scratch. *J. Mach. Learn. Res.* **2011**, *12*, 2493–2537.
11. Goudjil, M.; Koudil, M.; Bedda, M.; Ghoggali, N. A Novel Active Learning Method Using SVM for Text Classification. *Int. J. Automat. Comput.* **2018**, *15*, 290–298. [CrossRef]
12. Wu, K.; Gao, Z.; Peng, C.; Wen, X. Text Window Denoising Autoencoder: Building Deep Architecture for Chinese Word Segmentation. In Proceedings of the CCF International Conference on Natural Language Processing and Chinese Computing, Chongqing, China, 15–19 November 2013; Springer: Berlin/Heidelberg, Germany, 2013; pp. 1–12.
13. Zhang, Y.; Zheng, J.; Jiang, Y.; Huang, G.; Chen, R. A Text Sentiment Classification Modeling Method Based on Coordinated CNN-LSTM-Attention Model. *Chin. J. Electron.* **2019**, *28*, 120–126. [CrossRef]
14. Jing, R. A Self-attention Based LSTM Network for Text Classification. *J. Phys. Conf. Ser.* **2019**, *1207*, 12008–12012. [CrossRef]
15. Zhang, M.; Xiang, Y. A Key Sentences Based Convolution Neural Network for Text Sentiment Classification. *J. Phys. Conf. Ser.* **2019**, *1229*, 12062–12068.
16. Roy, S.S.; Rodrigues, N.; Taguchi, Y. Incremental Dilations using CNN for Brain Tumor Classification. *Appl. Sci.* **2020**, *10*, 4915. [CrossRef]
17. Zahid, M.; Ahmed, F.; Javaid, N.; Abbasi, R.A.; Zainab Kazmi, H.S.; Javaid, A.; Bilal, M.; Akbar, M.; Ilahi, M. Electricity Price and Load Forecasting using Enhanced Convolutional Neural Network and Enhanced Support Vector Regression in Smart Grids. *Electronics* **2019**, *8*, 122. [CrossRef]
18. Li, H.; Huang, J.; Ji, S. Bearing Fault Diagnosis with a Feature Fusion Method based on an Ensemble Convolutional Neural Network and Deep Neural Network. *Sensors* **2019**, *19*, 2034. [CrossRef]
19. Yang, Z.; Fan, K.; Lai, Y.; Gao, K.; Wang, Y. Short Texts Classification Through Reference Document Expansion. *Chin. J. Electron.* **2014**, *23*, 315–321.
20. Gao, Y.; Zhu, Z.; Riccaboni, M. Consistency and Trends of Technological Innovations: A Network Approach to the International Patent Classification Data. In Proceedings of the International Conference on Complex Networks and their Applications VI, Lyon, France, 29 November–1 December 2017; Springer: Cham, Switzerland, 2017; pp. 744–756.
21. Li, J.; Fong, S.; Zhuang, Y.; Khoury, R. Hierarchical Classification in Text Mining for Sentiment Analysis of Online News. *Soft Comput.* **2016**, *20*, 3411–3420. [CrossRef]
22. Wang, L.; Guo, X.; Wang, R. Automated Crowdturfing Attack in Chinese User Reviews. *J. Commun.* **2019**, *40*, 1–13.
23. Hughes, M.; Li, I.; Kotoulas, S.; Suzumura, T. Medical Text Classification Using Convolutional Neural Networks. *Stud. Health Technol. Inform.* **2017**, *235*, 246–250.
24. Tommasel, A.; Godoy, D. Short-Text Learning in Social Media: A Review. *Knowl. Eng. Rev.* **2019**, 38–49. [CrossRef]
25. Wang, H.; He, J.; Zhang, X.; Liu, S. A Short Text Classification Method based on N-gram and CNN. *Chin. J. Electron.* **2020**, *29*, 248–254. [CrossRef]
26. Mikolov, T.; Chen, K.; Corrado, G.; Dean, J. Efficient Estimation of Word Representations in Vector Space. *arXiv* **2013**, arXiv:1301.3781.
27. Blei, D.M.; Ng, A.Y.; Jordan, M.I. Latent Dirichlet Allocation. *J. Mach. Learn. Res.* **2003**, 993–1022.
28. Wei, W.; Nan, D.; Zhang, L.; Zhou, J.; Wang, L.; Tang, X. Short Text Data Model of Secondary Equipment Faults in Power Grids based on LDA Topic Model and Convolutional Neural Network. In Proceedings of the 2020 35th Youth Academic Annual Conference of Chinese Association of Automation (YAC), Zhanjiang, China, 16–18 October 2020; pp. 156–160.
29. Lu, X.; Zhou, M.; Wu, K. A Novel Fuzzy Logic-Based Text Classification Method for Tracking Rare Events on Twitter. *IEEE Trans. Syst. Man Cybern. Syst.* **2021**, *51*, 4324–4333. [CrossRef]
30. Fawcett, T. An Introduction to ROC Analysis. *Pattern Recognit. Lett.* **2006**, *27*, 861–874. [CrossRef]

Article

Short-Term Load Forecasting Model of Electric Vehicle Charging Load Based on MCCNN-TCN

Jiaan Zhang [1], Chenyu Liu [2,*] and Leijiao Ge [3,*]

[1] State Key Laboratory of Reliability and Intelligence of Electrical Equipment, Hebei University of Technology, Tianjin 300130, China; zhangjiaan@foxmail.com
[2] College of Artificial Intelligence and Data Science, Hebei University of Technology, Tianjin 300401, China
[3] Key Laboratory of Smart Grid of Ministry of Education, Tianjin University, Tianjin 300072, China
*   Correspondence: liuchenyux@163.com (C.L.); legendglj99@tju.edu.cn (L.G.)

Abstract: The large fluctuations in charging loads of electric vehicles (EVs) make short-term forecasting challenging. In order to improve the short-term load forecasting performance of EV charging load, a corresponding model-based multi-channel convolutional neural network and temporal convolutional network (MCCNN-TCN) are proposed. The multi-channel convolutional neural network (MCCNN) can extract the fluctuation characteristics of EV charging load at various time scales, while the temporal convolutional network (TCN) can build a time-series dependence between the fluctuation characteristics and the forecasted load. In addition, an additional BP network maps the selected meteorological and date features into a high-dimensional feature vector, which is spliced with the output of the TCN. According to experimental results employing urban charging station load data from a city in northern China, the proposed model is more accurate than artificial neural network (ANN), long short-term memory (LSTM), convolutional neural networks and long short-term memory (CNN-LSTM), and TCN models. The MCCNN-TCN model outperforms the ANN, LSTM, CNN-LSTM, and TCN by 14.09%, 25.13%, 27.32%, and 4.48%, respectively, in terms of the mean absolute percentage error.

Keywords: electric vehicle; short-term load forecasting; convolutional neural network; temporal convolutional network; climate factors; correlation analysis

Citation: Zhang, J.; Liu, C.; Ge, L. Short-Term Load Forecasting Model of Electric Vehicle Charging Load Based on MCCNN-TCN. *Energies* **2022**, *15*, 2633. https://doi.org/10.3390/en15072633

Academic Editor: Fabrice Locment

Received: 25 February 2022
Accepted: 1 April 2022
Published: 4 April 2022

Publisher's Note: MDPI stays neutral with regard to jurisdictional claims in published maps and institutional affiliations.

1. Introduction

The growth of the electric vehicle industry has captivated governments, automakers, and energy companies. EVs are seen as a viable solution to the depletion of fossil resources and rising pollution [1]. It is widely believed that the popularity of EVs can reduce greenhouse gas emissions (mainly carbon dioxide) [2]. Meanwhile, falling battery prices and government incentives will also promote rapid growth in the scale of EVs [3]. However, the increased charging demand resulting from the rapid development of EVs also poses various challenges to the grid. The EV charging load has a great impact on the stable operation of the distribution network [4], including the decline of power quality and the difficulty of optimizing and controlling the operation of the power grid [5,6]. The research on EV charging load forecasting is carried out not only to ensure the economical and stable operation of the power system [7] but also to support the development of EVs [8].

EV charging load forecasting approaches are now separated into probabilistic models, time series models, and machine learning models. The probabilistic modeling method establishes probabilistic models of residents' charging and travel behavior using statistical and queuing theory, followed by load forecasts using Monte Carlo simulation. Taylor J et al. [9] utilized the Monte Carlo method to establish a large-scale charging demand model, considering EV type, penetration rate, charging scenario, etc. In [10], it is assumed that the arrival time of EVs at the charging station follows Poisson distribution, and the charging load prediction is carried out based on queuing theory. With the deepening of research, the temporal and spatial distribution of EV charging load has attracted the interest of many researchers. Shun et al. [11] established a probabilistic model of the temporal and spatial distribution of EVs based on travel chains and Markov decision processes. Chen et al. [12] applied the OD matrix analysis method to plan the driving path of the logistics electric vehicle and solve the charging demand load value through the mixed-integer programming model. Xing et al. [13] proposed a data-driven EV charging load prediction method, which is based on Didi user travel data to establish a traffic network model, a vehicle spatiotemporal transfer model, and a resident travel probability model.

Currently, time series and machine learning algorithms are commonly employed to forecast EV charging load in the short term. The exponential smoothing model [14], the linear regression (LR) model [15], and the autoregressive integrated moving average (ARIMA) model [16] are the most often used time series models. While time series models have straightforward structures and require minimal training, they are incapable of capturing the nonlinear properties of load series. With the rapid advancement of artificial intelligence technology, intelligent algorithms such as artificial neural networks (ANNs) and deep neural networks are increasingly used to forecast EV charging load. The neural network has excellent power for feature extraction and the ability to form nonlinear mapping relationships [17], which effectively addresses the time series model's shortcomings. In [18], the SVR founded on an evolutionary algorithm is proposed for electric bus charging load forecasting. Yi et al. [19] proposed a multi-step EV load prediction model established on long short-term memory (LSTM), and the results suggest that the model is capable of accurately predicting sequence data. In [20], LSTM models show better performance and provide higher accuracy compared to the prediction results of ANNs. The gated recurrent unit (GRU) is a characteristic and efficient variant of LSTM. The GRU is characterized by making the network structure simpler. Zhu et al. [21] introduced GRU into short-term forecasting of EV charging load. In order to further improve the short-term load forecasting performance, some forecasting methods combined with LSTM and other recurrent neural networks (RNN) have also been proposed. Feng et al. [22] proposed an EV charging load prediction method based on a combination of the multivariate residual corrected grey model (EMGM) and LSTM network. Dabbaghjamanesh et al. [23] applied Q-Learning Technique based on ANN and RNN to improve the short-term prediction accuracy of EV charging load. The model based on LSTM and GRU is capable of learning long-term temporal correlations; however, due to the lack of convolution in the model, the feature extraction capability still has to be enhanced. Therefore, it is difficult for the above models to effectively utilize and extract the feature information in the EV charging load.

When confronted with this problem, approaches for extracting features are seen to be one of the most viable solutions. The convolutional neural networks (CNN) have excellent feature extraction [24], which is often used for feature extraction in short-term load forecasting. Li et al. [25] applied an evolutionary algorithm-optimized CNN model for EV charging load prediction. In addition, the CNN-LSTM model combining CNN and LSTM is often used in traditional short-term forecasting of power loads [26]. In the CNN-LSTM model, CNN extracts the feature information of load-related influencing factors, and LSTM is used to learn the temporal dependency between the feature information sequence extracted by CNN and the output [27]. Yan et al. [28] proposed a hybrid model based on CNN and LSTM to predict the short-term electricity load of a single household. However, most methods ignore the long-term temporal relationship of input variables, causing the load forecasting model to lack adequate prior knowledge.

Furthermore, EVs are abundant in urban areas, and EV users' travel behavior is influenced by many random factors, resulting in increasingly complicated fluctuations in the charging load of EVs. Given this problem, accurate forecasting by using a short-term load forecasting model on a single time scale is difficult [29]. Short-term load forecasting can be enhanced by decomposing the load into multiple intrinsic mode functions and then separately predicting and reconstructing the sub-model prediction results [30]. Wang et al. [31] proposed a "decomposition-predict-reconstruction" prediction model based on empirical mode decomposition (EMD) and LSTM, which effectively improved the accuracy of load prediction.

One-dimensional convolutional neural networks (1DCNN) can extract one-dimensional sequence features, commonly used to extract time series feature information. Wang et al. [32] utilized 1DCNN to extract the fusion features of bearing vibration signal and sound signal to realize bearing fault diagnosis. In [33], the influent load is first decomposed by EMD, and then 1DCNN extracts the latent features of each intrinsic mode function's periodic signal. However, although the 1DCNN model can achieve feature extraction at various time scales by adjusting the scope of the receptive field, it cannot extract the time series dependencies between time series data. With the advent of advanced TCN models that combine the advantages of CNN feature processing and RNN time-domain modeling, it is possible to extract time series dependencies between long intervals of historical data [34]. Yin et al. [35] proposed a feature fusion TCN structure that fuses model output features at multiple time delay scales. The TCN built on the convolutional network can process data in parallel on a large scale and has a faster computing speed than the RNN such as LSTM [36]. Although the signal decomposition method can obtain the components of EV charging load at various time scales, it still necessitates the selection and construction of low-dimensional features with a high degree of differentiation, which not only adds subjectivity and complexity to this identification method but also risks losing important information.

On the basis of the foregoing research, an EV charging load forecasting model based on the MCCNN-TCN is proposed in this paper. The MCCNN model can mine the fluctuation features of EV charging load at multi-time scales. The TCN model can establish the global time-series dependencies between the local time-series feature information at different time scales extracted by the MCCNN model. In addition, accurate load forecasting is frequently reliant on a thorough understanding of the elements that contribute to increasing or decreasing consumer demand [37]. The EV charging load is affected by numerous aspects, including weather temperature, date type, traffic conditions, user travel behavior, etc. [8]. Therefore, this paper introduces the maximum information coefficient (MIC) and Spearman rank correlation coefficient and proposes a similar day method based on weighted gray correlation analysis to screen historical loads. The main contributions of this paper are described as follows:

(1) The MIC was applied to eliminate input data redundancy and reduce the complexity of the model. The MIC was used to choose meteorological variables that have a substantial link with EV charging load. The selected meteorological variables were utilized as an input to both the prediction and comparable day selection models;

(2) A similar day selection model based on weighted grey relational analysis was proposed. The Spearman rank correlation coefficient of the week average daily load was used to calculate week type similarity. Then, by selecting meteorological variables obtained by MIC and week type similarity as the input, a similar day selection model based on weighted gray correlation analysis was used to choose a similar day load used as the forecasting model's input;

(3) An MCCNN-TCN model framework was built. Combining the multi-channel 1DCNN model with the TCN model can establish global temporal dependencies between time series features at multiple time scales, which effectively improves the prediction performance.

The remainder of this paper is organized as follows. In Section 2, a short-term EV charging load forecasting framework based on the MCCNN-TCN model is introduced. In Section 3, experiments are conducted with a real dataset of grid companies and compared with other models. In Section 4, the model proposed in this paper is analyzed compared to other state-of-the-art methods based on experimental results. In Section 5, the paper's conclusions and future research are given.

2. Materials and Methods

2.1. Selecting Similar Days

2.1.1. Screening of Meteorological Features Based on Maximum Information Coefficient

As a new type of electric load, EV charging load is not only related to residents' travel behavior but also affected by meteorological factors such as weather and temperature [38]. In order to lower the input size of the similar day model and forecast model, relevant meteorological features that strongly correlate with EV charging load must be selected [36]. At the same time, since meteorological features and EV charging load are both nonlinear time series, this paper uses MIC to examine the nonlinear relationship between each meteorological variable and EV charging load. Unlike other traditional correlation analysis methods, the benefit of MIC is that it does not require any assumptions about the data distribution and is acceptable for both linear and nonlinear data [39]. The MIC is calculated as follows [40].

For a binary dataset, D and $D \in R^2$, divide D into a grid of x rows and y columns. The obtained grid G based on different division methods forms set A. Find the maximum mutual information $\max I(D|G)$ in set A, conserve it as:

$$I^*(D, x, y) = \max_{G \in A} I(D|G) \tag{1}$$

where $D|G$ is the distribution of the binary data set D on the grid G.

The maximum normalized mutual information of the binary dataset D at different scales is formed into the feature matrix $M(D)$, and the elements of the feature matrix are defined as:

$$M(D)_{x,y} = \frac{I^*(D, x, y)}{\log_2 \min(x, y)} \tag{2}$$

The MIC is calculated by:

$$MIC(D) = \max_{rc < B(n)} \left\{ M(D)_{x,y} \right\} \tag{3}$$

where n indicates the size of the sample, $B(n)$ is a function about the size of the sample, and the constraint indicating the total number rc of squares of the grid G is less than $B(n)$, generally $B(n) = n^{0.6}$ [41]. A greater MIC value between the two variables indicates a stronger correlation.

2.1.2. Quantifying Week Type Similarity Based on Spearman Correlation Analysis

The characteristics of EV charging load in different months, seasons, and week kinds are investigated in this article to study the relationship between EV charging load and date types. The EV charging load has the maximum consumption level in December and the lowest in April, as shown in Figure A1 in the Appendix A. The consumption level of EV charging load in winter and fall is significantly higher than in spring and summer, and the load in winter represents a tendency of rising first and then reduce. In contrast, the load in summer has a fluctuating and rising trend, as shown in Figure A2 in the Appendix A. EV charging load consumption level is highest on Saturday and lowest on Monday, as shown in Appendix A Figure A3. In summary, it is critical to pay attention to the effect of date type on the charging load of EVs. In this paper, the date types were divided into season types and week types, and the similarity between week types under each season was established as the input of the similar day model. In order to avoid human subjective participation in setting the week types map value, using the average daily EV charging load between week types calculated the similarity between week types in this paper.

The data on electric vehicle charging load do not follow a normal distribution. Additionally, the Spearman coefficient does not require that the data remain normal [42]. As a result, this paper proposes utilizing the Spearman coefficient to quantify the similarity of week types. The week types under each season were divided into seven (Monday to

Sunday), and then the Spearman coefficient was calculated for the average daily load between the week types. The correlation value indicative is represented by F_{kg}^h, as in (4):

$$F_{kg}^h = 1 - \frac{6 \sum A_t^2}{n(n^2 - 1)} \quad t = 1, \ldots, 96 \tag{4}$$

where k and g represent the week type; h represents the season, $h = 1, 2, 3, 4$; n is the load sample number; and A_t indicates the difference of the position between the t-th daily load samples of week type k and week type g.

2.1.3. Similar Days Selection Model Based on Weighted Grey Correlation Analysis

When calculating the gray correlation, the traditional gray correlation analysis assigns the same weight to each feature, ignoring each influencing factor's difference [43]. Therefore, each influencing factor's weight is first analyzed based on the improved entropy weight method in this paper. Then the correlation degree between the forecasting day and history day is calculated based on the weighted grey correlation degree analysis.

According to the historical data, the entropy E_j of the j-th meteorological feature is calculated [44]:

$$\begin{cases} E_j = \alpha \cdot \sum\limits_{n}^{i=1} b_{ij} \ln b_{ij}, \quad j = 1, 2, \cdots, m \\ \alpha = -\frac{1}{\ln n} \\ b_{ij} = \frac{a_{ij}}{\sum\limits_{n}^{i=1} a_{ij}} \end{cases} \tag{5}$$

where n is the number of historical days, m indicates the dimension of the day feature; a_{ij} represents the value of the j-th feature of the i-th historical day. Additionally, if $b_{ij} = 0$, $b_{ij} \ln b_{ij} = 0$.

According to the entropy of each meteorological feature, the weight of the j-th day feature based on the improved entropy weight method is calculated as [45]:

$$w_j = \frac{\exp\left(\sum\limits_{t=1}^{m} E_t + 1 - E_j\right) - \exp(E_j)}{\sum\limits_{l=1}^{m}\left(\exp\left(\sum\limits_{t=1}^{m} E_t + 1 - E_l\right) - \exp(E_l)\right)} \tag{6}$$

The correlation coefficient of each day's feature is calculated using gray correlation analysis [18]. The following are the feature sequences of the forecasting and history days:

$$\begin{cases} X_d = [x_d(1), x_d(2), \cdots, x_d(m)] \\ X_{d-i} = [x_{d-i}(1), x_{d-i}(2), \cdots, x_{d-i}(m)] \end{cases} \tag{7}$$

where X_d represents the feature sequence of the forecasting day d, X_{d-i} represents the factor sequence of the history day $d - i$. The correlation coefficient of the j-th feature of X_d to X_{d-i} is:

$$\xi_d^{d-i}(j) = \frac{minmin_k|x_d(j) - x_{d-i}(j)| + \rho maxmax_i|x_d(j) - x_{d-i}(j)|}{|x_d(j) - x_{d-i}(j)| + \rho maxmax_{ij}|x_d(j) - x_{d-i}(j)|} \tag{8}$$

where $x_d(j)$ and $x_{d-i}(j)$ are the j-th feature of the forecasting day d and the history day $d - i$, respectively, ρ is the distinguishing coefficient and $\rho = 0.5$.

Based on calculating the grey correlation coefficients ξ of the factors and their weights w, the weighted grey correlation between forecast day d and historical day $d - i$ can be expressed as follows:

$$r_d^{d-i} = \sum\limits_{j=1}^{m} w_j \xi_d^{d-i}(j) \tag{9}$$

The first 14 days of the forecasting day are defined as a similar day rough set in this paper. Because the capacity of the similar day rough set is limited, it is not assumed that as the date distance increases, the similarity between the forecasting day and the historical day decreases. Furthermore, derived from the past EV charging load data, the average number of days with a Spearman's correlation coefficient larger than 0.4 between the forecasting day and each historic day in the similar day rough set is 3. In addition, the adjacent daily load is added to the similar day set to ensure time consistency between the forecasting day load and the historical day load. According to the above analysis, the size of the similar day set in this paper is 4.

2.2. Multi-Channel Convolutional Neural Network and Temporal Convolutional Network Model

Because the charging load of EVs is influenced by various factors, including weather conditions, residents' travel habits, and the traffic network, there is a high level of short-term volatility, making short-term load forecasting more complex. It was demonstrated that extracting the characteristics of EV charging load at various time scales is an effective strategy for improving prediction accuracy [31]. Different influencing factors affect the features of EV charging load at different time scales. In this regard, the paper proposes the MCCNN-TCN model framework. As illustrated in Figure 1, the model framework is divided into three layers: a multi-channel 1DCNN feature extraction layer, a multi-channel TCN layer, and an output layer. The model framework can extract EV charging load characteristics at various time scales and construct a worldwide time-series dependency between the historical and predicted day loads. The multi-channel 1DCNN is utilized as the gate of the MCCNN-TCN model to extract the local features of the input time series at different time scales. Deepening the TCN network can expand its receptive field, establishing the temporal dependencies between global features. The output layer's job is to create a nonlinear relationship between the forecasting load, meteorological and calendar features, and historical load. Sections 2.1.1 and 2.1.2 show that the meteorological and date factors impact the EV charging load, in addition to the influence of the historical load on the forecasting load. As a result, this paper combines the TCN model's output historical load feature vector with a high-dimensional feature vector derived from meteorological and date features. Then, it is input into a fully connected neural network. The fully connected neural network's output is forecasting day load.

The length of the 1DCNN layer's input feature map is sn, where s is the number of similar days and n is the number of daily load samples. The role of the multi-channel 1DCNN is to extract the features of a one-dimensional time series consisting of EV charging load sequences in similar daily sets at different time scales. The TCN layer takes the output of the multi-channel 1DCNN model as input and captures the global temporal dependencies at different time scales. The BP layer maps the feature composed of the meteorological factors simultaneously as the forecasting day load and the date type of forecasting day to the high-dimensional feature space. The high-dimensional feature vector obtained by integrating the BP model's output and the TCN model's output is used as the input of the fully connected layer in the output layer of the MCCNN-TCN.

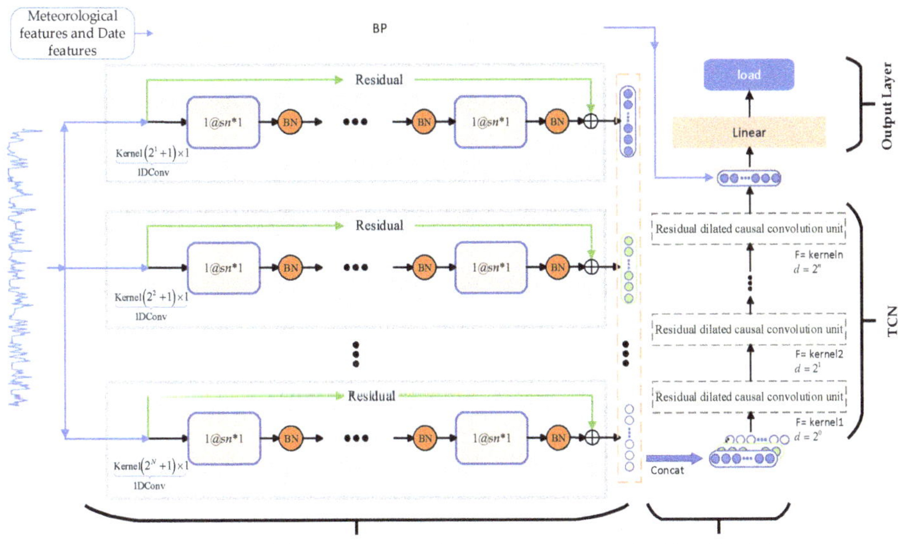

Figure 1. Multi-channel convolutional neural network and temporal convolutional network (Where, @ is preceded by the number of channels and followed by the output of the convolution layer).

2.2.1. Multi-Channel 1D Convolutional Network Model

CNN is a great neural network model that uses convolution kernels to extract essential information automatically [46]. Figure 2 shows the basic architecture of the 1DCNN, which can extract latent features in time series using multiple convolution kernels of the same weight. The same convolution kernel obtains a class of related features during the convolution process. Its mathematical model is described as [47]:

$$H_i = f(H_{i-1} \otimes W_i + b_i) \tag{10}$$

where H_i indicates the input of layer I; H_{i-1} indicates the output of layer $i-1$; W_i and b_i indicate the weight matrix and the corresponding bias vector of the convolution kernel of layer i, respectively; $\otimes$ indicates for convolution operation; and f indicates the activation function.

Following the convolution operations, the pooling layer uses data downsampling to downsample a huge matrix into a small one, reducing the amount of computation and avoiding overfitting. The pooling layer mathematical model is as follows:

$$H_i = down(H_{i-1}) \tag{11}$$

where H_{i-1} and H_i indicate the features before and after pooling, respectively, and "$down()$" indicates the pooling function.

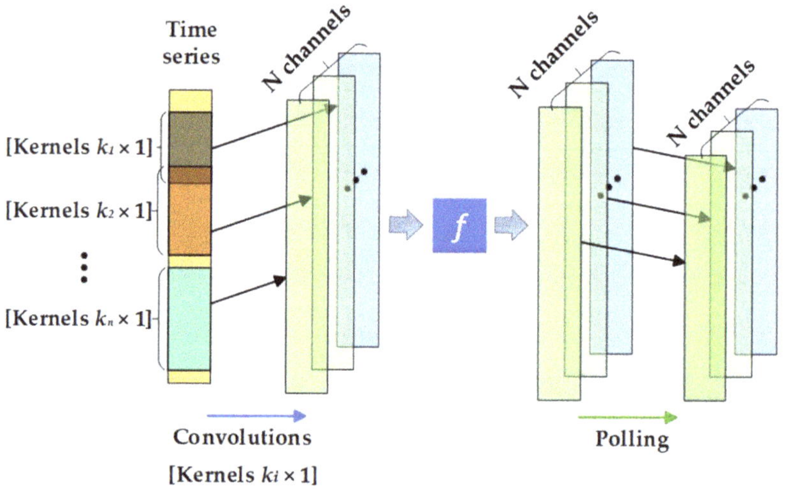

Figure 2. Structure of one-dimensional convolutional neural network.

As shown in Figure 3, the multi-channel 1DCNN is made up of numerous parallel 1D convolution blocks. The first convolutional layer of the multi-channel 1DCNN has a varied convolution kernel size. Long-term scale characteristics of EV charging load can be extracted using big convolution kernels. Short-time-scale characteristics of EV charging loads can be extracted using little convolution kernels. Rough features of EV charging load at different time scales are obtained after the first convolutional layer. This paper extracts detailed features by adding numerous convolutional layers with a convolution kernel of three to the initial convolutional layer to fully mine the detailed information under various EV charging load time scales. The first convolutional layer kernel size K of each channel is represented as follows:

$$K = 2^n + 1 \tag{12}$$

where $n \in (1, 2, 3, \ldots, N)$, N is the number of channels. The value of N depends on the length of the input layer time series.

Furthermore, earlier research has revealed that when the depth of the neural network increases, residual connections can effectively handle the problems of gradient disappearance and network overfitting [48]. As a result, each channel of the multi-channel 1DCNN is assigned a residual connection in this paper. The residual connection mathematical model is:

$$x_{l+1} = x_l + F(x_l, w_l) \tag{13}$$

where x_{l+1} is the output of layer $l + 1$, x_l is the input of layer l, and $F(x_l, w_l)$ is the residual of layer l.

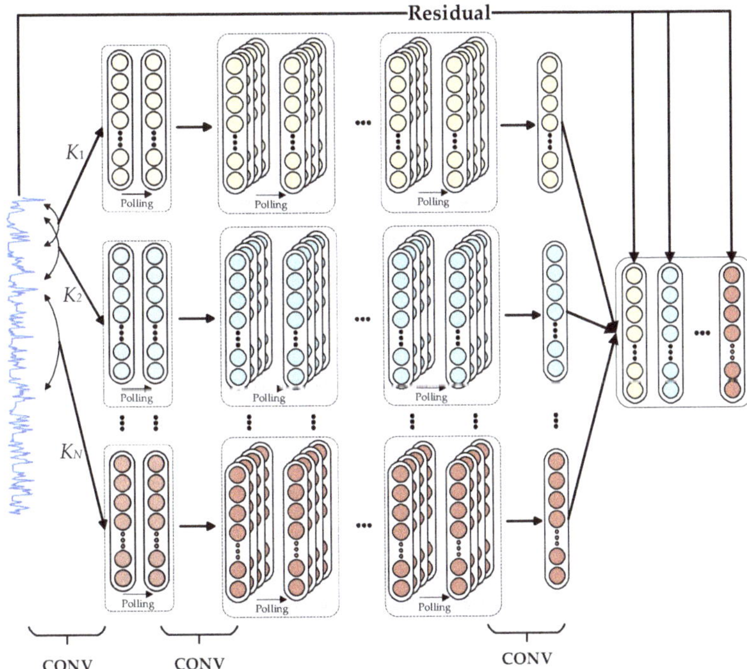

Figure 3. Multi-channel one-dimensional convolutional network.

2.2.2. Temporal Convolutional Network Model

The TCN developed by Bai et al. in 2018 is an algorithm for processing time series [49]. The TCN combines causal convolution, dilated convolution, and residual block to address the problem of extracting long-term time-series information.

The core of TCN is the residual dilated causal convolution unit (RDCCU), which consists of two rounds of dilated causal convolution with the same dilation factor, WeightNorm layer, activation function, Dropout layer, and residual connections formed by direct mapping of the input [35]. Multiple residual dilated causal convolutional units are connected to form a multi-layer TCN network structure, as shown in Figure 4.

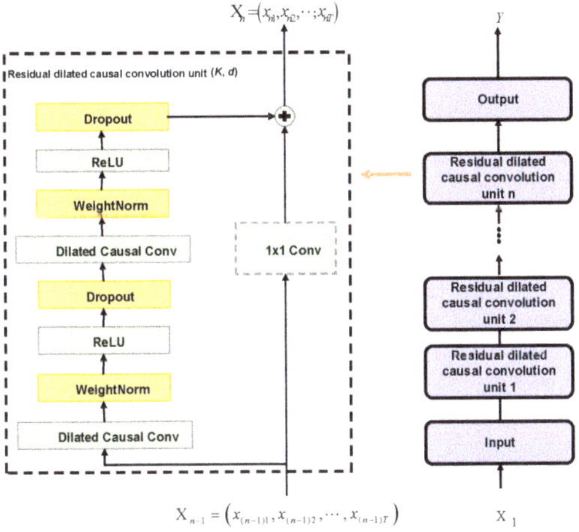

Figure 4. Connection of multiple residual dilated causal convolution units.

The fundamental core structure of the RDCCU is the dilated causal convolution [50], which is composed of causal convolution and dilated convolution [51]. The structure of the dilated causal convolution is shown in Figure 5.

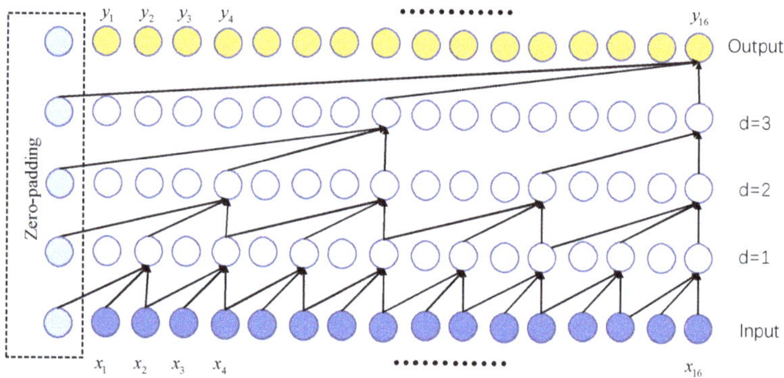

Figure 5. Schematic of dilated causal convolution.

Causal convolution refers to obtaining the output of time t through the convolution of elements at time t and earlier in the previous layer. It ensures that there will be no future information leakage, meeting the requirements of power load forecasting. Dilated convolution can expand the receptive field by increasing the dilation factor [52] and capture long enough historical information without increasing the depth of the model [53], which improves the efficiency of model training. Dilated convolution makes the input of the previous layer sampled at intervals, and the dilation factor d of each layer increases exponentially by 2, which can be described as:

$$l = \sum_{d=1}^{n} \left[(K-1) \cdot 2^d + 1 \right] \tag{14}$$

As illustrated in Figure 5, the kernel size of each dilated causal convolutional layer is 3. The dilation factor d grows from 1 to 4, which raises the effective history of neurons in the output layer from 3 to 15. In addition, to maintain the whole sequence information, each layer's output is zero-padded to match the number of input sequences. The mathematical model of dilated causal convolution is as follows [49]:

$$y(s) = (x_d^* f)(s) = \sum_{k-1}^{i=0} f(i) \cdot x_{s-d \cdot i} \tag{15}$$

where x is the input and y is the output.

Residual connections are a key structure of the RDCCU. The RDCCU is defined as follows [49]:

$$o = artivation(x + F(x)) \tag{16}$$

The output of the multi-channel 1DCNN is arranged in a T^*n two-dimensional data structure according to the channel direction and fed into the first RDCCU of the TCN model. The internal procedure of the RDCCU is shown in Figure 6. The width of the convolution kernel of the RDCCU corresponds to the number of input data channels. The number of output channels of this RDCCU is equal to the number of convolution kernels in the RDCCU. The output of the RDCCU is seamed in the channel direction and used as the input to the next RDCCU.

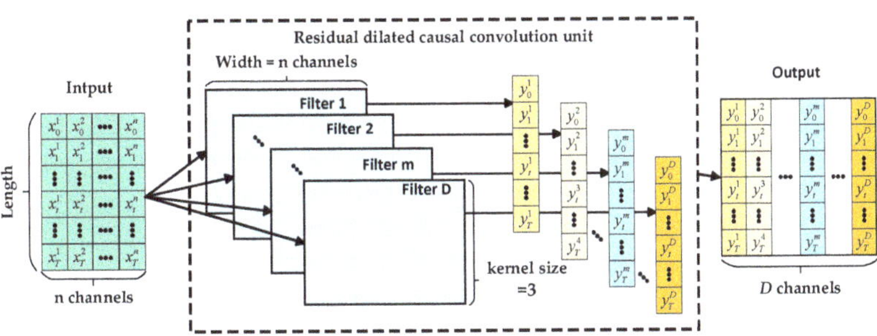

Figure 6. An illustration of the inputs and outputs of one residual dilated causal convolution unit.

3. Results

The subject of the study in the paper is EV charging load short-term forecasting in the urban area of a city in northern China. The dataset was data collected from 38 public DC charging stations in the city's urban area, from 1 January 2019 to 31 March 2020. The number of charging stations in residential, commercial, work and leisure areas is 8, 12, 11, and 7. These charging stations have 298 charging poles, each with a maximum charging power of 60 kW. The dataset included the active power of the charging poles, the transaction power, the charging start time and the charging end time, etc. The active power of the charging poles was sampled at 15 min intervals.

Meteorological data, which can be obtained from China Meteorological Data Network, include the temperature, humidity, precipitation, visibility, wind speed, and weather type. Among them, the temperature, humidity, and precipitation need to be interpolated by spline, and the purpose is to obtain the sampling value simultaneously with the load. Other data includes date type, season, etc.

All of the experimental models were run in the Python 3.6 programming environment, implemented under the Pytorch framework. The hardware used for the experiments was a PC with an Intel Core i7-10300H CPU, NVIDIA RTX 2060 GPU, and 32 GB of RAM.

3.1. Input Variables Selection and Processing

According to the investigation of influencing factors on EV charging load, these factors were divided into meteorological factors, date features, and similar daily load in this paper. Next, three types of features are selected and processed.

The MIC between each meteorological factor and EV charging load was calculated except for weather conditions. Table 1 shows the MIC and Pearson correlation coefficient between EV charging load and temperature, humidity, precipitation, visibility, and wind direction. As shown in Table 1, the EV charging load has a strong correlation with temperature, humidity, and rainfall but a weak correlation with visibility and wind speed. At the same time, the influence of weather conditions on the charging load of EVs cannot be ignored [25]. The min–max normalization was used to linearly transform the raw temperature, humidity, and rainfall data to $[0,1]$. The number of index mapping databases is referenced in Ref. [18]. In this paper, the mapping values were set to 0.1, 0.2, and 0.3 for the weather types sunny, cloudy and overcast, respectively, and 0.7, 0.1, and 1.5 for the weather types light rain or snow, rain or snow, and heavy rain or snow, respectively. Therefore, this paper selected weather type, temperature, humidity, and rainfall as the meteorological features that affect the EV charging load. Thus, this paper selected the temperature, humidity, rainfall, and weather conditions among meteorological factors as similar daily selection and prediction models.

Table 1. Correlation coefficient between electric vehicle charging load and meteorological factors.

	Temperature	Humidity	Precipitation	Visibility	Wind Speed
MIC	0.778	0.788	0.461	0.033	0.343
PCC	0.865	−0.881	−0.459	0.042	0.767

Since the month, season, and week type affect the EV charging load fluctuation characteristics, the season, month, day, week type, weekday, and holiday, selected as date features, were used as the input of the prediction model. Table 2 depicts the date features.

Table 2. Date feature factors.

Date Feature	Detailed Description
Season	1~4 represent spring, summer, fall, and winter
Month	1~12 represent January to December
Day	1~31 represents No. 1 to No. 31
Week	1~7 represents Monday to Sunday
Workday	0 represents a workday, 1 represents a weekend
Holiday	0 represents a non-holiday, 1 represents a holiday

Similar daily loads were obtained from the similar days model. The min–max normalization was adopted to constrain EV charging load to $[0, 1]$. After that, the forecasted load values were exponentiated to establish a nonlinear relationship between the exponentially mapped forecasted load values and the historical loads. It eliminates the lagging problem when the model takes the last moment of the input sequence as the forecasting load value.

3.2. Performance Evaluation

The paper considered the root mean square error (RMSE), the mean absolute error (MAE), and the mean absolute percentage error (MAPE) while assessing the performance of the forecasting model. These are the statistical metrics defined:

$$\text{RMSE} = \sqrt{\frac{\sum\limits_{N}^{i=1}\left(y_i - y_{fi}\right)^2}{N}} \tag{17}$$

$$\text{MAPE} = \sum\limits_{N}^{i=1}\left|\frac{y_{fi} - y_i}{y_i}\right| \times \frac{100}{N} \tag{18}$$

$$\text{MAE} = \frac{1}{N}\sum\limits_{N}^{i=1}\left|y_{fi} - y_i\right| \tag{19}$$

where N indicates the number of validation or testing instances. y_i and y_{fi} represents the actual load and forecasted load of the i-th instance, respectively.

Each statistical metric has different advantages and disadvantages. The RMSE evaluates the performance of a predictive model based on the mean absolute error of the deviation between predicted and actual loads. However, it is susceptible to outliers. In comparison to the RMSE, the MAE reflects the mean absolute error between forecasted and actual loads. It is more resilient to outliers than the RMSE but does not show the real degree of prediction bias. The MAPE is a forecast accuracy measure that considers the relative difference between forecasted and actual loads. However, the MAPE does not apply when the actual load is zero. Therefore, it is vital to employ multiple statistical metrics to assess the prediction performance.

3.3. Similar Daily Load Selection Based on Weighted Grey Correlation Analysis

The weather condition, temperature, humidity, rainfall, and week type are selected as daily features for the similar day in this paper. Since weather conditions and week type similarity are coarse-grained features, while temperature, humidity, and rainfall are fine-grained features, it is necessary to select the coarse-grained amounts of temperature, humidity, and rainfall. This paper selected daily maximum temperature, mean temperature, minimum temperature, as well as daily mean humidity and daily average rainfall as coarse-grained characteristics. Therefore, weather conditions, daily maximum temperature, daily average temperature, daily minimum temperature, humidity, rainfall, and week type similarity were selected as daily features. According to the selected day characteristics and the weighted gray correlation degree, a similar day set of the forecasting day was obtained.

Taking the EV charging load forecast on 15 December 2019 as an example, the weather forecast parameters on that day are shown in Table 3. Because the selected December belongs to winter, the week type similarity obtained by Spearman correlation analysis in this season is shown in Table 4.

Table 3. Forecasting day meteorological and date type parameters.

Forecasting Day	Week Type	Weather Condition	Maximum Temperature/°C	Minimum Temperature/°C	Mean Temperature/°C	Relative Humidity/%	Mean Rainfall/mm
15 December 2019	Sun	cloudy	0.1	−4.5	−2.4	51	0

Table 4. Values of winter day type similarity.

	Mon	Tues	Wed	Thurs	Fri	Sat	Sun
Mon	1						
Tues	0.8271	1					
Wed	0.8758	0.9082	1				
Thurs	0.7951	0.9044	0.9008	1			
Fri	0.8270	0.8485	0.7898	0.8670	1		
Sat	0.7800	0.8193	0.8665	0.7986	0.7299	1	
Sun	0.9113	0.8573	0.8897	0.7934	0.7698	0.7512	1

According to the historical meteorological data and week type before the forecast day (1 December 2019 to 14 December 2019), the weighted grey correlation degrees between the forecasting day and the historical days were calculated to obtain a similar day set. The results of a similar day set are shown in Table 5.

Table 5. Selection results of similar days.

Date	3 December 2019	6 December 2019	10 December 2019	14 December 2019
similarity	0.7219	0.7773	0.8122	0.6711

3.4. Validating the Multi-Channel Convolutional Neural Network and Temporal Convolution Network Model

3.4.1. Hyperparameters of the Multi-Channel Convolutional Neural Network and Temporal Convolution Network Model

From the similar day model results, it can be seen that the length of the similar day historical load sequence of the forecasting day is 384. In this paper, the number of channels of the multi-channel 1DCNN model was set to 4 to fully exploit the characteristics of EV charging load at different time scales. In the multi-channel 1DCNN model, the convolution stride in each channel was set to 1, and the activation function Tanh was selected to perform nonlinear mapping on the results after each convolution. The hyperparameters of the multi-channel 1DCNN model are shown in Table 6. The TCN model hyperparameters are shown in Table 7. The hyperparameters of the BP model and output layer are shown in Table 8. In this paper, meteorological features, date features, and similar daily loads were selected as input variables for the MCCNN-TCN model, as shown in Table 9.

Table 6. Layer architecture of the multi-channel 1D convolutional neural network-temporal convolution network.

Channel No.	Layer	Input	Output	Kernel	Kernel Number	Padding	Stride	Activation Function
C1	Input	$1 \times 384 \times 1$	$1 \times 384 \times 1$	-		-	-	-
	Residual Layer	$1 \times 384 \times 1$	$1 \times 384 \times 1$	-	-	-	-	-
	1D Conv1	$1 \times 384 \times 1$	$1 \times 384 \times 4$	1×3	4	1	1	Tanh
	1D Conv2	$1 \times 384 \times 4$	$1 \times 384 \times 1$	4×3	1	1	1	Tanh
	1D Conv3	$1 \times 384 \times 1$	$1 \times 384 \times 1$	1×3	1	1	1	Tanh
	Adding Layer	$1 \times 384 \times 1$	$1 \times 384 \times 1$	-	-	-	-	Tanh
C2	Input	$1 \times 384 \times 1$	$1 \times 384 \times 1$	-		-	-	-
	Residual Layer	$1 \times 384 \times 1$	$1 \times 384 \times 1$	-	-	-	-	-
	1D Conv1	$1 \times 384 \times 1$	$1 \times 384 \times 4$	1×5	4	2	1	Tanh
	1D Conv2	$1 \times 384 \times 4$	$1 \times 384 \times 1$	4×3	1	1	1	Tanh
	1D Conv3	$1 \times 384 \times 1$	$1 \times 384 \times 1$	1×3	1	1	1	Tanh
	Adding Layer	$1 \times 384 \times 1$	$1 \times 384 \times 1$	-	-	-	-	Tanh
C3	Input	$1 \times 384 \times 1$	$1 \times 384 \times 1$	-		-	-	-
	Residual Layer	$1 \times 384 \times 1$	$1 \times 384 \times 1$	-	-	-	-	-
	1D Conv1	$1 \times 384 \times 1$	$1 \times 384 \times 4$	1×9	4	4	1	Tanh
	1D Conv2	$1 \times 384 \times 4$	$1 \times 384 \times 1$	4×3	1	1	1	Tanh
	1D Conv3	$1 \times 384 \times 1$	$1 \times 384 \times 1$	1×3	1	1	1	Tanh
	Adding Layer	$1 \times 384 \times 1$	$1 \times 384 \times 1$	-	-	-	-	Tanh
C4	Input	$1 \times 384 \times 1$	$1 \times 384 \times 1$	-		-	-	-
	Residual Layer	$1 \times 384 \times 1$	$1 \times 384 \times 1$	-	-	-	-	-
	1D Conv1	$1 \times 384 \times 1$	$1 \times 384 \times 4$	1×17	4	8	1	Tanh
	1D Conv2	$1 \times 384 \times 4$	$1 \times 384 \times 1$	4×3	1	1	1	Tanh
	1D Conv3	$1 \times 384 \times 1$	$1 \times 384 \times 1$	1×3	1	1	1	Tanh
	Adding Layer	$1 \times 384 \times 1$	$1 \times 384 \times 1$	-	-	-	-	Tanh

Table 7. Layer architecture of the temporal convolutional network.

Layer	Input	Output	Kernel	Dilation	Dropout
Residual blocks 1	384×4	384×4	$4 \times 3 \times 4$	1	0.1
Residual blocks 2	384×4	384×2	$4 \times 3 \times 2$	2	0.1
Residual blocks 3	384×2	384×1	$2 \times 3 \times 1$	4	0.1

Table 8. Layer architecture of BP and Output layer.

Layer	Input	Output	Activation Function
BP	10	16	Sigmoid
Output layer	400	1	-

Table 9. Input variables description.

Type of Feature	Variables x	Detailed Description
Load features	$x_1 \sim x_{384}$	Historical load values on the 4 historical similar days
Meteorological features	$x_{385} \sim x_{388}$	Temperature, humidity, precipitation at the forecast time t and weather condition on the forecasting day
Date features	$x_{389} \sim x_{394}$	Season, month, day, week, workday, holiday on the forecast day

3.4.2. Comparative Analysis of Single-Channel and Multi-Channel Convolutional Neural Network and Temporal Convolution Network Model

On the same data set, compared with the prediction results of the single-channel 1DCNN-TCN model, the advanced nature of the MCCNN-TCN proposed in this paper was verified. Each single-channel 1DCNN-TCN and MCCNN-TCN had the same TCN structure, with the only distinction being the number of 1DCNN channels. The single-channel 1DCNN-TCN models were set as follows: Model 1: C1-TCN; Model 2: C2-TCN; Model 3: C3-TCN; Model 4: C4-TCN. Each single-channel 1DCNN-TCN model and MCCNN-TCN model, whose loss function is the MSE, were trained with the Adam optimizer, a learning rate of 0.001, and a batch size of 512.

From 1 June 2019 to 31 August 2019, the training set, validation set, and test set were selected according to the ratio of 8:1:1. Each model outputs a load forecast value at one time each time, and the one-day forecast value refers to the cyclic forecast load value at 96 times. The RMSE, MAPE, and MAE values of each single-channel 1DCNN-TCN and MCCNN TCN model on the test set are shown in Table 10.

Table 10. Prediction results of single-channel and multi-channel 1D convolutional neural network and temporal convolution network model.

Layer	RMSE/kW	MAE/kW	MAPE/%
C1-TCN	8.39	6.52	13.42
C2-TCN	9.26	7.32	15.57
C3-TCN	9.68	7.40	15.73
C4-TCN	9.75	7.49	15.88
MCCNN-TCN	7.62	5.79	11.50

From Table 10, it can be seen that the prediction performance of Model 1 to Model 4 decreases as the extracted time scale increases. This is due to the fact that the single-channel 1DCNN-TCN at the long-term scale loses the local short-term variation features of the EV charging load. The reason why the prediction performance of Model 1 is lower than that of the MCCNN-TCN model is that Model 1 lacks attention to the change trend features of EV charging load at a long-time scale. The advantage of the MCCNN-TCN model is that it can extract the local short-term change features and long-term change trend features of the EV charging load. Therefore, the RMSE, MAPE, and MAE values of the MCCNN-TCN model are lower than those of the single-channel 1DCNN-TCN models. It can be shown that extracting the multi-scale features of EV charging load can significantly improve the prediction accuracy.

3.4.3. Comparative Analysis of Different Forecasting Models

In order to evaluate the forecasting accuracy and superiority of the model proposed in this paper, ANN, LSTM, CNN-LSTM, and TCN prediction models, whose model structures are shown in Appendix B Figures A4–A7, were chosen for comparison. Table 11 shows the ANN, LSTM, and CNN-LSTM models' input. The TCN model's inputs are equal to those of the MCCNN-TCN model. The loss function of ANN, LSTM, CNN-LSTM, and TCN models is MSE. Meanwhile, ANN, LSTM, CNN-LSTM, and TCN models were trained with the Adam optimizer, with a learning rate of 0.001 and a batch size of 512. The dataset was selected between 1 January 2019 and 31 March 2020, with an 8:1:1 ratio for the training, validation, and test sets.

Table 11. Input variables description of ANN, LSTM, and CNN-LSTM models.

Type of Feature	Variable x	Detailed Description
Electrical features	$x_1 \sim x_{64}$	Historical load values from time t to $t + 16$ on historical similar days
	$x_{65} \sim x_{80}$	Historical load values at the time $t - 16$ to $t - 1$ on the forecasting day
Meteorological features	$x_{81} \sim x_{84}$	Temperature, humidity, precipitation at the forecast time t and weather conditions on the forecasting day
Date features	$x_{85} \sim x_{90}$	Season, month, day, week, workday, holiday on the forecast day

The forecasting load curve of the model mentioned above on the test set from 1 March to 7 March 2020 is shown in Figure 7. It can be seen from Figure 7 that the original load is an approximately constant value from 0:00 to 6:00 am every day. The forecasting value of this period, except for the BP model, the forecasting value of all models fluctuates and deviates from the actual value. Although the forecasting value of the ANN model remains constant, it deviates significantly from the actual value. The MCCNN-TCN model fluctuates less than other models and is proximate to the actual value. At the peak of the load curve, the predicted values of the LSTM, ANN, and CNN-LSTM models all deviate to a certain extent and lag significantly compared with the actual values. The TCN model has a significant deviation from the actual values. In comparison to other models, the changing trend of the MCCNN-TCN model is compatible with the actual situation, and the predicted value is more proximate to the actual value. In the rising stage of the load curve, the forecasting value of the MCCNN-TCN model can also maintain a trend similar to the actual value. By analyzing the forecast effect of each prediction model in three stages, it can be seen that the MCCNN-TCN model can improve the accuracy of the short-term load forecasting of EV charging load. This is because the MCCNN-TCN model can not only learn the variation law of EV load on a long timescale but also pay attention to the short-term fluctuation characteristics of EV charging load.

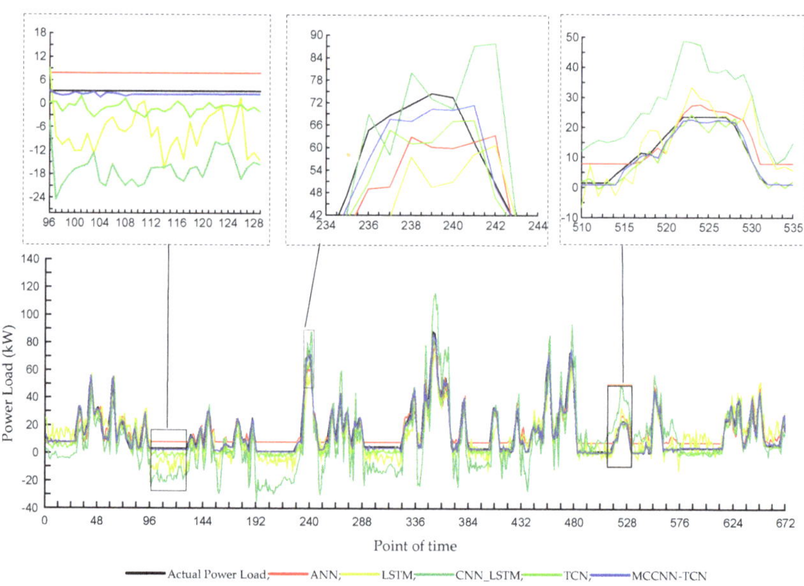

Figure 7. Comparison of forecasting results of load models in 7 days.

The RMSE, MAPE, and MAE of each model on the test set are shown in Table 12. It can be seen from Table 12 that the MAPE of the MCCNN-TCN model is 13.24%, which is 14.09%, 25.13%, 27.32%, and 4.48% higher than that of the ANN, LSTM, CNN-LSTM, and TCN models, respectively. The RMSE of the MCCNN-TCN model is 4.92 kW, which is also significantly less than that of other models. The absolute prediction error boxplots of the five models on the test dataset are shown in Figure 8. The wider the boxplot, the more spread out the prediction errors are. It can be seen from Figure 8 that the prediction error range of the MCCNN-TCN model is the narrowest while the LSTM is the widest, and the median absolute error of the MCCNN-TCN model is smaller than that of ANN, LSTM, CNN-LSTM, and TCN. From the prediction results, the MCCNN-TCN model is more effective than the ANN, LSTM, and CNN-LSTM models in complex fluctuation time series prediction.

Table 12. Prediction results of different models.

Layer	RMSE/kW	MAE/kW	MAPE/%
ANN	9.85	7.43	27.43
LSTM	12.16	9.59	38.47
CNN-LSTM	13.21	10.19	40.66
TCN	6.02	4.59	17.82
MCCNN-TCN	4.92	3.49	13.34

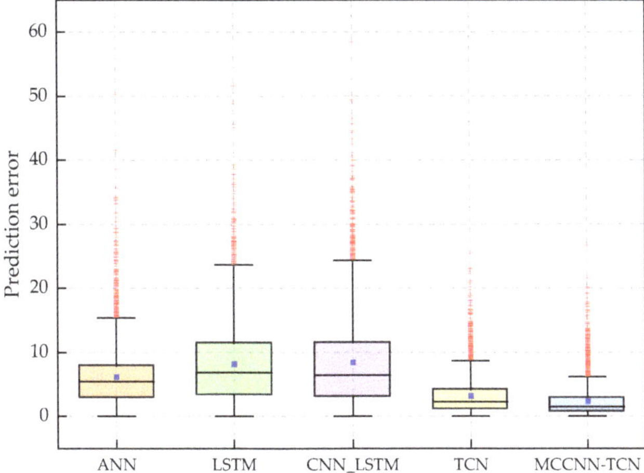

Figure 8. Box plot of absolute prediction errors for different methods.

In addition, it can be seen from Appendix A Figure A2 that in different seasons, the charging load of EVs will show different characteristics. Therefore, this means that the performance of the model proposed in this paper needs to be evaluated further during each season. According to the four seasons defined by meteorology, spring is from March 2019 to May 2019, summer is from June 2019 to August 2019, autumn is from September 2019 to November 2019, and winter is from December 2019 to February 2020. In this paper, each season's historical load and meteorological data are selected, respectively, and the training set, the verification set, and the test set are selected according to the ratio of 8:1:1. The prediction errors of different models on the test set of each season are presented in Table 13.

Table 13. Comparison of forecasting errors of models in each season.

	Spring			Summer			Fall			Winter		
	RMSE/kW	MAE/kW	MAPE/%	RMSE/kW	MAE/kW	MAPE/%	RMSE/kW	MAE/kW	MAPE/%	RMSE/kW	MAE/kW	MAPE/%
ANN	14.67	10.93	36.86	18.04	13.22	24.44	17.74	13.51	20.27	11.50	8.68	31.52
LSTM	13.09	9.88	32.22	20.41	14.73	28.04	18.34	14.10	24.80	11.95	9.41	35.17
CNN-LSTM	13.24	9.84	29.97	20.37	15.13	26.45	19.17	14.29	21.81	12.30	9.28	33.29
TCN	8.03	6.12	20.90	9.97	7.34	13.55	8.66	6.42	10.05	5.75	4.22	16.01
MCCNN-TCN	6.36	4.45	14.24	8.96	6.25	10.80	7.49	5.32	7.53	5.29	3.78	13.65

As shown in Table 13, by comparing the prediction results of the five models in each season, the advanced nature of the model proposed in this paper can be verified intuitively. Although the prediction performance of each prediction model is different in different seasons, the MCCNN-TCN model proposed in this paper has a significant decrease in MAPE, RMSE, and MAE compared with other models in each season. By taking the spring test set as an example, compared with other models, the MAPE of the MCCNN-TCN model decreased by 22.62%, 17.98%, 15.73%, and 6.66%, and the MAE decreased by 6.48, 5.43, 5.39, and 1.67, respectively. In addition, on the test set of each season, the RMSE, MAE, and MAPE of the MCCNN-TCN model and the TCN model are smaller than those of other models. However, since the TCN model does not have the characteristics of multi-time scale feature extraction, its RMSE, MAE, and MAPE in each season are higher than those of the MCCNN-TCN model. Additionally, the MCCNN-TCN model's mean absolute error is relatively concentrated and much lower than the other models under each season, as illustrated in Figure 9. Comparing the prediction results on the test set for each season demonstrates that the MCCNN-TCN model proposed in this paper has a stable prediction performance. This shows that the MCCNN-TCN model can adapt to the load forecasting demand of each season in a year and has good robustness and engineering application value.

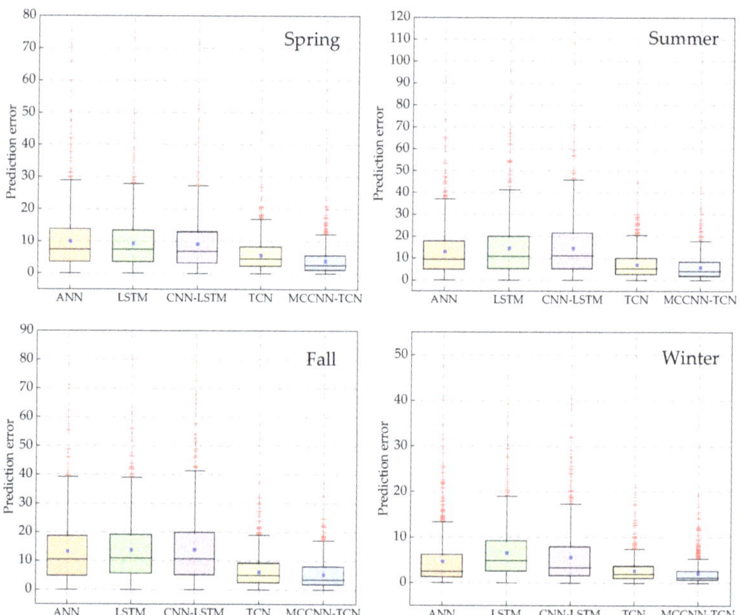

Figure 9. Box plot of absolute prediction errors for different methods in each season.

4. Discussion

By comparing with the single-channel 1DCNN-TCN model, it can be demonstrated that the method of extracting EV charging load feature information at different time scales by setting multiple parallel 1DCNN passes can significantly improve the short-term load prediction performance.

The results in Table 12 show that the MCCNN-TCN model can effectively improve short-term load prediction by using an approach that extracts EV charging load features at multiple scales and relies on TCN to establish long-time dependencies between features. The ANN model has the disadvantage of only establishing superficial nonlinear mapping relationships, which leads to a weaker ability to extract temporal correlations of EV charging loads. Recurrent neural network models such as LSTM have memory properties. They can learn long-term temporal correlations, but feature extraction is weak due to the lack of convolution in their models. This leads to its poor effectiveness in predicting EV charging loads characterized by substantial fluctuations over short periods. The TCN model has superior predictive capabilities over the LSTM and CNN-LSTM due to the availability of convolutional units for extracting shallow temporal features and establishing temporal dependencies. However, the TCN model can only extract features at a single scale, and therefore its prediction performance is poorer than that of the MCCNN-TCN. Further, the results in Table 13 show that the predictive performance of the MCCNN-TCN model proposed in this paper is stable and outperforms those of the comparison models under different seasons.

Combined with the above analysis, it can be seen that the EV charging load prediction model proposed in this paper has a high prediction accuracy. However, the model proposed in this paper relies on the accuracy of meteorological data and EV charging load data to achieve high accuracy prediction. Therefore, some problems need to be noted in the engineering application of this method. On the one hand, if there are deviations in the meteorological data measurement of the forecasting day, this will affect the selection of similar daily loads. This paper uses several meteorological and date factors as day features when selecting similar day loads. Additionally, the adjacent day loads of the forecasting day to be measured are also added to the similar day set, making the similar day selection model somewhat fault-tolerant. On the other hand, in the power system, there are disturbances in the power load data from the measurement system caused by errors in the electric power system, outliers due to data encoding errors, and EV charging start and end times falling between load sampling points. Suppose the deviation from the actual value is slight. In that case, the deviation from the actual value obtained from the prediction model will also be slight. Conversely, suppose there are significant deviations from the actual values. In that case, the actual values need to be estimated using data pre-processing techniques such as mean-fill, interpolation, and algorithmic mean filtering.

5. Conclusions

Due to the randomness of EV charging behavior, the short-term fluctuation characteristics of EV charging load are obvious in one day. In order to improve the load prediction accuracy, this paper proposes the MCCNN-TCN load model, which considers the multi-time scale characteristics of EV charging loads. The multi-channel 1DCNN model was used to extract the features of EV charging load at multiple time scales. The TCN model was used to establish global temporal dependencies between the features.

By considering the influence of various factors on the load, MIC and Spearman coefficient were used to reduce the meteorological feature dimension and establish the similarity of date types, respectively. Then, taking the selected meteorological features and the similarity of date types as the daily features, a similar day selection model based on the weighted grey correlation degree was established to select similar daily loads. The selected meteorological features, date features, and similar daily loads were used as the input of the MCCNN-TCN model.

From the comparative experiments of single-channel 1DCNN-TCN and MCCNN-TCN, it can be seen that MCCNN-TCN can improve the prediction accuracy of EV charging load. This shows that the prediction performance can be improved by extracting the feature information of time series at different time scales and establishing global time series dependencies.

According to the prediction results compared with ANN, LSTM, CNN-LSTM, and TCN models, compared with these models, due to the unique structure of the MCCNN-TCN network, it can learn the multi-scale features of the EV charging load time series and master the changing law of EV charging load.

The MCCNN-TCN network constructed in this paper also lacks the consideration of real-time electricity price factors. In the future, we can further consider the selection of richer feature data and take advantage of big data to improve the accuracy of load forecasting.

Author Contributions: Conceptualization, J.Z. and C.L.; methodology, C.L.; software, C.L.; validation, J.Z. and C.L.; formal analysis, J.Z. and L.G.; investigation, J.Z.; resources, J.Z. and L.G.; data curation, J.Z.; writing original draft preparation, C.L. and J.Z.; writing review and editing, J.Z., C.L. and L.G.; visualization, C.L.; supervision J.Z. and L.G.; project administration, J.Z.; funding acquisition, J.Z. All authors have read and agreed to the published version of the manuscript.

Funding: This research was funded by the State Key Laboratory of Reliability and Intelligence of Electrical Equipment (No. EERI_KF20200014), Hebei University of Technology.

Institutional Review Board Statement: Not applicable.

Informed Consent Statement: Not applicable.

Data Availability Statement: Not applicable.

Acknowledgments: We wish to thank Ping Zhang, Fei Li and Ning Liu for providing technical support.

Conflicts of Interest: The authors declare no conflict of interest.

Appendix A

Based on the EV charging load dataset used in Section 3 of the paper, the characteristics of EV charging load in different months, seasons, and week kinds are investigated. The box plot of EV charging load in each month is shown in Figure A1, and the average daily EV charging load curves for different seasons and different week kinds are shown in Figures A2 and A3, respectively.

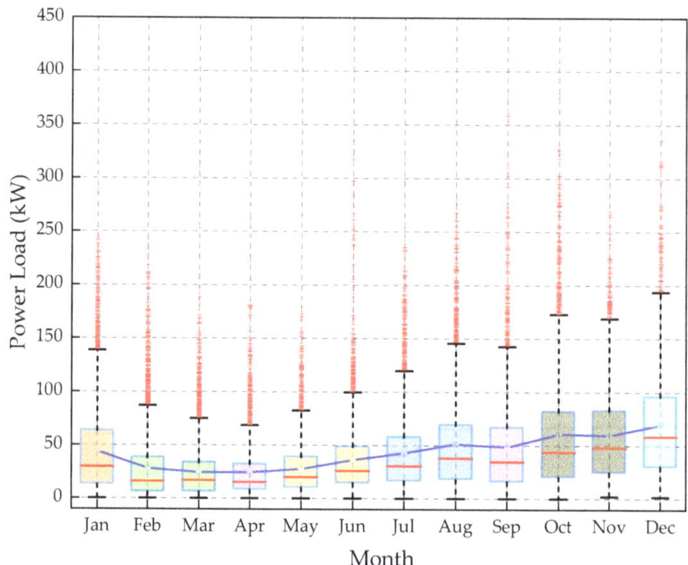

Figure A1. Average electric vehicle charging load per month.

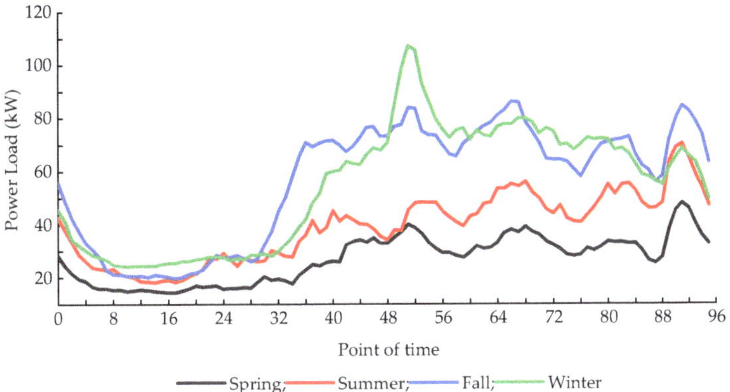

Figure A2. Average electric vehicle charging load for each season.

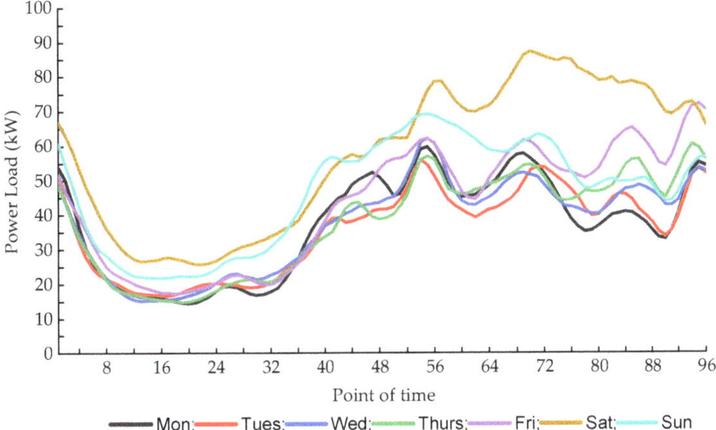

Figure A3. Average daily electric vehicle charging load for different weeks.

Appendix B

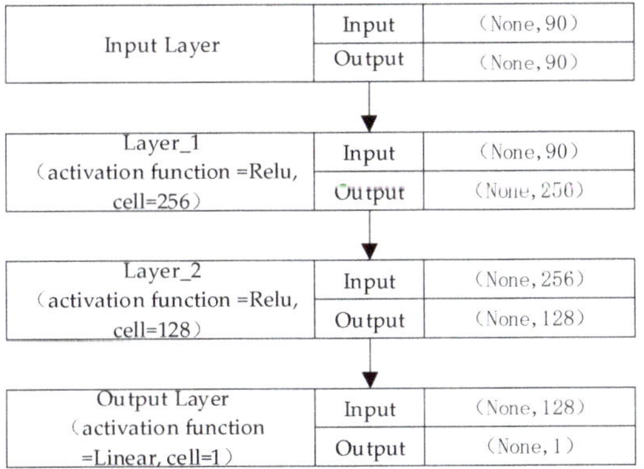

Input Layer	Input	（None, 90）
	Output	（None, 90）

Layer_1 （activation function =Relu, cell=256）	Input	（None, 90）
	Output	（None, 256）

Layer_2 （activation function =Relu, cell=128）	Input	（None, 256）
	Output	（None, 128）

Output Layer （activation function =Linear, cell=1）	Input	（None, 128）
	Output	（None, 1）

Figure A4. ANN model architecture.

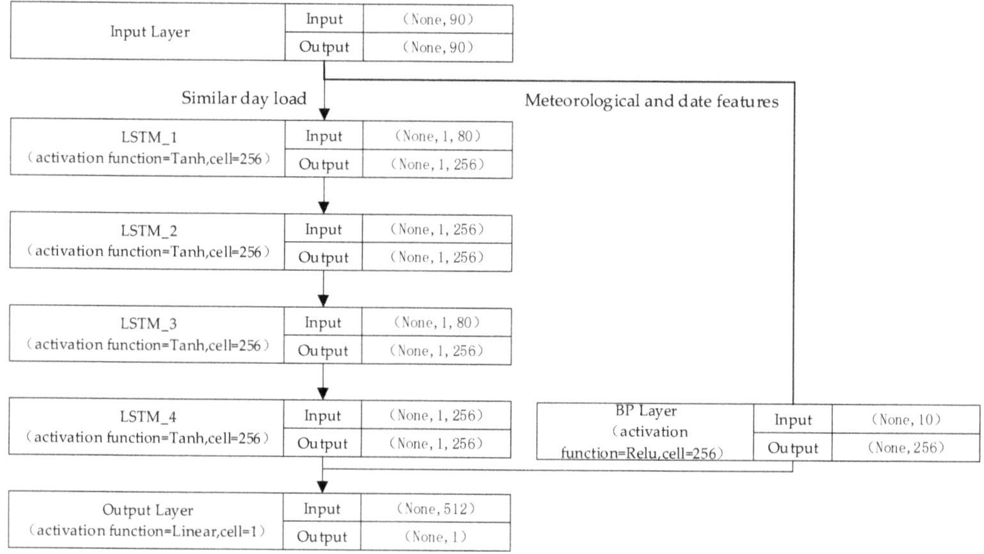

Figure A5. LSTM model architecture.

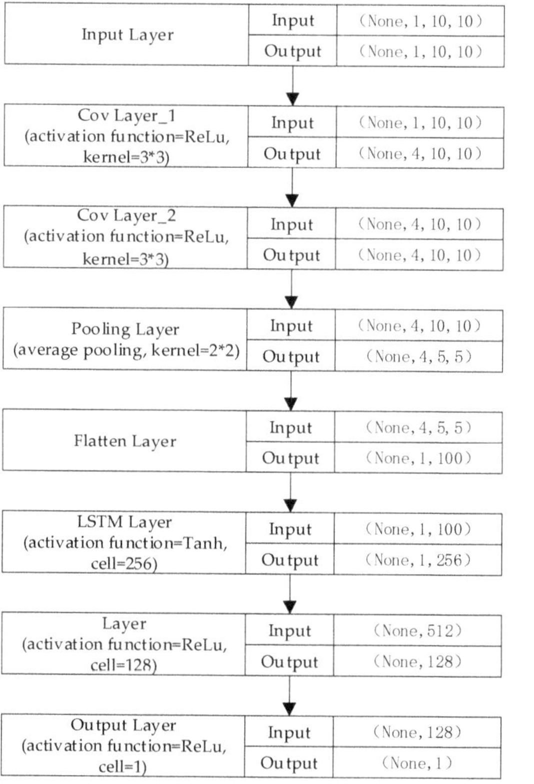

Figure A6. CNN-LSTM model architecture.

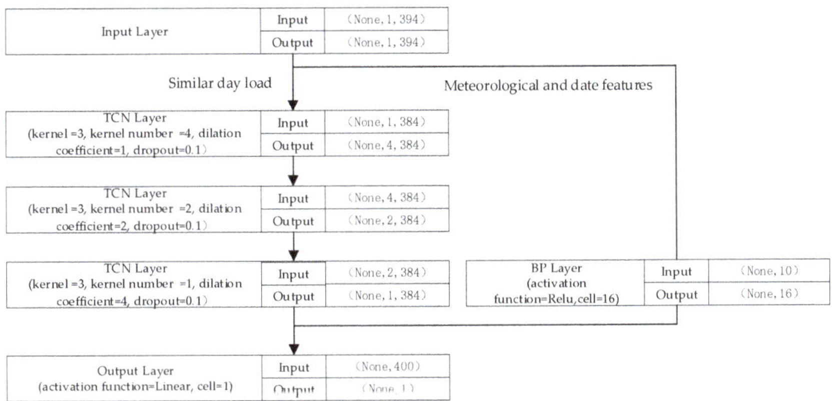

Figure A7. TCN model architecture.

References

1. Tribioli, L. Energy-Based Design of Powertrain for a Re-Engineered Post-Transmission Hybrid Electric Vehicle. *Energies* **2017**, *10*, 918. [CrossRef]
2. Habib, S.; Khan, M.M.; Abbas, F.; Sang, L.; Shahid, M.U.; Tang, H. A Comprehensive Study of Implemented International Standards, Technical Challenges, Impacts and Prospects for Electric Vehicles. *IEEE Access* **2018**, *6*, 13866–13890. [CrossRef]
3. Crozier, C.; Morstyn, T.; McCulloch, M. The opportunity for smart charging to mitigate the impact of electric vehicles on transmission and distribution systems. *Appl. Energy* **2020**, *268*, 114973. [CrossRef]
4. Lyu, L.; Yang, X.; Xiang, Y.; Liu, J.; Jawad, S.; Deng, R. Exploring high-penetration electric vehicles impact on urban power grid based on voltage stability analysis. *Energy* **2020**, *198*, 117301. [CrossRef]
5. Wang, H.; Yang, J.; Chen, Z.; Li, G.; Liang, J.; Ma, Y.; Dong, H.; Ji, H.; Feng, J. Optimal dispatch based on prediction of distributed electric heating storages in combined electricity and heat networks. *Appl. Energy* **2020**, *267*, 114879. [CrossRef]
6. Jiang, Y.; Yin, S. Recursive Total Principle Component Regression Based Fault Detection and Its Application to Vehicular Cyber-Physical Systems. *IEEE Trans. Ind. Inform.* **2018**, *14*, 1415–1423. [CrossRef]
7. Salah, F.; Ilg, J.P.; Flath, C.M.; Basse, H.; Dinther, C.V. Impact of electric vehicles on distribution substations: A Swiss case study. *Appl. Energy* **2015**, *137*, 88–96. [CrossRef]
8. Arias, M.B.; Bae, S. Electric vehicle charging demand forecasting model based on big data technologies. *Appl. Energy* **2016**, *183*, 327–339. [CrossRef]
9. Taylor, J.; Maitra, A.; Alexander, M.; Brooks, D.; Duvall, M. Evaluation of the impact of plug-in electric vehicle loading on distribution system operations. In Proceedings of the 2009 IEEE Power & Energy Society General Meeting, Calgary, AB, Canada, 26–30 July 2009.
10. Bae, S.; Kwasinski, A. Spatial and Temporal Model of Electric Vehicle Charging Demand. *IEEE Trans. Smart Grid.* **2012**, *3*, 394–403. [CrossRef]
11. Shun, T.; Kunyu, L.; Xiangning, X.; Jianfeng, W.; Yang, Y.; Jian, Z. Charging demand for electric vehicle based on stochastic analysis of trip chain. *IET Gener. Transm. Distrib.* **2016**, *10*, 2689–2698. [CrossRef]
12. Chen, T.; Zhang, B.; Pourbabak, H.; Kavousi-Fard, A.; Su, W. Optimal Routing and Charging of an Electric Vehicle Fleet for High-Efficiency Dynamic Transit Systems. *IEEE Trans. Smart Grid* **2018**, *9*, 3563–3572. [CrossRef]
13. Xing, Q.; Chen, Z.; Zhang, Z.; Huang, X.; Leng, Z.; Sun, K.; Chen, Y.; Wang, H. Charging Demand Forecasting Model for Electric Vehicles Based on Online Ride-Hailing Trip Data. *IEEE Access* **2019**, *7*, 137390–137409. [CrossRef]
14. Taylor, J.W. Short-Term Load Forecasting with Exponentially Weighted Methods. *IEEE Trans. Power Syst.* **2012**, *27*, 458–464. [CrossRef]
15. Hong, T.; Wilson, J.; Xie, J. Long Term Probabilistic Load Forecasting and Normalization with Hourly Information. *IEEE Trans. Smart Grid.* **2014**, *5*, 456–462. [CrossRef]
16. Zhang, J.; Wei, Y.; Li, D.; Tan, Z.; Zhou, J. Short term electricity load forecasting using a hybrid model. *Energy* **2018**, *158*, 774–781. [CrossRef]
17. Khodayar, M.; Liu, G.; Wang, J.; Mohammad, E.K. Deep learning in power systems research: A review. *CSEE J. Power Energy Syst.* **2021**, *7*, 209–220.
18. Xu, X.; Liu, W.; Xi, Z.; Zhao, T. Short-term load forecasting for the electric bus station based on GRA-DE-SVR. In Proceedings of the 2014 IEEE Innovative Smart Grid Technologies—Asia, Kuala Lumpur, Malaysia, 20–23 May 2014.
19. Yi, Z.; Liu, X.C.; Wei, R.; Chen, X.; Dai, J. Electric vehicle charging demand forecasting using deep learning model. *J. Intell. Transp. Syst.* **2021**, 1–14. [CrossRef]

20. Zhu, J.; Yang, Z.; Chang, Y.; Guo, Y.; Zhu, K.; Zhang, J. A novel LSTM based deep learning approach for multi-time scale electric vehicles charging load prediction. In Proceedings of the 2019 IEEE Innovative Smart Grid Technologies, Chengdu, China, 21–24 May 2019.

21. Zhu, J.; Yang, Z.; Guo, Y.; Zhang, J.; Yang, H. Short-Term Load Forecasting for Electric Vehicle Charging Stations Based on Deep Learning Approaches. *Appl. Sci.* **2019**, *9*, 1723. [CrossRef]

22. Feng, J.; Yang, J.; Li, Y.; Wang, H.; Ji, H.; Yang, W.; Wang, K. Load forecasting of electric vehicle charging station based on grey theory and neural network. *Energy Rep.* **2021**, *7*, 487–492. [CrossRef]

23. Dabbaghjamanesh, M.; Moeini, A.; Kavousi-Fard, A. Reinforcement Learning-Based Load Forecasting of Electric Vehicle Charging Station Using Q-Learning Technique. *IEEE Trans. Ind. Inform.* **2021**, *17*, 4229–4237. [CrossRef]

24. Krizhevsky, A.; Sutskever, I.; Hinton, G.E. ImageNet classification with deep convolutional neural networks. *Commun. ACM* **2017**, *60*, 84–90. [CrossRef]

25. Li, Y.; Huang, Y.; Zhang, M. Short-Term Load Forecasting for Electric Vehicle Charging Station Based on Niche Immunity Lion Algorithm and Convolutional Neural Network. *Energies* **2018**, *11*, 1253. [CrossRef]

26. Ozcanli, A.K.; Yaprakdal, F.; Baysal, M. Deep learning methods and applications for electrical power systems: A comprehensive review. *Int. J. Energy Res.* **2020**, *44*, 7136–7157. [CrossRef]

27. Shang, C.; Gao, J.; Liu, H.; Liu, F. Short-Term Load Forecasting Based on PSO-KFCM Daily Load Curve Clustering and CNN-LSTM Model. *IEEE Access* **2021**, *9*, 50344–50357. [CrossRef]

28. Yan, K.; Wang, X.; Du, Y.; Jin, N.; Huang, H.; Zhou, H. Multi-Step Short-Term Power Consumption Forecasting with a Hybrid Deep Learning Strategy. *Energies* **2018**, *11*, 3089. [CrossRef]

29. Wei, L. A Hybrid Model Based on ANFIS and Empirical Mode Decomposition for Stock Forecasting. *J. Econ. Bus. Manag.* **2015**, *3*, 356–359. [CrossRef]

30. Zheng, H.; Yuan, J.; Chen, L. Short-Term Load Forecasting Using EMD-LSTM Neural Networks with a Xgboost Algorithm for Feature Importance Evaluation. *Energies* **2017**, *10*, 1168. [CrossRef]

31. Wang, Y.; Gu, Y.; Ding, Z.; Li, S.; Wan, Y.; Hu, X. Charging Demand Forecasting of Electric Vehicle Based on Empirical Mode Decomposition-Fuzzy Entropy and Ensemble Learning. *Autom. Electr. Power Syst.* **2020**, *44*, 114–121.

32. Wang, X.; Mao, D.; Li, X. Bearing fault diagnosis based on vibro-acoustic data fusion and 1D-CNN network. *Measurement* **2021**, *173*, 108518. [CrossRef]

33. Heo, S.; Nam, K.; Loy-Benitez, J.; Yoo, C. Data-Driven Hybrid Model for Forecasting Wastewater Influent Loads Based on Multimodal and Ensemble Deep Learning. *IEEE Trans. Ind. Inform.* **2021**, *17*, 6925–6934. [CrossRef]

34. Wang, Y.; Chen, J.; Chen, X.; Zeng, X.; Kong, Y.; Sun, S.; Guo, Y.; Liu, Y. Short-Term Load Forecasting for Industrial Customers Based on TCN-LightGBM. *IEEE Trans. Power Syst.* **2021**, *36*, 1984–1997. [CrossRef]

35. Yin, L.; Xie, J. Multi-feature-scale fusion temporal convolution networks for metal temperature forecasting of ultra-supercritical coal-fired power plant reheater tubes. *Energy* **2022**, *238*, 121657. [CrossRef]

36. Tang, X.; Chen, H.; Xiang, W.; Yang, J.; Zou, M. Short-Term Load Forecasting Using Channel and Temporal Attention Based Temporal Convolutional Network. *Electr. Power Syst. Res.* **2022**, *205*, 107761. [CrossRef]

37. Mishra, M.; Nayak, J.; Naik, B.; Abraham, A. Deep learning in electrical utility industry: A comprehensive review of a decade of research. *Eng. Appl. Artif. Intell.* **2020**, *96*, 104000. [CrossRef]

38. Korolko, N.; Sahinoglu, Z.; Nikovski, D. Modeling and Forecasting Self-Similar Power Load Due to EV Fast Chargers. *IEEE Trans. Smart Grid* **2015**, *7*, 1620–1629. [CrossRef]

39. Zhang, J.; Liu, D.; Li, Z.; Han, X.; Liu, H.; Dong, C.; Wang, J.; Liu, C.; Xia, Y. Power prediction of a wind farm cluster based on spatiotemporal correlations. *Appl. Energy* **2021**, *302*, 117568. [CrossRef]

40. Ge, L.; Xian, Y.; Wang, Z.; Gao, B.; Chi, F.; Sun, K. Short-term Load Forecasting of Regional Distribution Network Based on Generalized Regression Neural Network Optimized by Grey Wolf Optimization Algorithm. *CSEE J. Power Energy Syst.* **2020**, *7*, 1093–1101.

41. Reshef, D.N.; Reshef, Y.A.; Finucane, H.K.; Grossman, S.R.; McVean, G.; Turnbaugh, P.J.; Lander, E.S.; Mitzenmacher, M.; Sabeti, P.C. Detecting Novel Associations in Large Data Sets. *Science* **2011**, *334*, 1518–1524. [CrossRef]

42. Nascimento, J.; Pinto, T.; Vale, Z. Day-ahead electricity market price forecasting using artificial neural network with spearman data correlation. In Proceedings of the 2019 IEEE Milan PowerTech, Milan, Italy, 23–27 June 2019.

43. Han, M.; Zhang, R.; Qiu, T.; Xu, M.; Ren, W. Multivariate Chaotic Time Series Prediction Based on Improved Grey Relational Analysis. *IEEE Trans. Syst. Man Cybern. Syst.* **2019**, *49*, 2144–2154. [CrossRef]

44. An, L.; Teng, Y.; Li, C.; Dan, Q. Combined Grey Model Based on Entropy Weight Method for Long-term Load Forecasting. In Proceedings of the 2020 5th Asia Conference on Power and Electrical Engineering, Chengdu, China, 4–7 June 2020.

45. Song, J.; He, C.; Li, X.; Liu, Z.; Tang, J.; Zhong, W. Daily load curve clustering method based on feature index dimension reduction and entropy weight method. *Autom. Electr. Power Syst.* **2019**, *43*, 65–76.

46. Abdel-Hamid, O.; Mohamed, A.; Jiang, H.; Deng, L.; Penn, G.; Yu, D. Convolutional Neural Networks for Speech Recognition. *IEEE ACM Trans. Audio Speech Lang. Process.* **2014**, *22*, 1533–1545. [CrossRef]

47. Zhang, W.; Peng, G.; Li, C.; Chen, Y.; Zhang, Z. A New Deep Learning Model for Fault Diagnosis with Good Anti-Noise and Domain Adaptation Ability on Raw Vibration Signals. *Sensors* **2017**, *17*, 425. [CrossRef] [PubMed]

48. He, K.; Zhang, X.; Ren, S.; Sun, J. Deep residual learning for image recognition. In Proceedings of the 2016 IEEE Conference on Computer Vision and Pattern Recognition, Las Vegas, NV, USA, 27–30 June 2016.
49. Bai, S.; Kolter, J.Z.; Koltun, V. An empirical evaluation of generic convolutional and recurrent networks for sequence modeling. In Proceedings of the AAAI Conference on Artificial Intelligence, New Orleans, LA, USA, 2–7 February 2018.
50. Lore, K.G.; Stoecklein, D.; Davies, M.; Ganapathysubramanian, B.; Sarkar, S. A deep learning framework for causal shape transformation. *Neural Netw.* **2018**, *98*, 305–317. [CrossRef] [PubMed]
51. Xu, B.; Ye, H.; Zheng, Y.; Wang, H.; Luwang, T.; Jiang, Y.G. Dense Dilated Network for Video Action Recognition. *IEEE Trans. Image Process.* **2019**, *28*, 4941–4953. [CrossRef] [PubMed]
52. Farha, Y.A.; Gall, J. MS-TCN: Multi-stage temporal convolutional network for action segmentation. In Proceedings of the 2019 IEEE/CVF Conference on Computer Vision and Pattern Recognition, Long Beach, CA, USA, 15–20 June 2019.
53. Li, Y.; Huang, J.B.; Ahuja, N.; Yang, M.H. Joint imagefdiltering with deep convolu-tional networks. *IEEE Trans. Pattern Anal. Mach. Intell.* **2019**, *41*, 1909–1923. [CrossRef] [PubMed]

Article

Two-Stage Energy Management Strategies of Sustainable Wind-PV-Hydrogen-Storage Microgrid Based on Receding Horizon Optimization

Jiarui Wang [1], Dexin Li [1], Xiangyu Lv [1], Xiangdong Meng [1], Jiajun Zhang [1], Tengfei Ma [2,*], Wei Pei [2] and Hao Xiao [2]

[1] State Grid Jilin Electric Power Research Institute, Changchun 130000, China; wangjiarui247@126.com (J.W.); lidexin0323@163.com (D.L.); caulxy@163.com (X.L.); mengxd22@sins.com (X.M.); 18362961072@163.com (J.Z.)

[2] Institute of Electrical Engineering, Chinese Academy of Sciences, Beijing 100190, China; peiwei@mail.iee.ac.cn (W.P.); xiaohao09@mail.iee.ac.cn (H.X.)

* Correspondence: flytengma@mail.iee.ac.cn

Abstract: Hydrogen and renewable electricity-based microgrid is considered to be a promising way to reduce carbon emissions, promote the consumption of renewable energies and improve the sustainability of the energy system. In view of the fact that the existing day-ahead optimal operation model ignores the uncertainties and fluctuations of renewable energies and loads, a two-stage energy management model is proposed for the sustainable wind-PV-hydrogen-storage microgrid based on receding horizon optimization to eliminate the adverse effects of their uncertainties and fluctuations. In the first stage, the day-ahead optimization is performed based on the predicted outpower of WT and PV, the predicted demands of power and hydrogen loads. In the second stage, the intra-day optimization is performed based on the actual data to trace the day-ahead operation schemes. Since the intra-day optimization can update the operation scheme based on the latest data of renewable energies and loads, the proposed two-stage management model is effective in eliminating the uncertain factors and maintaining the stability of the whole system. Simulations show that the proposed two-stage energy management model is robust and effective in coordinating the operation of the wind-PV-hydrogen-storage microgrid and eliminating the uncertainties and fluctuations of WT, PV and loads. In addition, the battery storage can reduce the operation cost, alleviate the fluctuations of the exchanged power with the power grid and improve the performance of the energy management model.

Keywords: sustainable wind-PV-hydrogen-storage microgrid; energy management; power-to-hydrogen; receding horizon optimization; storage

Citation: Wang, J.; Li, D.; Lv, X.; Meng, X.; Zhang, J.; Ma, T.; Pei, W.; Xiao, H. Two-Stage Energy Management Strategies of Sustainable Wind-PV-Hydrogen-Storage Microgrid Based on Receding Horizon Optimization. *Energies* **2022**, *15*, 2861. https://doi.org/10.3390/en15082861

Academic Editors: Leijiao Ge, Jun Yan, Yonghui Sun and Zhongguan Wang

Received: 1 March 2022
Accepted: 5 April 2022
Published: 14 April 2022

Publisher's Note: MDPI stays neutral with regard to jurisdictional claims in published maps and institutional affiliations.

1. Introduction

In order to protect the environment and cope with the energy crisis, the renewable energies, such as wind and solar, are being exploited in a more widespread way. However, the randomness and intermittency of renewable electricity are still challenging issues for the large-scale connection to the power grid [1].

Since hydrogen has the advantages of high-energy density, being environmentally friendly and easy storage, it has been regarded as a promising energy carrier and electricity storage medium to reduce carbon emission, improve the sustainability of the energy system, promote the consumption of renewable energies and alleviate their volatility [2]. Hydrogen can either be produced centrally from renewable electricity through electrolyzers situated close to wind or PV power plants and then transported to hydrogen consumers or can be directly generated on sites close to hydrogen consumers [3]. Though mass production in a central manner is more economical, the high transportation cost may erase the advantages of this hydrogen production mode. Consequently, the distributed hydrogen production mode based on a renewable energy microgrid is considered to be an effective way to reduce hydrogen production and transportation costs and promote the consumptions of distributed renewable energies [4]. The distributed hydrogen production mode based on a renewable energy microgrid has attracted more attention, and research is focused on the aspects of modeling, techno-economic analysis, cooperative operation and optimal planning, etc. For example, the accurate modeling method of the advanced alkaline electrolyzer system is proposed and demonstrated in [5]. The techno-economic feasibility of the production of hydrogen from the PV-wind microgrid has been evaluated in [6]. A cooperative operation method to increase profits for wind turbines and onsite hydrogen production and fueling stations has been proposed in [7]. A Nash-bargaining-based cooperative planning and operation method for a wind-hydrogen-heat multi-agent energy system has been proposed in [8]. In addition, optimal capacity planning of an isolated, batteryless, hydrogen-based microgrid is proposed in [9].

Due to the stochastic volatility of renewable electricity, such as wind power and PV, not only affecting the stable operation of the hydrogen production system, but also affecting hydrogen purity, the question of how to relieve the adverse effects of the uncertainties and volatility of the renewable energies is still a critical challenge and an open problem, which has drawn more and more attention. The energy management strategy based on model predictive controller or receding horizon optimization is considered to be one of the promising methods. For instance, an energy management strategy is proposed for a renewable hydrogen-based microgrid in [10], and both the long- and short-term optimal operation schedules are obtained by the model predictive controller. In [11], an energy management strategy based on the receding-horizon stochastic optimization method is proposed to increase renewable penetration and improve operational flexibility of the PV-hydrogen microgrids. In [12], a flexible weighted model predictive control energy management strategy is proposed for a multi-energy microgrid with the hydrogen energy storage system and the heat storage system. In [13], a real-time energy management method based on model predictive control is proposed for a microgrid composed of PV, battery, electrolyzer and fuel cell. In [14], in order to maximize the operational benefit of the microgrid and minimize the degradation causes of each storage system, energy management based on the model predictive control method is proposed. The energy management strategies proposed above all show good performance in relieving the uncertainties of renewable energies or loads. Furthermore, the energy storage as well as the demand response technologies is also helpful in mitigating the power fluctuations of the renewable energy microgrid. For example, in [15], the conventional operation strategy, demand response strategy for peak shaving, has been comparatively studied for grid-connected photovoltaic (PV)-hydrogen/battery systems and the battery storage has an important role in reducing the operation cost and mitigating the power fluctuation. In [16], the accurate model of a hybrid energy system including solar energy, lithium-ion battery and hydrogen is proposed; the coordinated operations of the short-term lithium-ion battery and long-term hydrogen storage show great advantage in keeping energy balanced and mitigating the power fluctuation of renewable energies. In [17], the advantages in reducing operation cost and relieving the intermittent use of a pumped-storage system with a dynamic tariff demand response strategy have been demonstrated in a system consisting of wind turbines, a photovoltaic array and a pumped hydro energy storage system. In order to improve the reliability and mitigate the stochastic volatility of wind farms, an optimal coordination operation and planning method of kinetic energy storage is proposed in [18]; simulation results show that the proposed method is effective in identifying the minimum capacity of kinetic energy storage and improving power supply reliability. Likewise, an optimization energy management method is proposed in [19] to reduce the operation cost of a wind power plant-flywheel energy storage system; simulation results show that the flywheel energy storage method is effective in relieving the stochastic fluctuation of wind power. In [20], a planning method is proposed to optimize the structure of the PV-wind-electrochemical storage system, and the energy storage system has been shown to play an important role in improving the power supply reliability.

These research results have laid a good foundation for the energy management problem of the renewable-energy-based microgrid. However, the energy management problem for the wind-PV-hydrogen-battery microgrid is still an open problem; the questions of how to mitigate the adverse effects of the stochastic and uncertain factors of renewable electricity and how to coordinate the operation of the whole system still need further investigation.

In order to alleviate the uncertainties and fluctuations of outpower of WT and PV, and the power and hydrogen demands, this paper proposes a two-stage energy management model for the sustainable wind-PV-hydrogen-storage microgrid based on receding horizon optimization, and the role of energy storage has also been explored. The main contributions are as follows.

(1) A two-stage energy management model based on receding horizon optimization is proposed to tackle the uncertainties and randomness of renewable energies and loads, as well as to minimize the operation cost.
(2) The day-ahead optimization is performed to minimize the overall operation cost, while the intra-day optimization model is carried out to trace the day-ahead schemes and minimize the deviations of the intra-day and the day-ahead operation strategies.
(3) The roles of battery storage in reducing operation cost and improving the performance of the energy management model have been explored and demonstrated.

The remainder of this paper is organized as follows. Section 2 introduces the structures and the subsystem models of the sustainable wind-PV-hydrogen-storage microgrid. Section 3 proposes the two-stage energy management model. Section 4 presents the simulation and result analysis. At last, Section 5 draws the conclusion.

2. The Sustainable Wind-PV-Hydrogen-Storage Microgrid

Figure 1 illustrates the sustainable wind-PV-hydrogen-storage (WPHS) microgrid. It is mainly composed of wind turbines (WT), photovoltaics (PV), battery storage and the power-to-hydrogen (P2H) subsystem. The WPHS microgrid is responsible for meeting the hydrogen demands and power demands of the end users. The sustainable WPHS microgrid is connected to the upstream power grid, and the renewable electricity is mainly consumed locally to produce hydrogen and meet the power loads. Bilateral power exchange with the power grid is supported, the surplus power can be fed back to the power grid to make profit and the insufficient electricity can also be purchased from the power grid. Therefore, the WPHS microgrid comprises a high proportion of renewable energy systems, which can realize the sustainability of energy supply. The models of WT, PV, battery storage and the power-to-hydrogen subsystem are as follows.

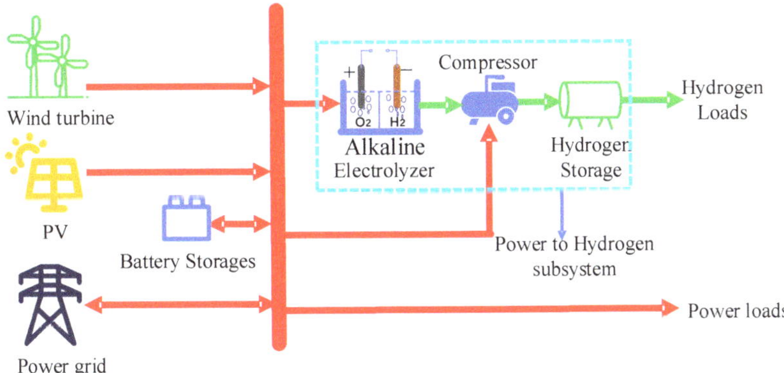

Figure 1. The schematic of wind-PV-hydrogen-storage microgrid.

2.1. The Wind Turbine Model

The outpower of wind turbine can be expressed as the function of wind speed [15].

$$P_{WT}^t = \begin{cases} 0 & v_t \leq v_{in}, v_t \geq v_{out} \\ \frac{v_t - v_{in}}{v_r - v_{in}} P_{RWT} & v_{in} \leq v_t \leq v_r \\ P_{RWT} & v_r \leq v_t \leq v_{out} \end{cases} \tag{1}$$

where P_{WT}^t is the outpower of wind turbine at time slot t; v_t is the wind speed at time slot t; v_{in} and v_{out} are cut-in and cut-out wind speed, respectively; v_r is the rated wind speed of wind turbine; P_{RWT} is the rated power of wind turbine.

2.2. The PV Model

The outpower of PV panels can be expressed as the function of solar radiation intensity and the cell temperature [21].

$$P_{PV}^t = N_{PV} \cdot P_{rSTC} \cdot \frac{I_t}{I_{STC}} [1 + 0.005 \cdot (T_t - 25)] \tag{2}$$

where P_{PV}^t is the outpower of PV array; N_{PV} is the number of PV panes; I_{STC} is the standard irradiance, 1000 W/m^2; P_{rSTC} is the rated power of each PV panel at standard test conditions (cell temperature is 25 °C, irradiance is 1000 W/m^2); I_t and T_t are irradiance and cell temperature (which approximates to the ambient temperature) at time slot t.

2.3. The Battery Storage Model

The battery storages are helpful in alleviating the volatility of renewable energies. Let E_{bat}^t be the energy stored in the batteries at time slot t; E_{bat}^{min} and E_{bat}^{max} denote the minimum and maximum capacity of battery storages, respectively. Let $P_{bat,c}^t$ and $P_{bat,d}^t$ denote the charging and discharging power, respectively, and let $P_{bat,c}^{max}$ and $P_{bat,d}^{max}$ denote the maximum values of charging and discharging power, respectively. Then, the battery storage model can be formulated as follows [22].

$$\begin{cases} E_{bat}^t = E_{bat}^{t-1} + \left(P_{bat,c}^t e_{bat,c} - \frac{P_{bat,d}^t}{e_{bat,d}} \right) \Delta t \\ 0 \leq P_{bat,c}^t \leq u_{bat}^t \cdot P_{bat,c}^{max} \\ 0 \leq P_{bat,d}^t \leq (1 - u_{bat}^t) \cdot P_{bat,d}^{max} \\ E_{bat}^{min} \leq E_{bat}^t \leq E_{bat}^{max} \\ E_{bat}^T = E_{bat}^1 \end{cases} \tag{3}$$

where the first equation of Equation (3) denotes the stored energy variation during time interval Δt before and after charging or discharging. The second and third items of Equation (3) indicate that the charging and discharging power cannot exceed their maximums. u_{bat}^t is a binary variable to avoid charging and discharging power simultaneously. The fourth item of Equation (3) denotes that the stored energy should be constrained between the minimum and maximum capacity. The last item of Equation (3) shows that the stored energy at the end of the dispatch period is to be equal to its initial value.

2.4. The Power-to-Hydrogen Subsystem Model

The power-to-hydrogen production system mainly consists of alkaline electrolyzer, hydrogen compressor and hydrogen storage tank.

2.4.1. The Model of Electrolyzer

Currently the alkaline electrolyzer (AE) and proton exchange membrane (PEM) are two major ways to produce hydrogen from electricity. The alkaline electrolyzer technology is more mature and economic, and was thus chosen to produce hydrogen in this paper. Since the start and response speed of the electrolyzer is quick [16], the ramp up/down constraints are assumed to be satisfied in this paper. The mass of hydrogen production of AE is approximately linear to the consumed power [23].

$$\begin{cases} m^t_{H_2} = \eta_{H_2} P^t_{el} \cdot \Delta t \\ 0 \le P^t_{el} \le P^{max}_{el} \end{cases} \tag{4}$$

where $m^t_{H_2}$ is the hydrogen mass produced at time slot t, kg; η_{H_2} is hydrogen production rate, $kg/kW \cdot h$; P^t_{el} denotes the power consumed by electrolyzer at time slot t, kW; Δt is the time step; P^{max}_{el} is the maximum power of electrolyzer.

2.4.2. The Model of Hydrogen Compressor

A hydrogen compressor is used to compress the hydrogen into high-pressure hydrogen. The power consumption of the hydrogen compressor can be expressed as follows [24]:

$$\begin{cases} P^t_{com} = \dfrac{C_{H_2} m^t_{com} T_{in} \kappa}{\eta_{com}(\kappa-1)} \left[\left(\dfrac{P_{out}}{P_{in}} \right)^{\frac{\kappa-1}{\kappa}} - 1 \right] \\ 0 \le P^t_{com} \le P^{max}_{com} \end{cases} \tag{5}$$

where P^t_{com} is the electric power consumed by compressor at time t; C_{H_2} is the specific heat of hydrogen at constant pressure, $14.304 \ kJ/kg \cdot K$; m^t_{com} is the hydrogen flow rate through compressor at time t, kg/s; T_{in} is the inlet hydrogen temperature (293 K); η_{com} is the efficiency of compressor (0.7); κ is the isentropic exponent of hydrogen (1.4); P^{max}_{com} is the maximum power of hydrogen compressor.

2.4.3. The Model of Hydrogen Storage Tank

The compressed hydrogen is stored in the hydrogen storage tank. The pressure of the hydrogen tank can be formulated as follows [25].

$$\begin{cases} M^{t+1}_{H_2} = M^t_{H_2} + m^t_{H_2} - L^t_{H_2} \\ M^{min}_{H_2} \le M^t_{H_2} \le M^{max}_{H_2} \\ M^{min}_{H_2} = \gamma^{min} C^R_{tank}, \ M^{max}_{H_2} = \gamma^{max} C^R_{tank} \\ M^0_{H_2} = M^T_{H_2} = \gamma^0 C^R_{tank}, \end{cases} \tag{6}$$

where $M^t_{H_2}$ is the stored hydrogen mass in the hydrogen tank at time slot t, kg; $L^t_{H_2}$ is the hydrogen load at time slot t, kg; C^R_{tank} is the capacity of hydrogen tank, kg; γ^{min} and γ^{max} denote the minimum and maximum ratio of the rated capacity of hydrogen tank; $M^0_{H_2}$ and $M^T_{H_2}$ are the stored hydrogen at the initial and end time slot, respectively.

3. The Two-Stage Energy Management Model

The randomness and uncertainty of the outpower of WT and PV will affect the stable operation of the whole system and may reduce the hydrogen purity. As illustrated in Figure 2, in order to alleviate these adverse effects, a two-stage energy management model based on receding horizon optimization is proposed for the wind-PV-hydrogen-storage microgrid. In the first stage of the energy management model, the day-ahead optimization is performed to minimize the total operational cost based on the predicted outpower of WT and PV, as well as the predicted power and hydrogen demands, the time-of-use price, the feed-in tariff and the operation constraints of the whole system. In the second stage of the energy management model, the intra-day optimization model based on the receding horizon optimization is executed to eliminate the power fluctuations caused by the forecast errors. The specific day-ahead optimization and intra-day optimization models will be formulated in the following subsections.

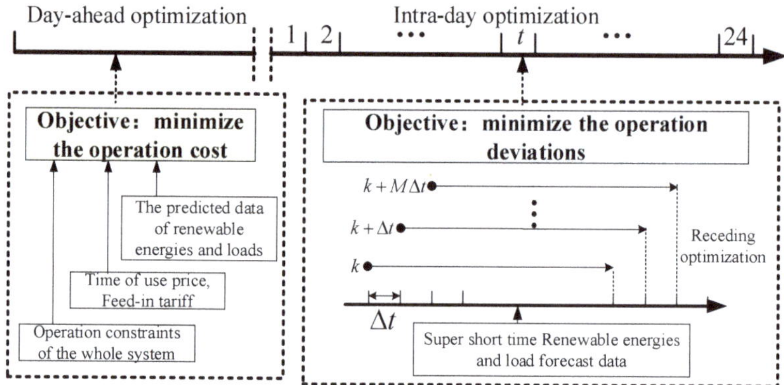

Figure 2. The schematic of the two-stage energy management model.

3.1. The Day-Ahead Optimization Model

The objective of the day-ahead operation is to minimize the comprehensive operation cost C_{DAC}, which is composed of the operational and maintenance costs of PV (C_{PV}) and WT (C_{WT}); the degradation costs of batteries (C_{bat}) and electrolyzer (C_{el}); and the net energy cost (C_e).

The operational and maintenance costs of PV and WT are formulated as the functions of their output power.

$$\begin{cases} C_{PV} = \sum_{t=1}^{T} \lambda_{PV} P_{PV}^t \Delta t \\ C_{WT} = \sum_{t=1}^{T} \lambda_{WT} P_{WT}^t \Delta t \end{cases} \tag{7}$$

where T is the total number of time slots, λ_{PV} and λ_{WT} are maintenance cost coefficients of PV and WT, respectively; their values are assumed to be 0.005 ¥/kWh and 0.0045 ¥/kWh, respectively [26]. P_{PV}^t and P_{WT}^t are output power of PV and WT at time slot t, respectively.

The degradation of energy storage is caused by charging and discharging, as well as the depth of discharge. Refs. [27,28] have shown that the degradation density function of the state of charge (SoC) is almost flat between minimum and maximum of Soc. Thus, as in the model in [29–31], the amortized battery degradation cost C_{bat} can be computed by the power of discharging and charging (Equation (8)), while the degradation cost of battery considering the depth of discharge can be found in [17,32].

$$C_{bat} = \sum_{t=1}^{T} \lambda_{bat} \left(P_{bat,c}^t + P_{bat,d}^t \right) \Delta t \tag{8}$$

where $P_{bat,c}^t$ and $P_{bat,d}^t$ are the charging and discharging power of battery, respectively. λ_{bat} denotes the degradation cost coefficient.

Similarly, the amortized degradation cost C_{el} of electrolyzer can be expressed as follows.

$$C_{el} = \sum_{t=1}^{T} \lambda_{el} P_{el}^t \Delta t \tag{9}$$

where P_{el}^t and λ_{el} are the consumed power and degradation cost coefficient of electrolyzer, respectively.

The microgrid is allowed to buy electricity from the utility grid when its electricity is insufficient and it may sell power back to the grid when its power is surplus. Then, the net energy cost is formulated as the electricity purchasing cost minus the revenue from selling electricity.

$$C_e = \sum_{t=1}^{T} \left(\pi_b P_b^t - \pi_s P_s^t \right) \Delta t \tag{10}$$

where π_b and π_s denote the electricity buying and selling prices, respectively. P_b^t and P_s^t are the power purchased from and sold to the utility grid, respectively.

Then, the objective of the day-ahead optimization model can be expressed as follows.

$$\min C_{DAC} = \min(C_{PV} + C_{WT} + C_{bat} + C_{el} + C_e) \tag{11}$$

The power balance should be satisfied at each time slot.

$$P_{PV}^t + P_{WT}^t + P_b^t + P_{bat,d}^t = P_{bat,c}^t + P_{el}^t + P_{com}^t + P_s^t + P_{load}^t \tag{12}$$

where P_b^t and P_s^t are the power buying from and selling to the power grid at time slot t, respectively. P_{load}^t is the predicted power load at time slot t.

The buying and selling power cannot happen simultaneously. Let $P_{grid}^{\max}$ denote the maximum power allowed when selling to or buying from the power grid, and χ_{bs}^t denote the binary variable; then, the constraints of power exchanged with the power grid can be formulated as follows.

$$\begin{cases} 0 \leq P_b^t \leq \chi_{bs}^t P_{grid}^{\max} \\ 0 \leq P_s^t \leq \left(1 - \chi_{bs}^t\right) P_{grid}^{\max} \end{cases} \tag{13}$$

Furthermore, the operation constraints of WT, PV, battery storage and the power-to-hydrogen subsystem should be satisfied. Then, the day-ahead optimization model in compact form can be expressed as follows.

$$\min C_{DAC} = \min \left\{ \begin{array}{l} \sum_{t=1}^{T} \lambda_{PV} P_{PV}^t + \sum_{t=1}^{T} \lambda_{WT} P_{WT}^t + \sum_{t=1}^{T} \lambda_{bat} \left(P_{bat,c}^t + P_{bat,d}^t \right) \\ + \sum_{t=1}^{T} \lambda_{el} P_{el}^t + \sum_{t=1}^{T} \left(\pi_b P_b^t - \pi_s P_s^t \right) \end{array} \right\} \tag{14}$$

$$s.t. \ (1) - (6), (12) - (13)$$

3.2. The Intra-Day Optimization Model

According to the day-ahead operation schemes, the intra-day optimization model will be performed to minimize the operation errors based on the ultra-short-term prediction data of WT, PV, power and hydrogen demands. The intra-day optimization model is built based on the receding horizon optimization, which is an effective method to tackle the uncertainty and volatility of renewable energies and loads. The main idea of the receding horizon operation is illustrated in Figure 3. It mainly contains the following three steps [3]. (1) Take the day-ahead operation schemes as set points; at the time slot k, solve the intra-day operation strategies during the receding horizon based on the real-time predicted values of renewable energy generation, power loads and hydrogen loads. (2) From the first step, obtain the operation strategies over the k-th to the $(k + M - 1)$-th time slot. Only the operation strategies at the k-th slot are implemented. (3) Move to the $k + 1$-th time slot, update the prediction data and repeat the first two steps. It is obvious that by means of receding horizon optimization, the operation strategies are updated step-by-step to alleviate the impacts of the uncertainty and volatility of renewable energies and loads.

The objective of the intra-day energy management is to trace the day-ahead operation schemes based on the updated predicted data for the output power of PV and WT, the power loads and hydrogen loads. Take the day-ahead operation strategies as set points and the objection function intra-day energy management can be expressed as follows.

$$\min C_{IDC} = \min \left\{ \sum_{t=k}^{k+M-1} \left[\begin{array}{l} w_1 \left(P^t_{grid} - \hat{P}^t_{grid} \right)^2 + w_2 \left(P^t_{el} - \hat{P}^t_{el} \right)^2 \\ + w_3 \left(P^t_{com} - \hat{P}^t_{com} \right)^2 + w_4 \left(P^t_{bat} - \hat{P}^t_{bat} \right)^2 \end{array} \right] \right\}$$

$$s.t. \left\{ \begin{array}{l} (1) - (6), (12) - (13) \\ \Delta P^t_{grid} = \left| P^t_{grid} - \hat{P}^t_{grid} \right| \leq \Delta P^{max}_{grid} \\ \Delta P^t_{el} = \left| P^t_{el} - \hat{P}^t_{el} \right| \leq \Delta P^{max}_{el} \\ \Delta P^t_{com} = \left| P^t_{com} - \hat{P}^t_{com} \right| \leq \Delta P^{max}_{com} \\ \Delta P^t_{bat} = \left| P^t_{bat} - \hat{P}^t_{bat} \right| \leq \Delta P^{max}_{bat} \end{array} \right. \tag{15}$$

where $P^t_{grid} = P^t_b - P^t_s$, $P^t_{bat} = P^t_{bat,c} - P^t_{bat,d}$, $\hat{P}^t_{grid}$, $\hat{P}^t_{el}$, $\hat{P}^t_{com}$ and $\hat{P}^t_{bat}$ denote the day-ahead operation schemes of buying power, electrolyzer, compressor and battery storage. w_1, w_2, w_3 and w_4 are weight factors; they can be optimized based on their significance. In this paper, they are assumed to have equal weights. ΔP^{max}_{grid}, ΔP^{max}_{el}, ΔP^{max}_{com} and ΔP^{max}_{bat} are the admissible maximum errors of exchanged power, the power of electrolyzer, the power of compressor and the net charging power of battery storage, respectively. The objective function (15) of the intra-day operation model is to minimize the operation deviation between the intra-day strategies and the day-ahead strategies. The constraints are the operation limitations of each component and the power exchange with the power grid. The decision variables are the operation strategies of each component and the power exchange with the power grid; the decision variable vector is $x = \left[P^t_{grid}, P^t_{el}, P^t_{com}, P^t_{bat} \right]$. More details about the input and output variables of the two-stage energy management model can be found in Appendix A.

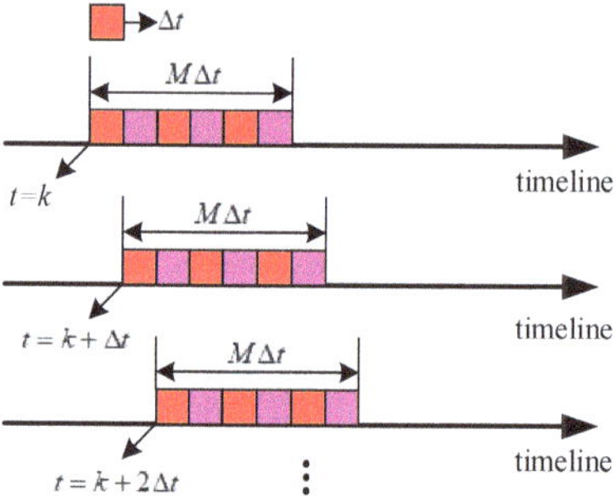

Figure 3. Schematic of receding horizon optimization.

4. Numerical Analysis

4.1. Basic Parameter Settings

The microgrid in Figure 1 is taken as an example to demonstrate the proposed two-stage energy management method. In the first stage, the day-ahead optimization is performed based on the predicted outpower of WT and PV, the predicted power and hydrogen loads. In the second stage, the intra-day optimization is performed based on the actual data. Without loss of generality, the actual data are assumed to be the sum of predicted data and the forecast error. Assume that the day-ahead forecast errors of wind power, PV, power and hydrogen demands follow standard normal distribution. The standard deviation for the day-ahead forecast errors of wind power, PV, power and hydrogen demands is set as 25%, 20%, 15% and 15% of their day-ahead forecast data, respectively. In fact, the ultra-short-term prediction data of WT, PV, power and hydrogen demands can be predicted by the long short-term memory (LSTM), neural network or other artificial intelligence methods [33]. In this paper, the prediction cycle is 1 h, and the control cycle is 30 min and the receding horizon optimization is performed once per 5 min. Therefore, the intra-day optimization model will be executed 288 times during the 24 h.

The predicted and actual output power of WT and PV is shown in Figures 4 and 5 [34], respectively. The predicted and actual power load and hydrogen load are illustrated in Figures 6 and 7 [7,34], respectively. Table 1 gives the power prices of the power grid [35]; the buying price is time-of-use price and the feed-in price is fixed price. The other parameters of the micro are given in Table 2 [34,35]. The maximum of the deviations for P_{grid}^t, $\hat{P}_{el}^t$, $\hat{P}_{com}^t$ and $\hat{P}_{bat}^t$ are 200 kW.

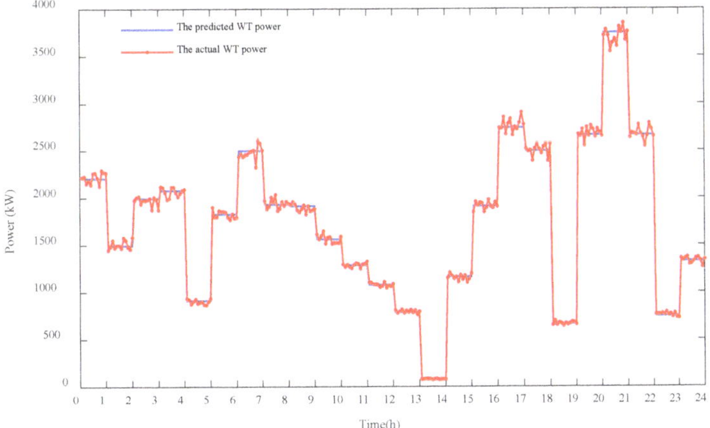

Figure 4. The predicted and actual outpower of WT.

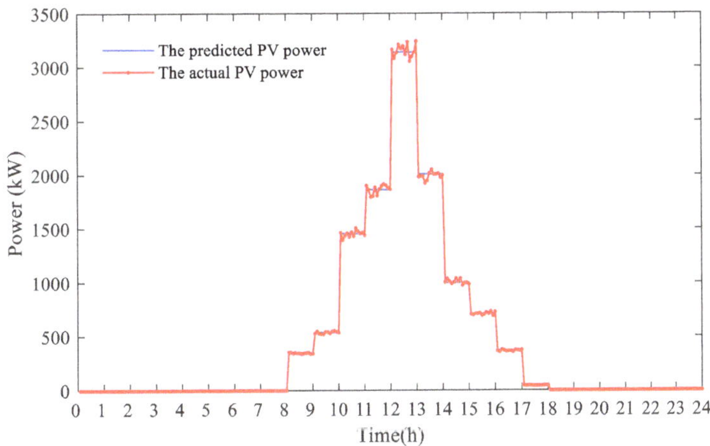

Figure 5. The predicted and actual outpower of PV.

Table 1. Power prices (¥/kWh).

Time Slots	Buying Price	Selling Price
01:00–07:00, 23:00–24:00	0.3376	0.4
12:00–14:00, 19:00–22:00	0.8654	0.4
08:00–11:00, 15:00–18:00	0.5980	0.4

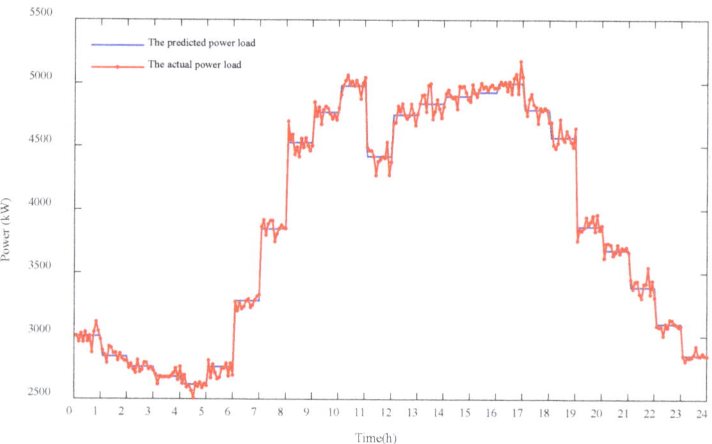

Figure 6. The predicted and actual power load.

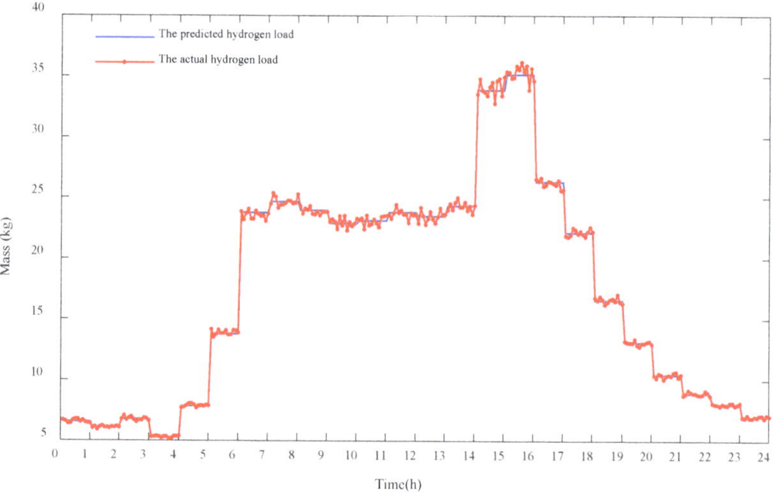

Figure 7. The predicted and actual hydrogen load.

Table 2. The parameters of the WPHS microgrid.

η_{H_2}	0.0192	η_{com}	0.7	S_B^{max}	5700 kWh	ΔP_{grid}^{max}	200 kW
P_{el}^{max}	5000 kW	κ	1.4	S_B^{min}	600 kWh	ΔP_{el}^{max}	100 kW
P_{grid}^{max}	6000 kW	$m_{H_2}^{min}$	0 kg	$S_B^{t=0}$	600 kWh	ΔP_{com}^{max}	10 kW
R_{H_2}	14.304	$m_{H_2}^{max}$	1000 kg	$P_{bat,c}^{max}$	2100 kW	ΔP_{bat}^{max}	200 kW
T_{in}	40 °C	P_{com}^{max}	500 kW	$P_{bat,d}^{max}$	2400 kW		

4.2. The Analysis and Discussions of the Simulation Results

Figures 8–11 show the day-ahead schemes and the intra-day operation strategies of buying or selling power, charging and discharging power of battery storage, electrolyzer and compressor, respectively. Figures 12 and 13 illustrate the storage states of battery storage and hydrogen tank, respectively.

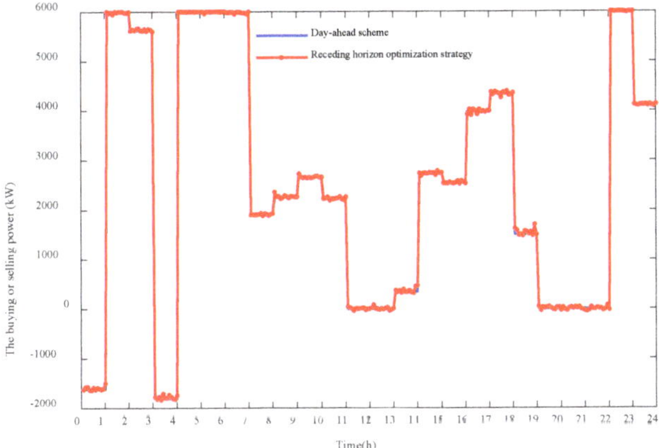

Figure 8. The exchanged power with power grid.

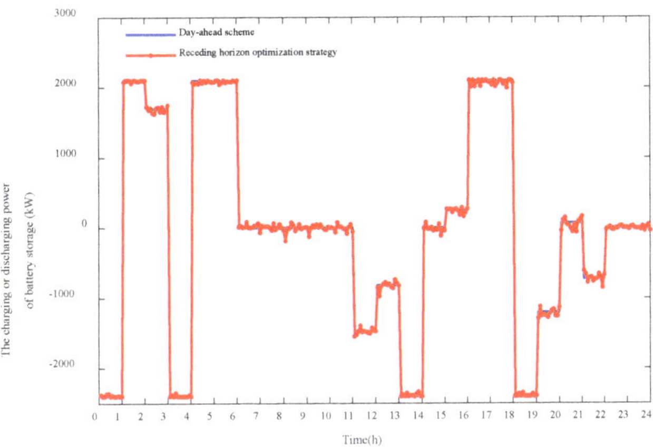

Figure 9. The charging and discharging power of battery storage.

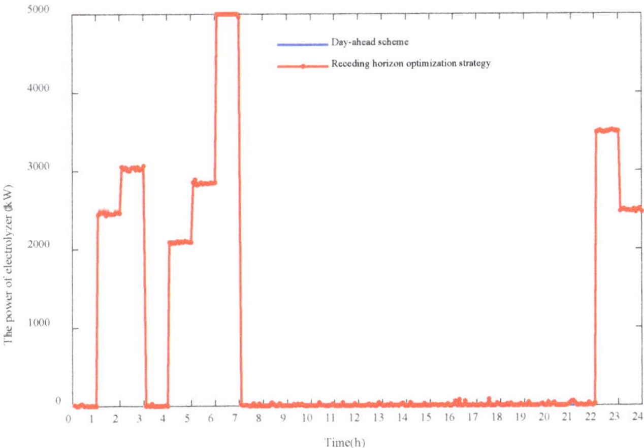

Figure 10. The power of electrolyzer.

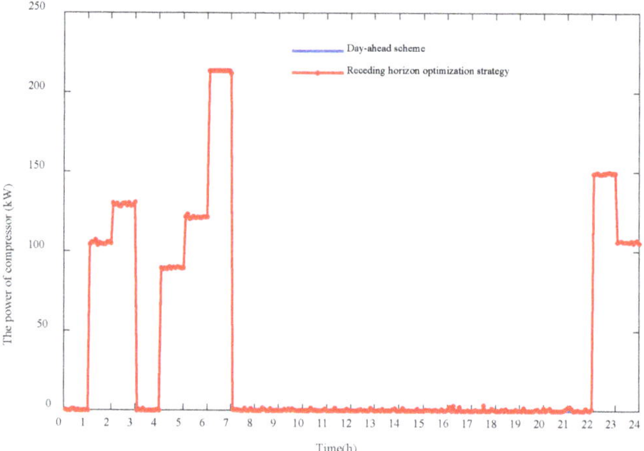

Figure 11. The power of compressor.

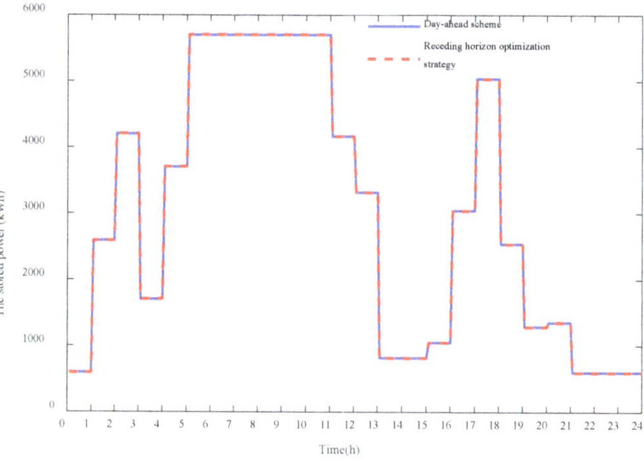

Figure 12. The storage state of battery storage.

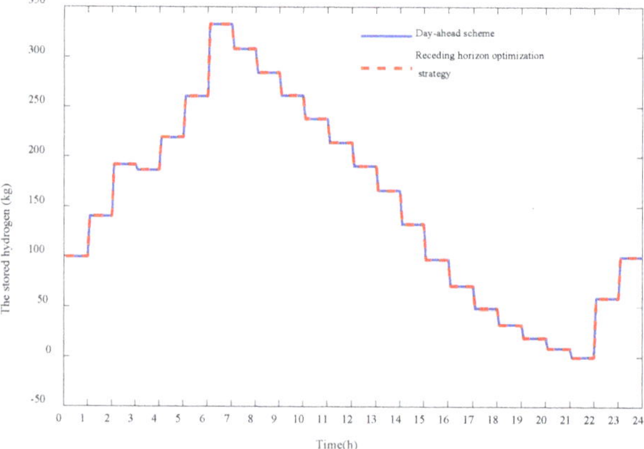

Figure 13. The storage state of hydrogen tank.

4.2.1. The Day-Ahead Simulation Results

The simulation results of Figures 8 and 9 show that the sustainable WPHS microgrid buys more electricity from the power grid during the valley periods and flat periods than the peak periods. This is because the buying power price is low during the valley and flat periods; in order to reduce the operation cost, the microgrid buys more electricity to meet the demands of power loads, produce hydrogen or charge the batteries. During the peak periods, the required power of the WPHS microgrid is mainly met by the battery storage, the WT and PV. In addition, it can be seen from Figures 10, 11 and 13 that the hydrogen is produced and stored in the tank during the valley periods, and the tank discharges hydrogen during the flat and peak periods to satisfy hydrogen demand. Figures 9 and 12 show that the battery storage mainly stores the electricity during the valley or flat periods and discharges power during the peak periods to reduce the operation cost. Therefore, the day-ahead optimization can effectively coordinate the operation of the WT, PV, battery storage and power-to-hydrogen subsystems, and realize high-efficiency operation.

4.2.2. The Intra-Day Simulation Results

Figures 8–13 show that the intra-day operation strategies are effective in tracing the day-ahead operation schemes and eliminating the effects of the volatility of renewable energies, power and hydrogen loads. Furthermore, the intra-day operation strategies of battery storage and hydrogen tank can completely trace their day-ahead states. The maximum deviations of exchanged power, the power of electrolyzer, the power of compressor and the power of battery storage are 199.45 kW, 81.34 kW, 3.48 kW and 191.62 kW, respectively. They all satisfy their maximum error constraints. Therefore, the intra-day optimization model is able to improve the operation stability of the WPHS microgrid and eliminate the adverse influence of the fluctuations of WT, PV, power and hydrogen demands.

4.2.3. The Simulation Results of WPHS Microgrid without Battery Storage

In this section, the sustainable WPHS microgrid in Figure 1 without battery storage is taken as the comparative microgrid (WPH microgrid) to demonstrate the roles of battery storages. Figures 14–17 illustrate the day-ahead schemes and the intra-day operation strategies of buying or selling power, electrolyzer and compressor, respectively. Figure 17 illustrates the storage states of hydrogen tank. It can be seen that the proposed two-stage energy management model is robust and effective in coordinating the operation of the sustainable WPH microgrid, and intra-day receding horizon optimization strategies can effectively trace the day-ahead schemes. The operation costs for the microgrid with and without battery storage are 27,727 CNY and 31,815 CNY, respectively. The battery storage can reduce the operation cost dramatically by 12.85%. Furthermore, the maximum of the deviation of the receding horizon optimization strategy and the day-ahead scheme is 202.0123 kW and 231.5762 kW for the microgrid with and without battery storage, respectively. This deviation is reduced by 12.77% when the battery storage is considered. Therefore, the battery storage can also alleviate the fluctuations of the exchanged power with power grid and improve the performance of the intra-day optimization model.

Remark 1. *Though other methods, such as the scenario-based stochastic programming method and robust optimization [36], can also tackle the uncertainties, the former needs the probability distribution of uncertain factors and a huge number of scenario simulations, which may be a heavy burden. While the robust optimization can incorporate the uncertainties with a range without underlying probability distributions, and the optimal solutions in the worst case can be obtained, however, these solutions are very conservative [36]. The two-stage energy management method needs neither probability distribution nor huge scenario simulations; the robust solutions can be obtained based on the updated predicted data.*

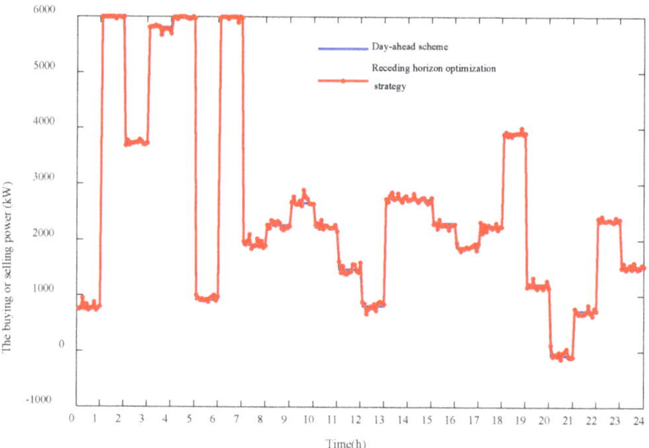

Figure 14. The exchanged power with power grid of WPH microgrid.

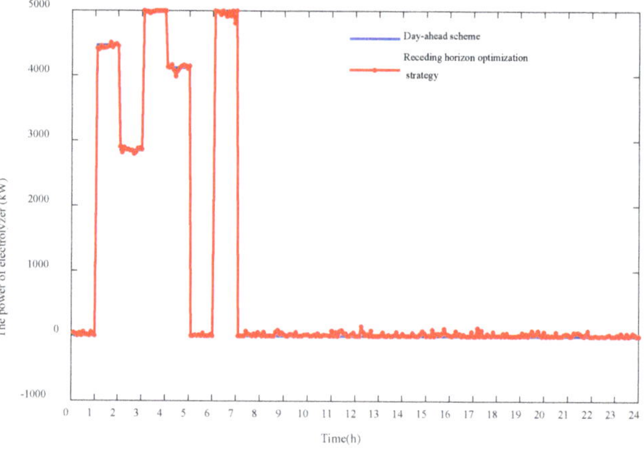

Figure 15. The power of electrolyzer of WPH microgrid.

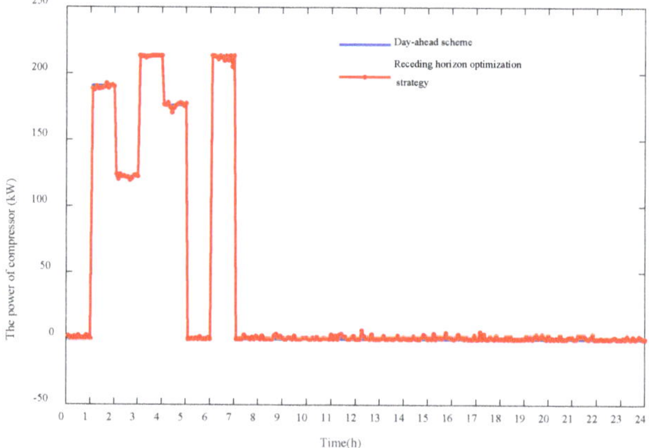

Figure 16. The power of compressor of WPH microgrid.

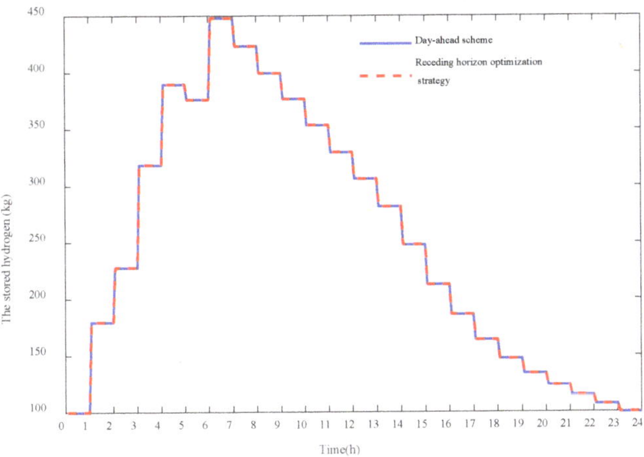

Figure 17. The storage state of hydrogen tank of WPH microgrid.

5. Conclusions

A two-stage energy management model is proposed for the sustainable wind-PV-hydrogen-storage microgrid based on receding horizon optimization. In the first stage, the day-ahead optimization is performed based on the predicted outpower of WT and PV, the predicted demands of power and hydrogen loads. In the second stage, the intra-day optimization is performed based on the actual data to trace the day-ahead operation schemes. The following conclusions are drawn.

(1) The proposed two-stage optimization is effective in managing the operation of the micro and eliminating the uncertainties and fluctuations of WT, PV and loads. The day-ahead optimization can effectively coordinate the operations of the WT, PV, battery storage and power-to-hydrogen subsystems, and realize the high-efficiency operations. The intra-day optimization model is able to improve the operation stability of the WPHS microgrid and eliminate the adverse influence of the fluctuations of WT, PV, power and hydrogen demands.

(2) The proposed two-stage energy management model is robust and effective in coordinating the operation of the sustainable WHP microgrid, and intra-day receding horizon optimization strategies can effectively trace the day-ahead schemes. In addition, the battery storage can reduce the operation cost dramatically by 12.85%, as well as alleviate the fluctuations of the exchanged power with the power grid, and the maximum deviation of the exchanged power between the day-ahead and intra-day strategies is reduced by 12.77% when the battery storage is considered.

Furthermore, in the future work, more accurate models of each component, including consideration of the startup cost and ramp time of the green hydrogen system will be considered. The demand side management issue is another interesting topic, which can be integrated in the two-stage energy management model. The mean efficiency of the whole process of the system can also be discussed and analyzed in the future work.

Author Contributions: Conceptualization, W.P.; methodology, H.X.; investigation, T.M. and J.W.; resources, D.L.; writing—original draft preparation, J.W. and T.M.; writing—review and editing, X.L.; supervision, X.M.; project administration, J.Z.; funding acquisition, J.W. All authors have read and agreed to the published version of the manuscript.

Funding: This research was funded by the Science and Technology Project of State Grid JILIN Electric Power Supply Company (NO. B7234220002J).

Institutional Review Board Statement: Not applicable.

Informed Consent Statement: Not applicable.

Conflicts of Interest: The authors declare no conflict of interest.

Abbreviations

PV	Photovoltaic		WT	Wind turbine
WPHS	Wind-PV-hydrogen-storage		WPH	Wind-PV-hydrogen

Parameters and variables of wind turbine model

P_{WT}^t	Outpower of WT at time slot t		P_{RWT}	Rated power of WT
v_t	Wind speed at time slot t		v_{in}	Cut-in wind speed
v_{out}	Cut-out wind speed		v_r	Rated wind speed of wind turbine

Parameters and variables of PV model

P_{PV}^t	Outpower of PV array		N_{PV}	Number of PV panes
I_{STC}	Standard irradiance		P_{rSTC}	Rated power of each PV panel at standard test conditions
I_t	Irradiance at time slot t		T_t	Temperature at time slot t

Parameters and variables of battery storage model

E_{bat}^t	Energy stored in the batteries at time slot t		E_{bat}^{min}	Minimum capacity of battery storages
E_{bat}^{max}	Maximum capacity of battery storages		$P_{bat,c}^t$	Charging power at time slot t
$P_{bat,d}^t$	Discharging power at time slot t		$P_{bat,c}^{max}$	Maximum charging power
$P_{bat,d}^{max}$	Maximum discharging power		u_{bat}^t	Binary variable

Parametersand variables of power-to-hydrogen system

η_{H_2}	Hydrogen production rate		P_{el}^{max}	Maximum power of electrolyzer
$m_{H_2}^t$	Hydrogen mass-produced at time slot t		P_{el}^t	Power consumed by electrolyzer at time slot t
C_{H_2}	Specific heat of hydrogen at constant pressure		T_{in}	Inlet hydrogen temperature
η_{com}	Efficiency of compressor		P_{out}/P_{in}	Compression ratio of hydrogen
P_{com}^{max}	Maximum power of compressor		m_{com}^t	Hydrogen flow rate through compressor at time
κ	Isentropic exponent of hydrogen		$M_{H_2}^t$	Stored hydrogen mass in the hydrogen tank at time slot t
$L_{H_2}^t$	Hydrogen load at time slot t		C_{tank}^R	Capacity of hydrogen tank
γ^{min}	Minimum ratio of the rated capacity of hydrogen tank		γ^{max}	Maximum ratio of the rated capacity of hydrogen tank

Variables of the two-stage energy management model

C_{DAC}	Day-ahead comprehensive operation cost		C_{PV}	Operational and maintenance costs of PV
C_{WT}	Operational and maintenance costs of WT		C_{bat}	Degradation costs of battery storage
C_{el}	Degradation costs of electrolyzer		C_e	Net energy cost
λ_{PV}	Maintenance cost coefficient of PV		λ_{WT}	Maintenance cost coefficient of WT
λ_{bat}	Degradation cost coefficient of battery storage		λ_{el}	Degradation cost coefficient of electrolyzer
P_b^t	Buying power from the power grid at time slot t		P_s^t	Selling power to the power grid at time slot t
P_{load}^t	Predicted power load at time slot t		X_{bs}^t	Binary variable
$P_{H_2,f_s}^{t,0}$	Hydrogen production at time slot t		$P_{el,f_s}^{t,0}$	Power consumed by electrolyzer device at time slot t

Appendix A

Take the exchanged power with power grid, the power of electrolyzer, the power of compressor, the charging/discharging power of battery storage, the power storage state of battery storage and the hydrogen storage state of hydrogen tank to constitute state vector $x(k) = \begin{bmatrix} P_{grid}(k) & P_{el}(k) & P_{com}(k) & P_{bat}(k) & E_{bat}(k) & M_{H_2}(k) \end{bmatrix}^T$; take the increment power of electrolyzer, the increment power of compressor and the increment charging/discharging power of battery storage to constitute the control variables $u(k) = \begin{bmatrix} \Delta P_{el}(k) & \Delta P_{com}(k) & \Delta P_{bat}(k) \end{bmatrix}^T$; take the increment power of the ultra-short-term predicted power of wind turbine, PV, power load and hydrogen load as disturbance input vector $r(k) = \begin{bmatrix} \Delta P_{WT}(k) & \Delta P_{PV}(k) & \Delta P_{load}(k) & \Delta L_{H_2}(k) \end{bmatrix}^T$; take the exchanged power with power grid, the power of electrolyzer, the power of compressor and the charging/discharging power of battery storage as the output variable vector $y(k) = \begin{bmatrix} P_{grid}(k) & P_{el}(k) & P_{com}(k) & P_{bat}(k) \end{bmatrix}^T$; then the multi-input and multi-output state space model can be formulated in the following matrix form [34]:

$$
\begin{aligned}
x(k+\Delta t) &=
\begin{bmatrix}
P_{grid}(k+\Delta t) \\
P_{el}(k+\Delta t) \\
P_{com}(k+\Delta t) \\
P_{bat}(k+\Delta t) \\
E_{bat}(k+\Delta t) \\
M_{H_2}(k+\Delta t)
\end{bmatrix}
=
\begin{bmatrix}
1 & 0 & 0 & 0 & 0 & 0 \\
0 & 1 & 0 & 0 & 0 & 0 \\
0 & 0 & 1 & 0 & 0 & 0 \\
0 & 0 & 0 & 1 & 0 & 0 \\
0 & 0 & 0 & 0 & 1 & 0 \\
0 & 0 & 0 & 0 & 0 & 1
\end{bmatrix}
\begin{bmatrix}
P_{grid}(k) \\
P_{el}(k) \\
P_{com}(k) \\
P_{bat}(k) \\
E_{bat}(k) \\
M_{H_2}(k)
\end{bmatrix} \\
&+
\begin{bmatrix}
-1 & -1 & -1 \\
1 & 0 & 0 \\
0 & 1 & 0 \\
0 & 0 & 1 \\
0 & 0 & \eta_{bs} \\
\eta_{H_2} & 0 & 0
\end{bmatrix}
\begin{bmatrix}
\Delta P_{el}(k) \\
\Delta P_{com}(k) \\
\Delta P_{bat}(k)
\end{bmatrix}
+
\begin{bmatrix}
1 & 1 & -1 & 0 \\
0 & 0 & 0 & 0 \\
0 & 0 & 0 & 0 \\
0 & 0 & 0 & 0 \\
0 & 0 & 0 & 0 \\
1 & 1 & 1 & -1
\end{bmatrix}
\begin{bmatrix}
\Delta P_{WT}(k) \\
\Delta P_{PV}(k) \\
\Delta P_{load}(k) \\
\Delta L_{H_2}(k)
\end{bmatrix}
\end{aligned}
\tag{A1}
$$

$$
y(k) =
\begin{bmatrix}
P_{grid}(k) \\
P_{el}(k) \\
P_{com}(k) \\
P_{bat}(k)
\end{bmatrix}
=
\begin{bmatrix}
1 & 0 & 0 & 0 & 0 & 0 \\
0 & 1 & 0 & 0 & 0 & 0 \\
0 & 0 & 1 & 0 & 0 & 0 \\
0 & 0 & 0 & 1 & 0 & 0
\end{bmatrix}
\begin{bmatrix}
P_{grid}(k) \\
P_{el}(k) \\
P_{com}(k) \\
P_{bat}(k) \\
E_{bat}(k) \\
M_{H_2}(k)
\end{bmatrix}
\tag{A2}
$$

References

1. Lux, B.; Pfluger, B. A supply curve of electricity-based hydrogen in a decarbonized European energy system in 2050. *Appl. Energy* **2020**, *269*, 115011. [CrossRef]
2. Rabiee, A.; Keane, A.; Soroudi, A. Technical barriers for harnessing the green hydrogen: A power system perspective. *Renew. Energy* **2021**, *163*, 1580–1587. [CrossRef]
3. El-Taweel, N.A.; Khani, H.; Farag, H.E.Z. Hydrogen Storage Optimal Scheduling for Fuel Supply and Capacity-Based Demand Response Program under Dynamic Hydrogen Pricing. *IEEE Trans. Smart Grid* **2019**, *10*, 4531–4542. [CrossRef]
4. Wu, X.; Qi, S.; Wang, Z.; Duan, C.; Wang, X.; Li, F. Optimal scheduling for microgrids with hydrogen fueling stations considering uncertainty using data-driven approach. *Appl. Energy* **2019**, *253*, 113568. [CrossRef]
5. Ulleberg, Ø. Modeling of advanced alkaline electrolyzers: A system simulation approach—ScienceDirect. *Int. J. Hydrogen Energy* **2003**, *28*, 21–33. [CrossRef]
6. Bernal-Agustín, J.L.; Dufo-López, R. Techno-economical optimization of the production of hydrogen from PV-Wind systems connected to the electrical grid. *Renew. Energy* **2010**, *35*, 747–758. [CrossRef]
7. Wu, X.; Li, H.; Wang, X.; Zhao, W. Cooperative Operation for Wind Turbines and Hydrogen Fueling Stations with On-Site Hydrogen Production. *IEEE Trans. Sustain Energy* **2020**, *11*, 2775–2789. [CrossRef]
8. Ma, T.; Pei, W.; Deng, W.; Xiao, H.; Yang, Y.; Tang, C. A Nash bargaining-based cooperative planning and operation method for wind-hydrogen-heat multi-agent energy system. *Energy* **2022**, *239*, 122435. [CrossRef]
9. Mohseni, S.; Brent, A.C.; Burmester, D. A comparison of metaheuristics for the optimal capacity planning of an isolated, battery-less, hydrogen-based micro-grid. *Appl. Energy* **2020**, *259*, 114224. [CrossRef]
10. Petrollese, M.; Valverde, L.; Cocco, D.; Cau, G.; Guerra, J. Real-time integration of optimal generation scheduling with MPC for the energy management of a renewable hydrogen-based microgrid. *Appl. Energy* **2016**, *166*, 96–106. [CrossRef]
11. Appino, R.R.; Wang, H.; Ordiano, J.G.; Faulwasser, T.; Mikut, R.; Hagenmeyer, V.; Mancarella, P. Energy-based stochastic MPC for integrated electricity-hydrogen VPP in real-time markets. *Electr. Power Syst. Res.* **2021**, *195*, 106738. [CrossRef]
12. Shi, M.; Wang, H.; Lyu, C.; Xie, P.; Xu, Z.; Jia, Y. A hybrid model of energy scheduling for integrated multi-energy microgrid with hydrogen and heat storage system. *Energy Rep.* **2021**, *7*, 357–368. [CrossRef]
13. Li, Q.; Member, S.; Zou, X.; Pu, Y.; Member, S.; Chen, W. A real-time energy management method for electric-hydrogen hybrid energy storage microgrid based on DP-MPC. *CSEE J. Power Energy Syst.* **2020**. [CrossRef]
14. Garcia-Torres, F.; Bordons, C. Optimal Economical Schedule of Hydrogen-Based Microgrids With Hybrid Storage Using Model Predictive Control. *IEEE Trans. Ind. Electron.* **2015**, *62*, 5195–5207. [CrossRef]
15. Zhang, Y.; Campana, P.E.; Lundblad, A.; Yan, J. Comparative study of hydrogen storage and battery storage in grid connected photovoltaic system: Storage sizing and rule-based operation. *Appl. Energy* **2017**, *201*, 397–411. [CrossRef]
16. Möller, M.C.; Krauter, S. Hybrid Energy System Model in Matlab/Simulink Based on Solar Energy, Lithium-Ion Battery and Hydrogen. *Energies* **2022**, *15*, 2201. [CrossRef]
17. Alotaibi, M.A.; Eltamaly, A.M. A Smart Strategy for Sizing of Hybrid Renewable Energy System to Supply Remote Loads in Saudi Arabia. *Energies* **2021**, *14*, 7069. [CrossRef]
18. Tomczewski, A.; Kasprzyk, L. Optimisation of the Structure of a Wind Farm—Kinetic Energy Storage for Improving the Reliability of Electricity Supplies. *Appl. Sci.* **2018**, *8*, 1439. [CrossRef]

19. Tomczewski, A.; Kasprzyk, L.; Nadolny, Z. Reduction of Power Production Costs in a Wind Power Plant–Flywheel Energy Storage System Arrangement. *Energies* **2019**, *12*, 1942. [CrossRef]
20. Kasprzyk, L.; Tomczewski, A.; Pietracho, R.; Mielcarek, A.; Nadolny, Z.; Tomczewski, K.; Trzmiel, G.; Alemany, J. Optimization of a PV-Wind Hybrid Power Supply Structure with Electrochemical Storage Intended for Supplying a Load with Known Characteristics. *Energies* **2020**, *13*, 6143. [CrossRef]
21. Soroudi, A.; Aien, M.; Ehsan, M. A probabilistic modeling of photo voltaic modules and wind power generation impact on distribution networks. *IEEE Syst. J.* **2012**, *6*, 254–259. [CrossRef]
22. Xu, L.; Ruan, X.; Mao, C.; Zhang, B.; Luo, Y. An improved optimal sizing method for wind-solar-battery hybrid power system. *IEEE Trans. Sustain. Energy* **2013**, *4*, 774–785.
23. Korpas, M.; Holen, A.T. Operation planning of hydrogen storage connected to wind power operating in a power market. *IEEE Trans. Energy Convers.* **2006**, *21*, 742–749. [CrossRef]
24. Gökçek, M.; Kale, C. Techno-economical evaluation of a hydrogen refuelling station powered by Wind-PV hybrid power system: A case study for İzmir-Çeşme. *Int. J. Hydrogen Energy* **2018**, *43*, 10615–10625. [CrossRef]
25. El-Taweel, N.A.; Khani, H.; Farag, H.E.Z. Optimal Sizing and Scheduling of LOHC-Based Generation and Storage Plants for Concurrent Services to Transportation Sector and Ancillary Services Market. *IEEE Trans. Sustain. Energy* **2020**, *11*, 1381–1393. [CrossRef]
26. Maleki, A.; Hafeznia, H.; Rosen, M.A.; Pourfayaz, F. Optimization of a grid-connected hybrid solar-wind-hydrogen CHP system for residential applications by efficient metaheuristic approaches. *Appl. Therm. Eng.* **2017**, *123*, 1263–1277. [CrossRef]
27. Han, S.; Han, S.; Aki, H. A practical battery wear model for electric vehicle charging applications. *Appl. Energy* **2014**, *113*, 1100–1108. [CrossRef]
28. Kim, K.; Choi, Y.; Kim, H. Data-driven battery degradation model leveraging average degradation function fitting. *Electron. Lett.* **2017**, *53*, 102–104. [CrossRef]
29. Kim, H.; Lee, J.; Bahrami, S.; Wong, V.W.S. Direct Energy Trading of Microgrids in Distribution Energy Market. *IEEE Trans. Power Syst.* **2020**, *35*, 639–651. [CrossRef]
30. Urgaonkar, R.; Urgaonkar, B.; Neely, M.J.; Sivasubramaniam, A. Optimal power cost management using stored energy in data centers. In Proceedings of the ACM SIGMETRICS, San Jose, CA, USA, 7–11 June 2011; pp. 221–232.
31. Wang, H.; Huang, J. Incentivizing Energy Trading for Interconnected Microgrids. *IEEE Trans. Smart Grid* **2018**, *9*, 2647–2657. [CrossRef]
32. Eltamaly, A.M.; Alotaibi, M.A.; Elsheikh, W.A.; Alolah, A.I.; Ahmed, M.A. Novel Demand Side-Management Strategy for Smart Grid Concepts Applications in Hybrid Renewable Energy Systems. In Proceedings of the 2022 4th International Youth Conference on Radio Electronics, Electrical and Power Engineering (REEPE), Moscow, Russian, 17–19 March 2022; pp. 1–7. [CrossRef]
33. Xiao, H.; Pei, W.; Li, K. Multi-time Scale Coordinated Optimal Dispatch of Microgrid Based on Model Predictive Control. *Autom. Electr. Power Syst.* **2016**, *40*, 7–14.
34. Ma, T.; Pei, W.; Xiao, H.; Li, D.; Lyu, X.; Hou, K. Cooperative Operation Method for Wind-solar-hydrogen Multi-agent Energy System Based on Nash Bargaining Theory. *Proceeding CSEE* **2021**, *41*, 25–39.
35. Du, Y.; Pei, W.; Chen, N.; Ge, X.; Xiao, H. Real-time microgrid economic dispatch based on model predictive control strategy. *J. Mod. Power Syst. Clean Energy* **2017**, *5*, 787–796. [CrossRef]
36. Zheng, Q.P.; Wang, J.; Liu, A.L. Stochastic Optimization for Unit Commitment—A Review. *IEEE Trans. Power Syst.* **2015**, *30*, 1913–1924. [CrossRef]

Article

A DC Series Arc Fault Detection Method Based on a Lightweight Convolutional Neural Network Used in Photovoltaic System

Yao Wang [1,2,*], **Cuiyan Bai** [1,2], **Xiaopeng Qian** [3], **Wanting Liu** [1,2], **Chen Zhu** [4] **and Leijiao Ge** [5]

1   State Key Laboratory of Reliability and Intelligence of Electrical Equipment, Hebei University of Technology, Tianjin 300132, China; 202021401054@stu.hebut.edu.cn (C.B.); 181579@stu.hebut.edu.cn (W.L.)
2   Key Laboratory of Electromagnetic Field and Electrical Apparatus Reliability of Hebei Province, Hebei University of Technology, Tianjin 300132, China
3   Zhejiang High and Low Voltage Electric Equipment Quality Inspection Center, Yueqing 325604, China; dubucgualbcu34@gmail.com
4   State Grid Beijing Electric Power Co., Ltd., Fangshan Power Supply Branch, Beijing 102400, China; chen.zhu.sg@gmail.com
5   School of Electrical and Information Engineering, Tianjin University, Tianjin 300072, China; legendglj99@tju.edu.cn
*   Correspondence: wangyao@hebut.edu.cn; Tel.: +86-22-6020-4409

Citation: Wang, Y.; Bai, C.; Qian, X.; Liu, W.; Zhu, C.; Ge, L. A DC Series Arc Fault Detection Method Based on a Lightweight Convolutional Neural Network Used in Photovoltaic System. *Energies* **2022**, *15*, 2877. https://doi.org/10.3390/en15082877

Academic Editor: Luis Hernández-Callejo

Received: 1 April 2022
Accepted: 12 April 2022
Published: 14 April 2022

Publisher's Note: MDPI stays neutral with regard to jurisdictional claims in published maps and institutional affiliations.

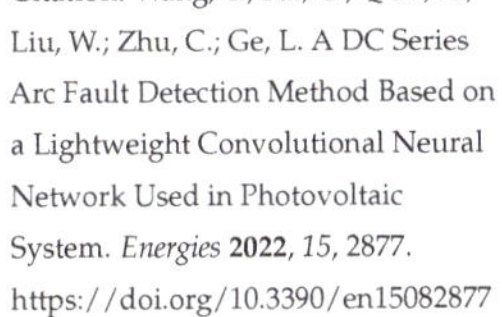

Abstract: Although photovoltaic (PV) systems play an essential role in distributed generation systems, they also suffer from serious safety concerns due to DC series arc faults. This paper proposes a lightweight convolutional neural network-based method for detecting DC series arc fault in PV systems to solve this issue. An experimental platform according to UL1699B is built, and current data ranging from 3 A to 25 A is collected. Moreover, test conditions, including PV inverter startup and irradiance mutation, are also considered to evaluate the robustness of the proposed method. Before fault detection, the current data is preprocessed with power spectrum estimation. The lightweight convolutional neural network has a lower computational burden for its fewer parameters, which can be ready for embedded microprocessor-based edge applications. Compared to similar lightweight convolutional network models such as Efficientnet-B0, B2, and B3, the Efficientnet-B1 model shows the highest accuracy of 96.16% for arc fault detection. Furthermore, an attention mechanism is combined with the Efficientnet-B1 to make the algorithm more focused on arc features, which can help the algorithm reduce unnecessary computation. The test results show that the detection accuracy of the proposed method can be up to 98.81% under all test conditions, which is higher than that of general networks.

Keywords: photovoltaic (PV) system; DC series arc fault; power spectrum estimation; attentional mechanism; lightweight convolutional neural network

1. Introduction

With the frequent occurrence of climate changes caused by global warming [1], environmental problems have attracted more and more attention. In order to reduce carbon dioxide emissions, the use of fossil energy is limited and green energy is more and more widely used. Solar energy is a kind of green energy that adds no pollution to the environment. Photovoltaic (PV) systems can convert solar energy into electric energy for people to use conveniently [2]; they play an essential role in distributed generation systems [3], so they are widely used in households and other places where solar energy is plentiful [4]. However, arc faults on the DC side of a PV system may cause severe electrical fires due to the high temperature above 5000 °C, which may ignite surrounding combustible material [5].

Due to the harmfulness of the DC arc faults in PV systems, in 2011 the National Electrical Code (NEC) required that roof PV systems with DC voltage higher than 80 V must be equipped with series DC arc fault circuit breakers. In 2014, this requirement was applied to all types of photovoltaic systems to reduce the fire hazard caused by the DC arc fault [6]. Moreover, the location of arc faults is stochastic, and arc current may be disturbed by high-frequency noise due to the pulse width modulated (PWM) control of the PV inverter [7], which makes it challenging to detect DC arc faults.

DC arc faults in PV systems can be categorized as parallel arc faults and series arc faults [8,9]. Parallel arc faults are generally caused by line-to-line and line-to-ground short circuit faults. The current amplitude of parallel arc faults can be larger than the current amplitude in a normal state, and can be easily detected by current changes [10]. Poor connection of wires or insulation deterioration can result in a series arc fault. Conversely, the current for a series arc fault does not increase due to the limitations of the series load, and the current is more likely to be affected or even masked by the noise from the series load, which makes the detection of series arc faults more challenging than it is for parallel arc faults [11,12]. This research focused on series arc faults.

In order to solve the problem of DC arc fault detection, many scholars have proposed different detection methods. A series of physical characteristics occurs in the process of arc faults, such as arc light and electromagnetic radiation. Murakami et al. [13] used a high-speed camera to observe the light emission of arcs. Yue et al. [14] detected arc fault using intermittent discharges or sparks occurring before series DC arc faults. In [6,15], a method based on high-frequency components of electromagnetic radiation was used to detect DC arc faults. Using physical characteristics to detect an arc fault is not complicated. However, the arc location in a PV system is stochastic, so it is difficult to judge the arc location and detect arc faults accurately.

Since arc current is independent of arc fault location, it is the most common parameter for arc fault detection. Arc current usually has the characteristics of transient and stochastic changes, which can be detected by different time domain methods, frequency domain methods, and time-frequency domain methods. In [16,17], circuit current data were used to identify arc faults by the time-domain method. Park et al. [18] used the time domain method to detect an arc fault initially, then used the frequency domain method to ensure the accuracy of the detection. Gu et al. [19] proposed a method based on fast Fourier transform (FFT) to detect arc faults. However, FFT does not reflect the time domain information, so it is impossible to determine the exact time of arc occurrence [20]. Liu et al. [21] proposed a method combining the time domain and the time-frequency domain to analyze circuit current and PV-side voltage, which improved the anti-interference ability of arc fault detection. Wang et al. [22] and Chen et al. [23] used wavelet transform to analyze arc signals with multiple resolutions in the time-frequency domain. These methods are superior to the methods of detecting the physical characteristics of DC arc faults and are not affected by the location of the arc fault. They are simple and easy to realize, at low cost. However, the thresholds used to judge normal states and arc fault states are set artificially and need to be adjusted, due to different PV systems' current and voltage complexities [24]. Therefore, more efficient arc fault detection methods are required.

In recent years, some scholars have explored artificial intelligence methods in arc fault detection. The neural network has become the first choice because of its robust feature learning and detail recognition ability. Li et al. [25] proposed an arc fault detection method based on a back propagation (BP) neural network, and the accuracy was 95.23%. Yang et al. [26] proposed a temporal domain visualization convolutional neural network (TDV-CNN) method. The current data was filtered and converted into gray images as the input of the CNN, and the accuracy of arc fault detection was 98.7%. Lu et al. [27] proposed domain adaptation combined with a deep convolutional generative adversarial network (DA-DCGAN) to detect DC arc faults. The PV loop current data were converted into a 2D matrix as the input of DA-DCGAN. Pedersen et al. [28] used a radial basis network to detect DC arc faults. The network's inputs were vectors that simplified the processing steps of the input data. Other neural networks can also be applied to arc fault detection. The neural network methods have high accuracy and do not need to set the threshold artificially. To further improve the accuracy of arc fault detection, the depth scaling, width scaling, and resolution scaling of the network need to be increased. However, if the three scales are added together, this dramatically increases the requirement for computer computing resources. Therefore, most existing methods add one network scaling to improve accuracy.

Although the existing AI-based arc fault detection methods have achieved good accuracy, higher than 95%, the accuracy needs to be further improved to reduce fire risk. Moreover, existing methods have not considered the situation of high current value and the influence of normal operations, such as PV inverter startup and irradiance mutation, on arc fault detection; therefore, the robustness of the methods needs to be improved. Furthermore, the number of model parameters of the existing methods is vast. The computational burden is too enormous for industrial embedded microcontrollers to implement.

In this paper, we propose a lightweight convolutional neural network-based method for detecting DC series arc faults in PV systems.

The main contributions of this paper are as follows:

1. Since the actual DC arc faults of PV systems are very stochastic, it is difficult to capture a large amount of DC series arc fault data directly for algorithm research of arc fault detection. Therefore, this paper establishes a DC arc fault experimental platform for an arc fault detection device (AFDD) installed within the inverter in the UL 1699B standard and analyzes DC series arc faults under different current values from 3 A to 25 A. Moreover, PV inverter startup and irradiance mutation are also considered, to evaluate the robustness of the algorithm.
2. Due to the complex working conditions of PV systems, it is essential to find the apparent characteristics of the current signal for arc fault detection. This paper takes the DC series arc current signal in PV systems as the research object and analyzes its power spectrum characteristics. The AR model of the DC current signal is established to obtain the power spectrum images, by exploring the principles of commonly used power spectrum estimation methods. The results show that the power spectrum images of current data in normal states and arc fault states have apparent differences.
3. An algorithm based on a lightweight convolutional neural network is proposed to detect DC series arc faults in PV systems. The gray images of the power spectrum of the DC current data are fed into the network model, and the detection accuracy of the proposed method is 98.81%, which is higher than the accuracies of GoogLeNet, AlexNet, and existing general networks. This algorithm, with fewer model parameters, has a low computational burden, provides better performance during the running process, and is feasible to run in an embedded microprocessor.

2. Data Collection and Analysis

2.1. Arc Fault Experiment Platform

Since the actual DC arc faults in PV systems have stochastic characteristics, it is challenging to directly capture a large amount of DC arc current data for the arc fault detection algorithm. Therefore, a DC arc fault experimental platform is established to generate DC series arc faults under different working conditions for collecting current data.

The experimental platform mainly includes a PV string, an arc fault generator, signal acquisition devices, and a PV inverter. The UL1699B standard includes four application examples for different AFDD installation positions. The first case, in which an AFDD is installed within the inverter, was used in this experimental platform. A GOODWE GW36K-MT three-phase inverter was used as the load, and the AFDD was installed within the inverter. The UL1699B standard indicates that the PV simulator can replace the actual PV string. Therefore, the ITECH IT6018C PV simulator replaced the PV string to make the experiments more convenient and diversified. The voltage range was 0–1500 V, and the current range was 0–40 A. The ITECH IT6018C PV can simulate the I-V curve under various weather conditions, such as irradiance. In accordance with the UL1699B standard, two circuit forms were used: (1) the circuit of one PV string for a centralized power inverter; (2) the circuit of two PV strings for a centralized power inverter. Figure 1 shows the circuit of the two PV strings for a centralized power inverter.

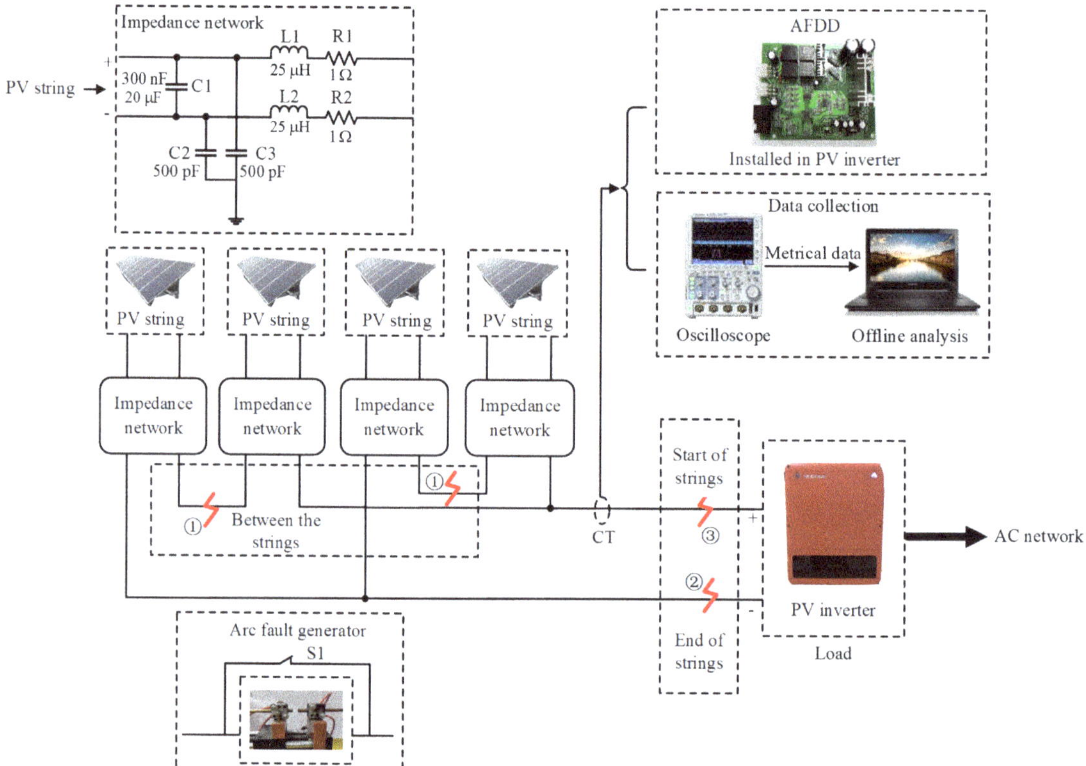

Figure 1. DC series arc fault experimental platform.

The different locations of arc faults have different effects on the DC side current of PV systems. In accordance with the UL 1699B standard, an arc generator was added to the circuit for simulating arc faults, as shown in ①, ②, and ③ in Figure 1. They are between the PV strings, at the end of the PV strings, and at the start of the PV strings. The arc generator was integrated into the system and combined with the system to generate a series arc fault.

In order to simulate the parasitic capacitance and inductance generated by the long line (80 m) between the AFDD and the PV string in PV systems, an impedance network module was added to the circuit to simulate the high-frequency characteristics of the PV system. The impedance network parameters shown in Figure 1 were set in accordance with the UL 1699B standard. When C1 was set to two parameters for testing—300 nF and 20 μF, respectively—the arc fault was the most serious, so each situation had to be tested. The standard stipulates that a decoupling network should be added in front of the impedance network to control the output capacitance of the PV simulator and simulate the DC characteristics of the PV system. The decoupling network is shown in Figure 2. According to the UL1699B standard, when I_{mpp} = 3 A, R3 = R4 = 27 Ω, and when I_{mpp} = 16 A, R3 = R4 = 4.5 Ω. According to the IEC 63027 standard, when I_{mpp} = 25 A, R3 = R4 = 2.5 Ω.

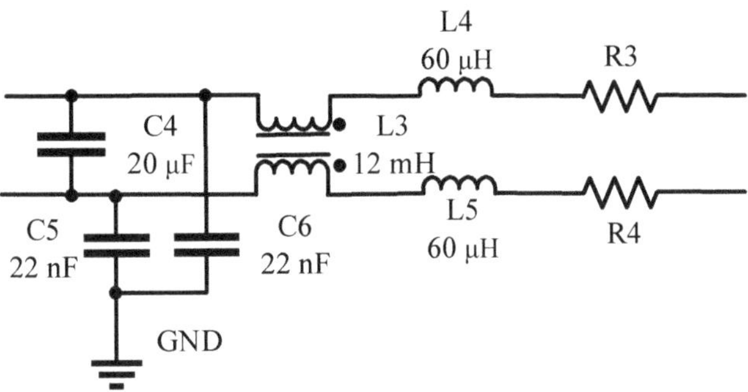

Figure 2. Decoupling network.

2.2. Different Operating Conditions and Power Spectra of Current Data

2.2.1. Different Operating Conditions in Experiments

Different operating conditions were used for data collection to verify the generalization ability of the algorithm. In this experiment, we selected three various tests from the UL 1699B and IEC63027 standards, as shown in Table 1. In order to simulate the worst arc fault situation, the impedance network component C1 was set to two parameters for testing, 300 nF and 20 µF, respectively. Each test, as shown in Table 1, was performed at the three arc fault locations shown in Figure 1 to verify the reliability of the algorithm. The minimum I_{arc} represents a realistic arc event with one or two strings at low irradiance, and I_{mpp}, V_{mpp} represent current and voltage in the maximum power point, respectively. V_{oc} represents open-circuit voltage. The PV simulator can set the four parameters. A stepping motor controller can set the gap and arcing speed. In this experiment, we added two situations, as shown in Table 1: (1) a PV inverter startup, and (2) irradiance mutation, which causes current mutation. These two situations, which belong to the normal state, were tested to verify the robustness of the algorithm.

Table 1. Different operating conditions of DC series arc fault experiments.

Test No.	Description	Minimum I_{arc}/A	I_{mpp}/A	V_{mpp}/V	V_{oc}/V	Gap/mm	Arcing Speed/(mm/s)
1	Arcing test	2.5	3.0	312.0	480.0	0.8	2.5
2	Arcing test	14.0	16.0	318.0	490.0	1.1	5
3	Arcing test	22.5	25.0	318.0	490.0	2.5	5
4	PV inverter startup	2.5 14.0 22.5	3.0 16.0 25.0	312.0 318.0 318.0	480.0 490.0 490.0	/ [1] / /	/ / /
5	Irradiance mutation	Mutation from 1.25 to 2.5 Mutation from 2.5 to 1.25 Mutation from 7.0 to 14.0 Mutation from 14.0 to 7.0 Mutation from 11.25 to 22.5 Mutation from 22.5 to 11.25	Mutation from 1.5 to 3.0 Mutation from 3.0 to 1.5 Mutation from 8.0 to 16.0 Mutation from 16.0 to 8.0 Mutation from 12.5 to 25.0 Mutation from 25.0 to 12.5	312.0 312.0 318.0 318.0 318.0 318.0	480.0 480.0 490.0 490.0 490.0 490.0	/ / / / / /	/ / / / / /

[1] The symbol "/" indicates that the test excludes this variable.

The DC arc current in PV systems presents the characteristics of stochastic high-frequency burrs in the time domain. In contrast, the frequency spectrum amplitude increases slightly in a specific frequency band (such as 40–100 kHz) in the frequency domain. The high-frequency noise of a similar frequency band will be generated when the PV inverter is in the PWM state, and its frequency spectrum amplitude is the same as or even higher than the arc current signal. Therefore, it is difficult to distinguish between the normal state and the arc fault state according to the amplitude difference. However, the PWM noise generated by power electronic devices has regularity due to periodic modulation and system inertia, so the current signals under different working conditions can be distinguished by analyzing the power spectrum. The power spectrum can describe the stochastic signal, which defines the power of the current signal as a function of frequency, and it is susceptible to the change of the signal. It can essentially reflect the objective law of signal change. The process of solving the power spectrum is called power spectrum estimation. Modern power spectrum estimation methods mainly include parametric model spectrum estimation and nonparametric model spectrum estimation. Compared with parametric model spectrum estimation, nonparametric model spectrum estimation has better spectrum estimation performance. However, it requires a large amount of calculation and model complexity, which present challenges in meeting the real-time requirements of DC arc fault detection in practical applications. Therefore, the power spectrum estimation method of the parametric model, with less calculation, is selected. The general power spectrum estimation methods of the parametric model include the autoregressive (AR) model and the autoregressive moving average (ARMA) model.

2.2.2. AR Model

The time series $x(n)$ of the p-order AR model is obtained by the superposition of the signal value at the first p moments and the white noise, and the calculation formula is

$$x(n) = - \sum_{m=1}^{p} a_m x(n-m) + w(n) \tag{1}$$

In Formula (1), a_m is the coefficient of the corresponding time series data, and w is the Gaussian white noise with mean value 0 and variance σ^2.

The system transfer function expression of the p-order AR model is

$$H(z) = \frac{1}{1 + \sum_{m=1}^{p} a_m z^{-m}} \tag{2}$$

According to Equation (2), the AR model is an all-pole model, which can directly reflect the peak distribution in the power spectrum. The Fourier transform processes the transfer function in Equation (2) to obtain the power spectrum calculation, as shown in Equation (3):

$$\widetilde{S}_x(\omega) = \frac{k^2}{\left|1 + \sum_{m=1}^{p} a_m e^{-j\omega m}\right|^2} \tag{3}$$

2.2.3. ARMA Model

The time series calculation formula of the (p, q) order ARMA model is

$$x(n) = \sum_{i=0}^{q} b_i w(n-i) - \sum_{m=0}^{p} a_m x(n-m) \tag{4}$$

According to Equation (4), the system transfer function of the ARMA model is

$$H(z) = \frac{\sum_{i=0}^{q} b_i z^{-1}}{\sum_{m=0}^{p} a_m z^{-m}} \tag{5}$$

According to Equation (5), the ARMA model is a zero-pole model, which can directly reflect the peak and valley distribution in the power spectrum. The Fourier transform processes the transfer function in Equation (5) to obtain the power spectrum calculation, as shown in Equation (6):

$$\widetilde{S}_x(\omega) = \frac{k^2 \left| \sum\limits_{i=0}^{q} b_i e^{-j\omega i} \right|^2}{\left| \sum\limits_{m=0}^{p} a_m e^{-j\omega m} \right|^2} \tag{6}$$

The AR model has a simpler structure and fewer calculations than the ARMA model. Therefore, the AR model is selected as the power spectrum estimation model. After the AR model of the DC current signal is established, the model parameters need to be calculated.

2.2.4. The Selection of Optimal Parameters in the AR Model

It can be seen from Equation (3) that the prediction accuracy of the power spectrum depends on the coefficient a_m and the order p, so choosing a suitable model parameters calculation method is necessary. Commonly used calculation methods for model parameters include the Levinson-Durbin algorithm and Burg algorithm. In this paper, the Burg algorithm was selected as the parameter calculation method for the current signal AR model of the PV system for research, because it has the minimum sum of total mean square error. The calculation process is as follows.

Assuming n sample data $x(1), x(2) \ldots, x(n)$, initialize the forward prediction error e^f and the backward prediction error e^b, where $n = 1, 2, 3, \ldots, N$.

$$\begin{cases} e_0^f(n) = e_0^b(n) = x(n) \\ e_m^f(n) = e_{m-1}^f(n) + k_m e_{m-1}^b(n-1) \\ e_m^b(n) = e_{m-1}^b(n-1) + k_m e_{m-1}^f(n) \end{cases} \tag{7}$$

In Equation (7), k_m is the reflection coefficient. The forward and backward prediction error power ε is defined as:

$$\varepsilon = \sum_{n=m}^{N-1} \left[e_m^f(n)^2 + e_m^b(n)^2 \right] \tag{8}$$

To minimize the error power ε, make $\frac{\partial \varepsilon}{\partial k_m} = 0$; the reflection coefficient k_m is calculated by Equations (7) and (8):

$$k_m = - \frac{2 \sum\limits_{n=m}^{N-1} \left[e_{m-1}^f(n) \right] \left[e_{m-1}^b(n-1) \right]}{\sum\limits_{n=m}^{N-1} \left\{ \left[e_{m-1}^f(n) \right]^2 + \left[e_{m-1}^b(n-1) \right]^2 \right\}} \tag{9}$$

In Equation (9), $m = 1, 2, 3, \ldots, p$. Since the reflection coefficient k_m is an unbiased estimation of the partial correlation coefficient, the autocovariance function R_{xx} of order from 0 to p, which is related to the parameter, can be derived from the Yule-Walker formula:

$$R_{xx}(m) = \begin{cases} -\sum\limits_{k=1}^{p} a_m(l) R_{xx}(m-l), m > 0 \\ -\sum\limits_{k=1}^{p} a_m(l) R_{xx}(m-l) + \sigma_p^2, m = 0 \\ R_{xx}(-m), m < 0 \end{cases} \tag{10}$$

In Equation (10), $l = 1, 2, \ldots, m - 1$. The following Equation (11) can be obtained by cycle calculation.

$$\begin{cases} \Delta_m = R_{xx}(m) + \sum\limits_{l=1}^{m-1} a_{m-1}(l) R_{xx}(m-i) \\ c_m = -\Delta_m / \sigma_{m-1} \\ a_m(l) = a_{m-1}(l) + c_m a_{m-1}(m-l), l = 1, 2, \ldots, m-1 \end{cases} \tag{11}$$

In Equation (9), the reflection coefficient k_m can be used as the estimated value of c_m. The Levinson recurrence Formula (12) can be obtained by substituting it into Equation (11). The AR model coefficient a_m is calculated according to the recurrence relationship:

$$\begin{cases} a_m(m) = k_m \\ a_m(l) = a_{m-1}(l) + k_m a_{m-1}(m-l) \end{cases} \tag{12}$$

In Equation (12), $l = 1, 2, \ldots, m - 1$. After the calculation, add 1 to the value of m and repeat the above steps until $m = p$.

After using the Burg algorithm to obtain the AR model coefficient a_m, it is necessary to determine the optimal order p of the model. If the order is not selected correctly, the estimation results will be inconsistent with reality. Using the Akaike information criterion (AIC) to fit the asymptotic unbiased estimation of the difference between the AR model and truth-value, the best order of the model can be determined when the model is unknown. The smaller the AIC value, the better the fitting effect of the model.

The general form of the AIC criterion is:

$$\text{AIC} = -lnL + 2k \tag{13}$$

where k is the number of parameters and L is the likelihood function. Assuming that the number of current samples is N and SSR is the sum of squares of residuals, Equation (13) can be converted to:

$$\text{AIC} = Nln(\frac{SSR}{N}) + 2k \tag{14}$$

Equation (14) is applied to the order determination of the AR model. k represents the order p, N is the number of samples, and $\frac{SSR}{N}$ is the variance of the prediction error of the AR model, which can be replaced by σ_p^2; then Equation (14) is converted to:

$$\text{AIC}(p) = ln\sigma_p{}^2 + \frac{2p}{N} \tag{15}$$

In Equation (15), σ_p^2 can be calculated by the reflection coefficient k_p in the Burg algorithm by Equation (9), and the calculation formula is:

$$\sigma_p{}^2 = (1 - \left| k_{p-1} \right|^2)\sigma_{p-1}{}^2 \tag{16}$$

In order to obtain the optimal order of the AR model, the above arc fault experimental platform was used to collect eighteen groups of DC side current data by tests no. 1, no. 2, and no. 3 with a 250 kHz sampling rate. The arc current is disordered and stochastic, and it influences the calculation result, so the current in the normal state was selected for calculating AIC values and analysis. The time window of each group of data was 10 ms. Thus, each time a window had 2500 samples, which ensured the validity of the calculation, and the samples were not very large. The order p was from 1 to 20, and the AIC values corresponding to different orders could be obtained according to Equations (15) and (16). The results are shown in Figure 3.

It can be seen from Figure 3 that when the order $p = 12$, the AIC value of the current data was the smallest. When the order p increased, the AIC value changed indistinctly and had a slightly increasing trend. Therefore, the optimal order of the DC current signal AR model was $p = 12$.

The Burg algorithm was used to solve the 12-order AR model coefficient of the current signal, and the expression of the transfer function is:

$$H(z) = \cfrac{1}{1 + \sum\limits_{m=1}^{12} a_m z^{-m}} \tag{17}$$

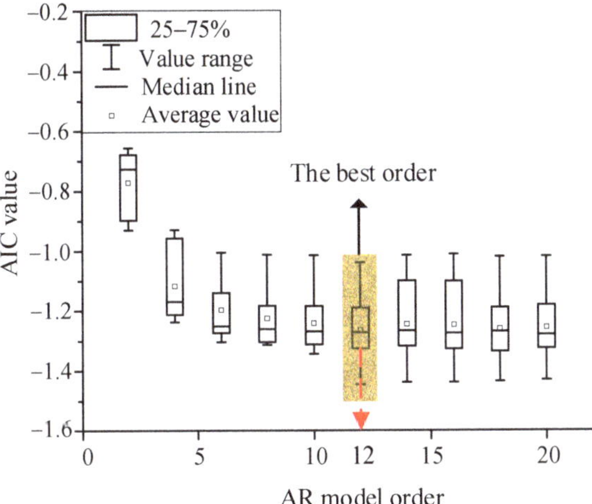

Figure 3. AIC values of current data.

According to Equation (17), the power spectrum estimation expression of the 12-order AR model of the current signal is calculated as:

$$\widetilde{S}_x(\omega) = \frac{k^2}{\left|1 + \sum\limits_{i=1}^{12} a_i e^{-j\omega i}\right|^2} \tag{18}$$

When using the AR model to calculate the power spectrum of the PV system current data, it is necessary to select a suitable time window scale to enlarge the difference in the power spectrum of the current signal under different time windows. In particular, the difference can reflect the changing characteristics in arc current, which is significantly different from the normal state. Since the correlation coefficient can reflect the relationship between two variables, the correlation coefficients of the power spectrum under different time windows were calculated by three groups of current data in test no. 1 separately. The characteristics of tests no. 2 and no. 3 were similar to those of test no. 1, and the variance was also calculated. The larger the variance value, the more pronounced the power spectrum difference in different time windows. It can be seen from Table 2 that when 10 ms and 17 ms time windows were selected for power spectrum estimation, the variance of the correlation coefficient in the arc fault state was considerable.

Table 2. The variance of correlation coefficient of arc fault states' and normal states' power spectrum values under different time window scales.

Time Window Scale/ms	Arc Fault State	Normal State
1	0.05939	0.01268
4	0.06125	0.00185
6	0.07528	0.00265
8	0.06233	0.00195
10	0.24262	0.00140
12	0.02257	0.00085
15	0.18921	0.00110
17	0.37401	0.00151
20	0.21711	0.00806

In contrast, the variance of the normal state correlation coefficient was much smaller than that of the arc fault state. However, the 17 ms time window was too long to process data quickly. Therefore, 10 ms was selected as the time window scale for calculating the DC current power spectrum.

After the time window scale was determined, the power spectrum of current signals was drawn for comparative analysis. Six groups of current data were selected by tests no. 1, no. 2, and no. 3, the 12-order AR model was established, and the power spectrum was calculated. One of the results of test no. 1 is shown in Figure 4.

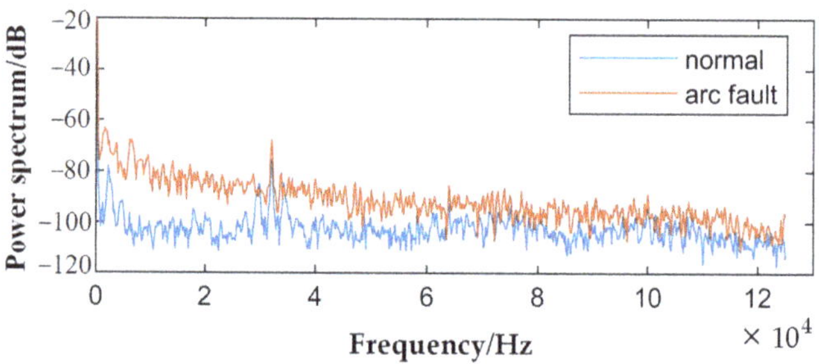

Figure 4. Comparison of power spectrum between the normal state and the arc fault state.

In Figure 4, the orange line represents the power spectrum of the current in the arc fault state. With the increase of frequency, the power spectrum values decreased gradually. The values of the low-frequency part were significantly higher than those of the high-frequency part. The blue line represents the power spectrum of the current in the normal state. The power spectrum values were basically unchanged with the frequency increase, except for 0–10 kHz. In addition, the power spectrum values of arc fault were higher than those of the normal state. The spike at 32 kHz was due to the noise interference of the PV inverter. Therefore, the power spectrum was significantly different between the arc fault state and the normal state, and could be used as the neural network input to detect arc fault.

2.3. Data Processing and Creating the Dataset

The power spectrum of the DC current under the normal state and the arc fault state were different, so it could be used as the input of the neural network model for training. We used the experimental platform to collect the current data of the tests shown in Table 1. Tests no. 1 to no. 3 contained eighteen groups of data that included the normal state and the arc fault state. Test no. 4 contained three groups of data, and test no. 5 contained six groups of data. Both test no. 4 and test no. 5 belonged to the normal state.

The original data were split to extract arc fault data and normal data. Since neural network learning requires a large amount of data, the current data collected by the experimental platform were processed into a dataset, as input for the neural network. The dataset's format and size were unified to facilitate network training. In order to unify the size of the dataset, the classified data for the arc fault state and the normal state were processed into the same time scale, and the 10 ms sampling window was taken as the unit time window.

Since the dataset sampling rate was 250 kHz, the number of sampling points in the unit time window was 2500. The 12-order AR model was used to obtain the power spectrum data. According to the different range of power spectrum values under different working conditions, the values were normalized to map the data value between [0, 1]. The deviation standardization was used as the normalization method, and the equation was as follows:

$$x^* = \frac{x - \min}{\max - \min} \tag{19}$$

In Equation (19), max and min are the maximum and minimum values of power spectrum data and x^* is the normalized value. Each group's normalized power spectrum data has the same order of magnitude.

The normalized data were transferred into the two-dimensional image format. In order to improve the training efficiency of the model, the images were processed into the gray images shown in Figure 5. The resolution of the images was converted into 240 × 240 to meet the EfficientNet-B1 input requirement. The total number of images was 10,000 in the dataset, including 6000 images in the training set, 2000 in the validation set, and 2000 in the test set. After the data were processed into images, each group of data was labeled and divided into two types: arc fault and normal.

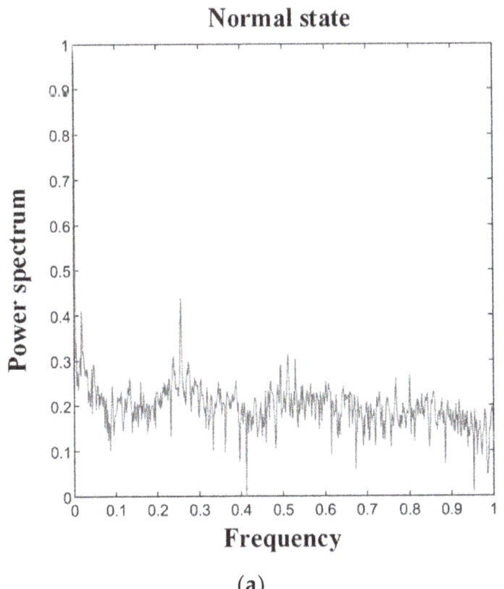

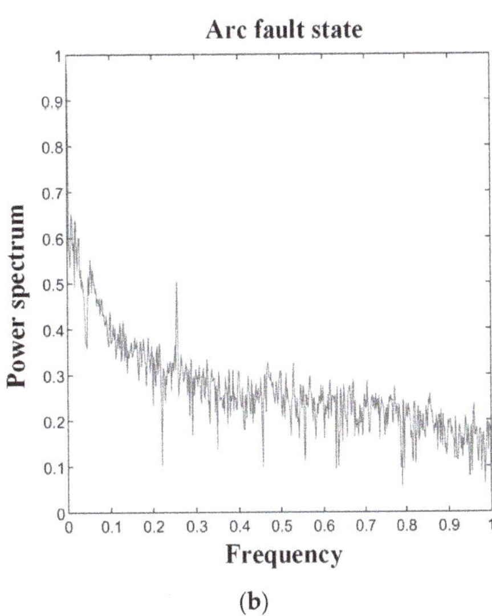

Figure 5. The inputs to the neural network model are two-dimensional images. (**a**) Normal state; (**b**) Arc fault state.

3. Methodology

Convolutional neural network (CNN) has emerged as a fundamental feature exaction program for applications in image tasks. However, the existence of multiple complex behaviors of arc current in PV systems makes some convolutional frameworks suboptimal for the arc fault detection task. Due to the complexity of the DC series arc fault current in PV systems, it is difficult to find a suitable set of CNN parameters, including depth, width, and resolution size, for effectively distinguishing between the arc fault state and the normal state using the current. Inspired by EfficientNet and the attention mechanism, this paper proposes a model based on a lightweight convolutional neural network with a channel and spatial attention mechanism for arc fault detection, and names it ArcDetectionNet (ADNet).

3.1. Lightweight Convolutional Backbone Network Structure

A lightweight convolutional backbone network structure, referring to the idea of EfficientNet, is shown in Figure 6. H, C, and W represent three dimensions of the convolutional neural network. First, we performed a 1 × 1 point-by-point convolution on the input data and changed the output channel dimension according to the expansion ratio. The global features were obtained in the channel dimension of the feature map, and then k × k depth convolution was carried out. Second, we performed an excitation operation on the output result. The 1 × 1 convolution result was multiplied by the activation ratio R, and the original channel dimension was restored at the end of the 1 × 1 point-by-point convolution.

Finally, the connection deactivation and the input jump connection were carried out. This structure is called mobile inverted bottleneck convolution (MBConv). Each convolution operation in this module is normalized and uses the swish activation function. The swish activation function equation is as follows:

$$f(x) = x \times sigmoid(\beta x) \tag{20}$$

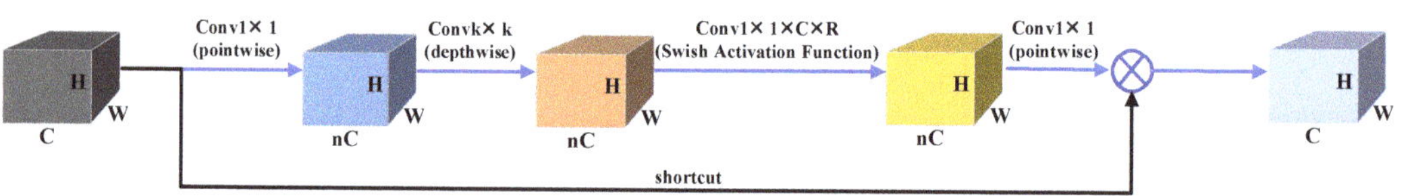

Figure 6. Lightweight convolutional backbone network structure.

In Equation (20), β is a constant or trainable parameter, which defaults to 1.

The effect of the swish function is better than that of the ReLU function on the deep network model. It has a lower bound without an upper bound, and it is smooth and non-monotonic. This method can make the model have stochastic depth, reduce the time required for model training, and improve model performance.

3.2. Arc Detection Attention Mechanism Module

The neural network uses the attention mechanism to generate different connection weights between layers and obtain the output of this layer, so it can focus on specific input characteristics, reduce the number of network operations, and improve network performance. This paper proposes an arc detection attention mechanism (ADAM) module. ADAM was calculated based on the channel and space dimensions for the feature map generated by the convolutional neural network. The calculation results were multiplied by the input data to carry out adaptive learning of features. Moreover, the module was designed for a convolutional neural network, which could be combined with various convolutional neural networks for end-to-end training. For example, we set the channel attention mechanism and then set the spatial attention mechanism after the channel attention mechanism. The structure of the ADAM module is shown in Figure 7.

As shown in Figure 7, the ADAM module extracts data features from two dimensions: channel and space. The channel attention mechanism performs pooling and convolution operations for the input data. The output data of the above processes are each channel's weight coefficient, and the weight coefficient is multiplied by the input data to weight and fuse the channels. The output features weighted by the channel attention mechanism are used as the input of the spatial attention mechanism module to weight the crucial regions in the spatial dimension.

The channel attention mechanism module and the spatial attention mechanism module are connected in serial. By changing the combination and position of the two modules, the optimal combination was selected to construct the ADNet model.

The ADAM module could be added at the front of the network, after the 3×3 convolution layer, or at the end of the network, after 16 MBConv modules. The optimal method was finally determined through the following experiments.

In addition to adding the ADAM module, it was also necessary to configure the other functions of the ADNet. The adaptive moment estimation algorithm was selected as the weight updating optimization algorithm, and the cross-entropy loss function was chosen as the loss function. The swish function was selected as an activation function.

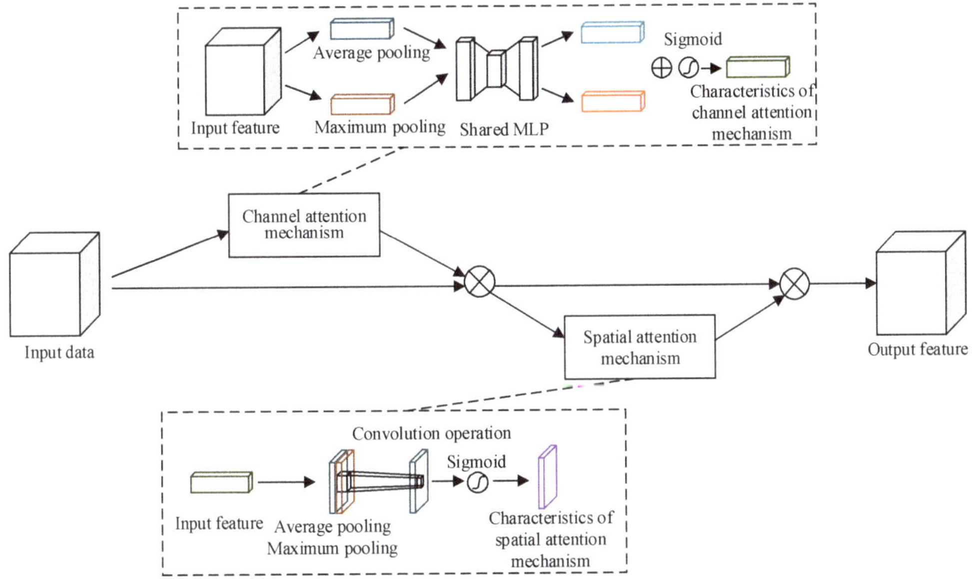

Figure 7. The structure of the ADAM module.

4. Experimental Results and Analysis

This section analyzes the experimental results to select the optimal structure of the proposed ADNet algorithm. The dataset included current data under the arc fault state and the normal state in tests no. 1 to no. 5, and the samples in the test set excluded those in the training and validation sets.

4.1. The Optimal Model Selection Based on EfficientNet

Since the ADNet network is based on the EfficientNet, and the EfficientNet model has eight models, the best model was selected at first. Among them, EfficientNet-B1~B7 are improved from the baseline model EfficientNet-B0. In order to get the most suitable network model, PyCharm software (JetBrains, Prague, Czech Republic) was used to build the program, and the environment was Python 3.7 (Guido van Rossum, Harlem, The Netherlands) and TensorFlow 2.4.0 (Google Brain, Mountain View, CA, USA). Due to the size of the dataset, a smaller network structure in the EfficientNet series networks was selected to reduce the number of parameters and unnecessary calculations for improving the training speed. The resolution of the input images becomes larger from EfficientNet-B0 to EfficientNet-B7, and the height and width of the output characteristic matrix of each layer structure will increase accordingly; the occupation of video memory will also increase. Therefore, the EfficientNet-B0–B3 of the EfficientNet series models were selected for training by the dataset. The model basic parameters and training results are shown in Table 3.

Table 3. EfficientNet-B0–B3 basic parameters and detection accuracy.

Network Model	Width	Depth	Resolution Ratio	Training Set Accuracy	Test Set Accuracy
B0	1.0	1.0	224×224	97.06%	95.32%
B1	1.0	1.1	240×240	99.32%	96.16%
B2	1.1	1.2	260×260	98.60%	95.65%
B3	1.2	1.4	300×300	97.29%	95.46%

It can be seen from Table 3 that the detection accuracy of the EfficientNet-B0~B3 networks can reach more than 95%. The EfficientNet-B1 has the highest detection accuracy, indicating that it is suitable for the DC series arc fault detection in PV systems. At the same time, it avoids the problem of reducing the calculation speed caused by the increasing network complexity, which is the advantage of the EfficientNet series model.

In order to further improve the generalization ability and accelerate the convergence speed of the ADNet, considering that not every part of the power spectrum image is equally important, the channel attention mechanism was used, and different convolution kernels were used to capture various features for channel weighted fusion. In addition, the judgment of whether the circuit has an arc fault mainly depends on some critical areas of the power spectrum image, and the characteristics of each part of the image cannot be treated equally. Therefore, the spatial attention mechanism was used to weight some important regions in space, to strengthen important information and suppress non-important information.

We continued with experimental verification to find the optimal ADNet model. The experimental results of different ADAM types used in the ADNet are shown in Table 4. In Table 4, C represents the channel attention mechanism, and S represents the spatial attention mechanism. Q represents putting the attention mechanism in the front of the network, which follows the 3×3 convolution layer, and H represents adding the attention mechanism to the end of the network, which follows the 16 MBConv modules.

Table 4. The ADNet detection accuracy in different ADAM types.

ADAM Type	Training Set Accuracy/%	Test Set Accuracy/%
CS–Q	99.92	98.36
CS–H	99.96	98.81
SC–Q	99.93	97.18
SC–H	99.95	98.38
S–Q	99.70	96.58
S–H	99.83	96.79
C–Q	99.82	97.37
C–H	99.90	97.32

According to Table 4, the ADNet model, compared with the original EfficientNet-B1 neural network model, improves the feature extraction ability of data samples and the accuracy of arc fault detection. Among the samples, the training set accuracy and test set accuracy of the improved CS-H model were the highest: the accuracy of arc fault detection of the training set was 99.96%, and that of the test set was 98.81%. Therefore, adding the channel attention mechanism first and then the spatial attention mechanism at the end of the network model can improve the model's detection accuracy. The ADAM module was more effective when applied to the deep layer of the network than when applied to the shallow layer of the network, because the characteristics of the deep layer of the network are more robust after multiple feature extractions. Thus, the ADNet model could capture some crucial features of power spectrum images with better robustness and performance after ADAM operation.

According to the above analysis, the EfficientNet-B1 and CS-H of the ADAM type were selected; the optimal structure of the ADNet model is shown in Figure 8.

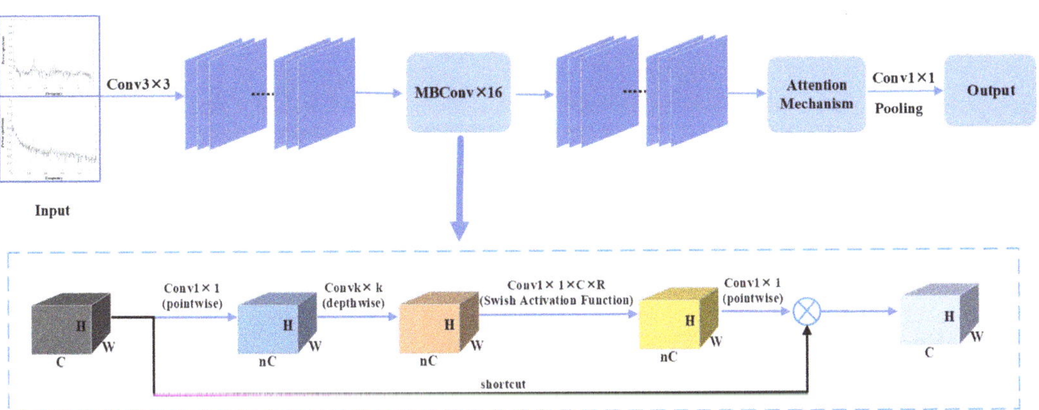

Figure 8. The ADNet model's optimal structure.

4.2. The Selection of ADNet Training Parameters

The learning rate directly affects the convergence state of the network model, which determines the step length of the weight iteration. The model will not converge when the learning rate is set too large. When the learning rate is set too small, the convergence speed of the model will become slower, and it will be unable to learn. The best initial learning rate usually uses the search method, which starts training the model from small to large. After many experiments, 0.001 was chosen as the learning rate of the network to accelerate the convergence speed and save the training time.

Batch size refers to the stochastic sample size used in the gradient descent algorithm, which affects the generalization performance of the convolutional neural network model. In a specific range, increasing the batch size will help the stability of convergence, improve the memory utilization rate, and speed up the processing speed of data volume. This paper set the batch size to 8 in many experiments with the ADNet model.

Since the ADNet model has a complex structure, dropout was used, and the dropout rate was set to 0.2 in many experiments for avoiding over-fitting, and the number of iterations was 120 times.

4.3. Influence of Different Current Values on Detection Results

In order to study the influence of different current values on the arc fault detection accuracy of the ADNet, we used 3 A, 16 A, and 25 A current data from the dataset to carry on experiments. Moreover, the PV inverter startup and irradiance mutation situations were considered the normal state to improve the robustness of the network. The results are shown in Table 5.

Table 5. The ADNet model's detection accuracy of different current values.

Current Value	Training Set Accuracy	Test Set Accuracy
3 A	100%	99.97%
16 A	99.86%	98.96%
25 A	99.68%	97.87%
overall	99.96%	98.81%

According to Table 5, with the increase of the current value, the accuracy of the training set and test set decreased gradually. By comparing Figures 4 and 9, it can be seen that with the increase of current values, the power spectrum values of current data also increased. Since the difference in power spectrum values between the high-frequency part and the low-frequency part decreased in the arc fault state, the power spectrum characteristics of arc fault and normal states were similar, which had a certain impact on arc fault detection. However, according to Figure 9, whether the original power spectrum or the normalized power spectrum was considered, the power spectrum values in the arc fault state were basically higher than those in the normal state, and arc fault could still be detected accurately by the ADNet model, as shown in Table 5. The ADNet model's detection accuracy was 98.81%, including three current levels, indicating that this method can detect arc fault accurately.

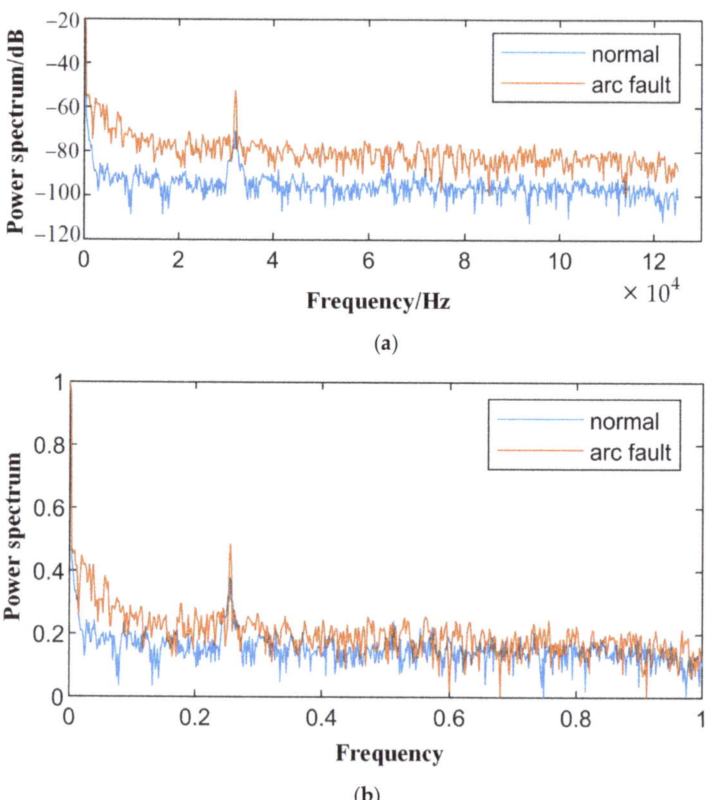

Figure 9. The power spectrum of 25 A current data. (**a**) Original power spectrum; (**b**) Normalized power spectrum.

4.4. Detection Accuracy of Different Existing Neural Networks

The existing research rarely used the power spectrum images as the input data for neural networks. Therefore, in order to verify whether the arc fault detection accuracy of the ADNet model is higher than that of the existing neural network models, we built GoogLeNet and AlexNet models to train and test with the same dataset as the ADNet model's and compared the accuracy of several existing arc fault detection networks. The results are shown in Table 6.

Table 6. The detection accuracy of different neural network models.

Model	Training Set Accuracy/%	Test Set Accuracy/%
GoogLeNet	96.37	96.23
AlexNet	96.91	96.83
BP neural network [25]	\	95.23
DA-DCGAN [27]	98.80	97.68
ADNet (ours)	99.96	98.81

According to Table 6, the detection accuracy of GoogLeNet, AlexNet, BP neural network, and DA-DCGAN is 96.23%, 96.83%, 95.23%, and 97.68%, respectively, and that of the ADNet model is 98.81%. Therefore, the arc fault detection accuracy of the ADNet model is higher than the other existing arc fault detection networks. The results indicate that the ADNet model has a better performance in arc fault detection.

4.5. Feasibility Analysis of Application in the Embedded Modules

The ADNet model can be used for edge applications based on embedded processors or modules of the arc fault detection equipment, such as Raspberry Pi, because: (1) The AR model-based data preprocessing method is employed to capture the arc features and remove un-sensitive parts of the power spectrum, which can help to reduce the amount of input data; (2) The ADNet model is based on EfficientNet-B1, a commonly-used lightweight convolutional neural network. Moreover, we used the attention mechanism to combine with the EfficientNet-B1, making the algorithm more concentrated on the arc features while ignoring the rest information. Specifically, spatial attention was used to locate the more sensitive part of the input signal, while channel attention was used to determine the more valuable channels or layers in the model [29]. Therefore, the proposed method can be further light-weighted with considerable detection accuracy; (3) Due to the above lightweight design and operation, the total parameters of the proposed ADNet model are only 6.58×10^6, which are less than those of other commonly used methods. Meanwhile, the detection accuracy was higher than that of others. Table 7 shows a detailed comparison.

Table 7. The comprehensive comparison of different neural networks.

Model	Total Parameters	Detection Accuracy	Computational Burden
GoogLeNet	10.31 M	96.23%	Medium
AlexNet	14.59 M	96.83%	Medium
Inception V3 [30]	23.63 M	94.10%	Large
Xception [30]	22.86 M	94.50%	Large
ResNet50 [31,32]	23.48 M	97.33%	Large
ADNet (ours)	6.58 M	94.10%	Small

The more model parameters, the greater the amount of calculation and the slower the running speed [33]. We compared the number of network model parameters with the built GoogLeNet, AlexNet, and several commonly used networks. As shown in Table 7, the total of the parameters was the sum of the model parameters. The total number of model parameters in GoogLeNet, AlexNet, Inception V3, Xception, and ResNet50, which are commonly used convolutional neural networks, are 10.31 M, 14.59 M, 23.63 M, 22.86 M, and 23.48 M, respectively. The quantity is too large, resulting in too much computation and slowing down the running speed. However, the total number of parameters in the ADNet model, which belongs to the lightweight convolutional neural network, is 6.58 M, which is lower than that of the above networks. The results show that the proposed method achieves the best detection accuracy, with minimum computational burden, due to the well-designed lightweight algorithm. Therefore, the ADNet model is ready for edge applications and can be implemented with embedded processors or modules, such as the Raspberry Pi 3B with a quad-core 1.2 GHz CPU and 1 GB RAM. This calls for further research in the future.

5. Conclusions

In this paper, we established an experimental platform, based on the UL1699B standard to collect DC current data in creating a dataset, which can obtain current data efficiently. The power spectrum image of current data can clearly distinguish the current in the normal state and the arc fault state. Therefore, it can be used as the input for the arc fault detection algorithm. In order to avoid the problem of excessive consumption of computing resources due to increasing algorithm complexity, this paper proposed a detection method of DC series arc faults in PV systems based on a lightweight convolutional neural network, which has fewer parameters and a low computational burden. The power spectrum images were normalized and converted into 240×240 gray images as the dataset. Compared with the EfficientNet series model, the EfficientNet-B1 was selected as the optimal network. The channel attention mechanism and the spatial attention mechanism were added to the deep layer of the EfficientNet-B1 to construct the ADNet model for improving the network's detection accuracy and making it more effective. This method considered the situations of PV inverter startup and irradiance mutation, enhancing the robustness of the network. The results showed that the accuracy of the training set was 99.96%, and that of the test set was 98.81%, which are higher than the accuracies of GoogLeNet, AlexNet, and other commonly used networks. According to the above analysis, this method can be used in PV systems to detect DC series arc faults accurately and to reduce arc fire hazards. Therefore, the safety of PV systems will be improved, and solar energy may be used sufficiently and stably.

Author Contributions: Conceptualization, Y.W.; methodology, C.B. and W.L.; software, C.B.; validation, C.B., C.Z. and X.Q.; formal analysis, L.G.; investigation, Y.W.; resources, Y.W. and X.Q.; data curation, C.B.; writing—original draft preparation, C.B., Y.W. and W.L.; writing—review and editing, Y.W. and C.B.; visualization, C.B., X.Q. and C.Z.; supervision, L.G.; project administration, Y.W.; funding acquisition, Y.W. All authors have read and agreed to the published version of the manuscript.

Funding: This research was funded by the Natural Science Foundation of Hebei Province, grant number E2020202204, the Key Laboratory of Special Machine and High Voltage Apparatus (Shenyang University of Technology), grant number KFKT202003, the Zhejiang Provincial Natural Science Foundation of China, grant number LGG20E070002, the National Natural Science Foundation of China, grant number 51807134, and the State Key Lab of Reliability and Intelligence of Electrical Equipment (Hebei University of Technology) Open Project Funding Projects, grant number EERI_KF20200014.

Institutional Review Board Statement: Not applicable.

Informed Consent Statement: Not applicable.

Data Availability Statement: Data are contained within the article. Data sharing is not applicable to this article.

Conflicts of Interest: The authors declare no conflict of interest.

References

1. Gu, J.; Lai, D.; Yang, M. A new DC series Arc fault detector for household photovoltaic systems. In Proceedings of the 2017 IEEE Power & Energy Society General Meeting, Chicago, IL, USA, 16–20 July 2017; pp. 1–5.
2. Ahmadi, M.; Samet, H.; Ghanbari, T. A New Method for Detecting Series Arc Fault in Photovoltaic Systems Based on the Blind-Source Separation. *IEEE Trans. Ind. Electron.* **2020**, *67*, 5041–5049. [CrossRef]
3. Ge, L.; Xian, Y.; Yan, J.; Wang, B.; Wang, Z. A Hybrid Model for Short-term PV Output Forecasting Based on PCA-GWO-GRNN. *J. Mod. Power Syst. Clean Energy* **2020**, *8*, 1268–1275. [CrossRef]
4. Ge, L.; Liu, J.; Wang, B.; Zhou, Y.; Yan, J.; Wang, M. Improved adaptive gray wolf genetic algorithm for photovoltaic intelligent edge terminal optimal configuration. *Comput. Electr. Eng.* **2021**, *95*, 107394–107409. [CrossRef]
5. Wang, Y.; Hou, L.; Paul, K.C.; Ban, Y.; Chen, C.; Zhao, T. ArcNet: Series AC Arc Fault Detection Based on Raw Current and Convolutional Neural Network. *IEEE Trans. Ind. Inform.* **2022**, *18*, 77–86. [CrossRef]
6. Xiong, Q.; Ji, S.; Zhu, L.; Zhong, L.; Liu, Y. A Novel DC Arc Fault Detection Method Based on Electromagnetic Radiation Signal. *IEEE Trans. Plasma Sci.* **2017**, *45*, 472–478. [CrossRef]

7. Liu, X. A Series Arc Fault Location Method for DC Distribution System Using Time Lag of Parallel Capacitor Current Pulses. In Proceedings of the 2018 IEEE International Power Modulator and High Voltage Conference (IPMHVC), Jackson, WY, USA, 3–7 June 2018; pp. 218–222.
8. Shekhar, A.; Ramirez-Elizondo, L.; Bandyopadhyay, S.; Mackay, L.; Bauera, P. Detection of Series Arcs Using Load Side Voltage Drop for Protection of Low Voltage DC Systems. *IEEE Trans. Smart Grid* **2018**, *9*, 6288–6297. [CrossRef]
9. Lu, S.; Phung, B.T.; Zhang, D. A comprehensive review on DC arc faults and their diagnosis methods in photovoltaic systems. *Renew. Sust. Energy Rev.* **2018**, *89*, 88–98. [CrossRef]
10. Xiong, Q.; Feng, X.; Gattozzi, A.L.; Liu, X.; Zheng, L.; Zhu, L.; Ji, S.; Hebner, R.E. Series Arc Fault Detection and Localization in DC Distribution System. *IEEE Trans. Instrum. Meas.* **2020**, *69*, 122–134. [CrossRef]
11. Dang, H.; Kim, J.; Kwak, S.; Choi, S. Series DC Arc Fault Detection Using Machine Learning Algorithms. *IEEE Access* **2021**, *9*, 133346–133364. [CrossRef]
12. Qu, N.; Wang, J.; Liu, J. An Arc Fault Detection Method Based on Current Amplitude Spectrum and Sparse Representation. *IEEE Trans. Instrum. Meas.* **2019**, *68*, 3785–3792. [CrossRef]
13. Murakami, M.; Ryonai, H.; Kubono, T.; Sekikawa, J. Properties of Short Arc Phenomena on AgCu Electrical Contact Pairs for Automotive Electronics Devices. In Proceedings of the Electrical Contacts—53rd IEEE Holm Conference on Electrical Contacts, Pittsburgh, PA, USA, 16–19 September 2007; pp. 146–150.
14. Yue, L.; Le, V.; Yang, Z.; Yao, X. A Novel Series Arc Fault Detection Method using Sparks in DC Microgrids with Buck Converter Interface. In Proceedings of the 2018 IEEE Energy Conversion Congress and Exposition (ECCE), Portland, OR, USA, 23–27 September 2018; pp. 492–496.
15. Zhao, S.; Wang, Y.; Niu, F.; Zhu, C.; Xu, Y.; Li, K. A Series DC Arc Fault Detection Method Based on Steady Pattern of High-Frequency Electromagnetic Radiation. *IEEE Trans. Plasma Sci.* **2019**, *47*, 4370–4377. [CrossRef]
16. Alam, M.K.; Khan, F.H.; Johnson, J.; Flicker, J. PV arc-fault detection using spread spectrum time domain reflectometry (SSTDR). In Proceedings of the 2014 IEEE Energy Conversion Congress and Exposition (ECCE), Pittsburgh, PA, USA, 14–18 September 2014; pp. 3294–3300.
17. Georgijevic, N.L.; Jankovic, M.V.; Srdic, S.; Radakovic, Z. The Detection of Series Arc Fault in Photovoltaic Systems Based on the Arc Current Entropy. *IEEE Trans. Power Electron.* **2016**, *31*, 5917–5930. [CrossRef]
18. Park, H.P.; Chae, S. DC Series Arc Fault Detection Algorithm for Distributed Energy Resources Using Arc Fault Impedance Modeling. *IEEE Access* **2020**, *8*, 179039–179046. [CrossRef]
19. Gu, J.; Lai, D.; Wang, J.; Huang, J.; Yang, M. Design of a DC Series Arc Fault Detector for Photovoltaic System Protection. *IEEE Trans. Ind. Appl.* **2019**, *55*, 2464–2471. [CrossRef]
20. Miao, W.; Xu, Q.; Lam, K.H.; Pong, P.W.T.; Poor, H.V. DC Arc-Fault Detection Based on Empirical Mode Decomposition of Arc Signatures and Support Vector Machine. *IEEE Sens. J.* **2021**, *21*, 7024–7033. [CrossRef]
21. Liu, S.; Dong, L.; Liao, X.; Cao, X.; Wang, X.; Wang, B. Application of the Variational Mode Decomposition-Based Time and Time-Frequency Domain Analysis on Series DC Arc Fault Detection of Photovoltaic Arrays. *IEEE Access* **2019**, *7*, 126177–126190. [CrossRef]
22. Wang, Z.; Balog, R.S. Arc Fault and Flash Signal Analysis in DC Distribution Systems Using Wavelet Transformation. *IEEE Trans. Smart Grid* **2015**, *6*, 1955–1963. [CrossRef]
23. Chen, S.; Li, X.; Meng, Y.; Xie, Z. Wavelet-based protection strategy for series arc faults interfered by multicomponent noise signals in grid-connected photovoltaic systems. *Sol. Energy* **2019**, *183*, 327–336. [CrossRef]
24. Li, K.; Zhao, S.; Wang, Y. A Planar Location Method for DC Arc Faults Using Dual Radiation Detection Points and DANN. *IEEE Trans. Instrum. Meas.* **2020**, *69*, 5478–5487. [CrossRef]
25. Li, T.; Xiong, Q.; Li, R.; Liu, H.; Ji, S.; Li, J. DC Arc Fault Risk Degree Evaluation Based on Back Propagation Neural Network. In Proceedings of the 2021 Power System and Green Energy Conference (PSGEC), Shanghai, China, 20–22 August 2021; pp. 655–659.
26. Yang, K.; Chu, R.; Zhang, R.; Xiao, J.; Tu, R. A Novel Methodology for Series Arc Fault Detection by Temporal Domain Visualization and Convolutional Neural Network. *Sensors* **2019**, *20*, 162. [CrossRef]
27. Lu, S.; Sirojan, T.; Phung, B.T.; Zhang, D.; Ambikairajah, E. DA-DCGAN: An Effective Methodology for DC Series Arc Fault Diagnosis in Photovoltaic Systems. *IEEE Access* **2019**, *7*, 45831–45840. [CrossRef]
28. Pedersen, E.; Rao, S.; Katoch, S.; Jaskie, K.; Spanias, A.; Tepedelenlioglu, C.; Kyriakides, E. PV Array Fault Detection using Radial Basis Networks. In Proceedings of the 2019 10th International Conference on Information, Intelligence, Systems and Applications (IISA), Patras, Greece, 15–17 July 2019; pp. 1–4.
29. Luo, Z.; Li, J.; Zhu, Y. A Deep Feature Fusion Network Based on Multiple Attention Mechanisms for Joint Iris-Periocular Biometric Recognition. *IEEE Signal Proc. Let.* **2021**, *28*, 1060–1064. [CrossRef]
30. Chollet, F. Xception: Deep Learning with Depthwise Separable Convolutions. In Proceedings of the 2017 IEEE Conference on Computer Vision and Pattern Recognition (CVPR), Honolulu, HI, USA, 21–26 July 2017; pp. 1800–1807.
31. Shen, J.; Liu, N.; Sun, H.; Zhou, H. Vehicle Detection in Aerial Images Based on Lightweight Deep Convolutional Network and Generative Adversarial Network. *IEEE Access* **2019**, *7*, 148119–148130. [CrossRef]

32. Mascarenhas, S.; Agarwal, M. A comparison between VGG16, VGG19 and ResNet50 architecture frameworks for Image Classification. In Proceedings of the 2021 International Conference on Disruptive Technologies for Multi-Disciplinary Research and Applications (CENTCON), Bengaluru, India, 19–21 November 2021; pp. 96–99.
33. Tao, Y.; Zongyang, Z.; Jun, Z.; Xinghua, C.; Fuqiang, Z. Low-altitude small-sized object detection using lightweight feature-enhanced convolutional neural network. *J. Syst. Eng. Electron.* **2021**, *32*, 841–853. [CrossRef]

Article

Multiple Power Supply Capacity Planning Research for New Power System Based on Situation Awareness

Dahu Li [1,2], Xiaoda Cheng [1], Leijiao Ge [3,*], Wentao Huang [1,*], Jun He [1] and Zhongwei He [4]

[1] Hubei Collaborative Innovation Center for High-Efficiency Utilization of Solar Energy,
     Hubei University of Technology, Wuhan 430068, China; li_dahu@126.com (D.L.); cxd_3939@163.com (X.C.);
     apm874@163.com (J.H.)
[2] State Grid Hubei Electric Power Co., Ltd., Wuhan 430077, China
[3] School of Electrical and Information Engineering, Tianjin University, Tianjin 300072, China
[4] Enshi Power Supply Company, State Grid Hubei Electric Power Co., Ltd., Enshi 445699, China;
     lizw19860727@126.com
*   Correspondence: legendglj99@tju.edu.cn (L.G.); 102010265@hbut.edu.cn (W.H.); Tel.: +86-013820176750 (L.G.)

Citation: Li, D.; Cheng, X.; Ge, L.; Huang, W.; He, J.; He, Z. Multiple Power Supply Capacity Planning Research for New Power System Based on Situation Awareness. *Energies* **2022**, *15*, 3298. https://doi.org/10.3390/en15093298

Academic Editors: Carlo Roselli and Paola Verde

Received: 16 February 2022
Accepted: 11 April 2022
Published: 30 April 2022

Publisher's Note: MDPI stays neutral with regard to jurisdictional claims in published maps and institutional affiliations.

Abstract: In the context of new power systems, reasonable capacity optimization of multiple power systems can not only reduce carbon emissions, but also improve system safety and stability. This paper proposes a situation awareness-based capacity optimization strategy for wind-photovoltaic-thermal power systems and establishes a bi-level model for system capacity optimization. The upper-level model considers environmental protection and economy, and carries out multi-objective optimization of the system capacity planning solution with the objectives of minimizing carbon emissions and total system cost over the whole life cycle of the system, further obtaining a set of capacity planning solutions based on the Pareto frontier. A Pareto optimal solution set decision method based on grey relativity analysis is proposed to quantitatively assess the comprehensive economic–environmental properties of the system. The capacity planning solutions obtained from the upper model are used as the input to the lower model. The lower model integrates system stability, environmental protection, and economy and further optimizes the set of capacity planning solutions obtained from the upper model with the objective of maximizing the inertia security region and the best comprehensive economic–environmental properties to obtain the optimal capacity planning scheme. The NSGA-II modified algorithm (improved NSGA-II algorithm based on dominant strength, INSGA2-DS) is used to solve the upper model, and the Cplex solver is called on to solve the lower model. Finally, the modified IEEE-39 node algorithm is used to verify that the optimized capacity planning scheme can effectively improve the system security and stability and reduce the carbon emissions and total system cost throughout the system life cycle.

Keywords: situation awareness; capacity configuration; wind-photovoltaic-thermal power system; carbon emission; multi-objective optimization; inertia security region

1. Introduction

In the context of the new power system, with the increased development of new energy generation, the proportion of wind power and photovoltaic integrated into the grid has been increasing year by year. The wind-photovoltaic-thermal power system can effectively bring into play the complementary characteristics and synergistic effects of different forms of energy, improve the level of new energy consumption within a certain range, and achieve the purpose of making full use of energy resources. Reasonable system capacity planning is the basis for the safe and stable operation of the system.

A great deal of research has been done by domestic and international scholars on the planning of power systems containing renewable energy. The paper [1] assesses the impact of regional and international renewable energy policy coordination on the economics, environmental performance, and planning outcomes of the North American power sector in the context of renewable energy policy coordination and identifies the need to integrate cost, emissions, trade, and infrastructure investment in future capacity planning decisions for renewable energy-containing power systems through a multi-model comparison analysis using multiple energy-economic models. The literature [2] proposes a modeling approach adapted to the planning of power systems containing renewable energy sources while dividing the modeling approach into four categories and concluding that the choice of model should depend on the purpose of the study as well as the system characteristics. It provides a reference value for the research on the planning of power systems containing renewable energy.

Most of the existing studies on multiple power supply planning have been considered in terms of economics and environmental protection. The literature [3] investigates the capacity configuration of scenic power generation systems by using data, such as network node voltage and scenic power output, through a nuclear limit learning machine method to find the solution that minimizes the total investment cost and network losses. The paper [4] takes into account the environmental factors and takes the CO_2 emission of the whole life cycle of the wind and solar power system as the optimization target to optimize the capacity allocation of the wind and solar power system. The literature [5] considers the construction and maintenance costs, energy wastage, and outage losses of wind-complementary micro-grids, and obtains an optimal allocation model that matches the meteorological conditions to optimize the capacity allocation of wind-complementary islanded microgrids. The literature [6] uses the lowest operating cost and the lowest system grid power supply rate as the optimization objectives for rational planning of the configuration of integrated energy systems, including wind power and photovoltaic power generation. A multi-objective optimization model with the objectives of minimizing total investment, node voltage exceedance probability, and undersupply probability was developed in [7]. An improved parallel elite non-dominated ranking genetic algorithm II is used to search for the Pareto optimal configuration solution for the optimal configuration of the wind and solar complementary system. The literature [8] investigates the optimization of wind capacity in power systems, considering system operation, economy, and reliability. The assessment of the economic aspects is obtained based on the social cost of the whole system, and the probabilistic method is used to assess the reliability of the system load loss probability. The planning problem of wind power capacity is solved through an opportunity-constrained planning approach. The literature [9,10] investigates the assessment of power system flexibility for the problem of planning systems with a high penetration of renewable energy sources, such as photovoltaic power. The concept of flexibility is reviewed and indicators for assessing the flexibility of power systems are summarised.

The above literature provides a reference for the study of multiple power supply planning in the context of new power systems but lacks consideration of system stability.

With the massive penetration of new energy sources, the inertia support of the system under active disturbance is severely weakened and the problem of inertia reduction cannot be ignored. Domestic and international new energy high percentage power grids have repeatedly experienced inertia shortages in operation, with significant frequency stability problems, thus exposing the system to the risk of large area cut-offs and load shedding.

At this stage, most of the research on system inertia presents an inertia assessment problem. The literature [11] describes the concept, characteristics, and assessment methods of the inertia security region. Literature [12] introduces the concept of minimum inertia demand for microgrids, establishes a minimum inertia demand assessment model, and proposes an optimal solution method. [13] proposes a method for estimating system inertia based on electromechanical oscillation parameters driven by stochastic data. Few studies have considered the impact on the optimal allocation of wind-photovoltaic-thermal power system capacity by taking the system inertia security region as an objective function.

Situation awareness is a technique for acquiring, understanding, and predicting the activities of elements that can cause changes in the system's situation [14]. Currently, situation awareness techniques are gradually being applied in the field of power systems [15].

Capacity planning involves finding the best capacity planning solution for building generation capacity subject to various economic and technical constraints. In the face of today's stricter environmental policies and increasing uncertainty in the power system, capacity planning studies need to be constantly innovated to meet new challenges [16]. The application of situation awareness methods to the study of multiple power sources in the context of new power systems is of great significance for the comprehensive awareness of system characteristics, in-depth understanding of system performance, effective prediction of system status, and significant improvement of grid operation efficiency.

This paper proposes a strategy for the capacity optimization of wind-photovoltaic-thermal power systems based on situation awareness, taking into account system economy, environmental protection, and stability. Situation awareness stage: data collection based on elements, such as equipment, meteorological environment, and users. Situation understanding stage: establishment of a bi-level model for system capacity optimization configuration. With the upper model taking the minimization of carbon emissions and the minimization of total system cost over the whole life cycle of the system as the optimization objective for the initial optimization of the system capacity configuration scheme, the Pareto frontier-based system capacity allocation scheme is obtained, and a grey relativity analysis-based Pareto optimal solution set evaluation method is proposed to quantitatively assess the integrated economic-environmental characteristics of the system. Using the upper model capacity configuration scheme as the input to the lower model, the lower model takes into account the stability, environmental protection, and economy of the system, and further optimizes the capacity planning scheme obtained from the upper model with the objective of maximizing inertia security region and the best the comprehensive economic–environmental properties to obtain the optimal capacity planning scheme. The upper model is solved using the INSGA2-DS algorithm and the lower model is solved using the Cplex solver. The data obtained in the situation awareness phase are used as the basis for understanding and evaluating the system state characteristics according to the bi-level model for optimal system capacity configuration. The situation prediction phase: the results of the capacity planning scheme are evaluated and analyzed to provide an effective basis for the relevant professionals to make decisions on the scheme. Finally, the effectiveness of the proposed strategies and algorithms is verified through a case study.

2. Wind-Photovoltaic-Thermal Power System Model for New Power System

2.1. Characteristics of Wind-Photovoltaic-Thermal Power Systems

With the proposal to build a new power system with wind power and photovoltaic as the main new energy sources, the proportion of new energy sources has increased significantly, gradually becoming the main power source. Thermal power is gradually transforming into a regulating, guaranteeing, and contingency power source. Wind, photovoltaic, and thermal power in the system can achieve complementarity on various time scales and guarantee total load demand.

From the perspective of system environmental protection, the whole life cycle of a power system generally includes four segments: manufacturing and installation, production and operation, operation and maintenance, and recycling and disposal. Wind power and photovoltaic power generation do not generate carbon emissions in the production and operation stages, and their carbon emissions are mainly concentrated in the remaining three stages. Thermal power generation generates carbon emissions in all four stages [17]. Thermal power generation, wind power generation, and photovoltaic power generation all produce carbon emissions during the whole life cycle of the system, but the carbon emission rate of thermal power generation is significantly greater than that of wind power generation and photovoltaic power generation. From the perspective of system stability, thermal power units can provide the rotational inertia required when the system is disturbed, which can effectively suppress the frequency fluctuations caused by faults in the system and is conducive to the frequency stability of the system. As wind turbines and photovoltaic battery units have power electronic characteristics, their transmission power is decoupled from the grid frequency and cannot provide inertia support to the system directly. Therefore, wind power and photovoltaic weaken the system after they replace thermal power units on a large scale to generate electricity. The level of inertia support is weakened by the large-scale replacement of thermal power units by wind power and photovoltaic, which affects system stability. From the perspective of system economics, the total cost of a thermal power plant consists of equipment investment costs, operating costs, replacement costs, and maintenance costs. Excluding the operating costs of wind power and photovoltaic power generation, the total cost of wind farms and photovoltaic power plants consists of equipment investment costs, replacement costs, and maintenance costs [18,19]. Figure 1 shows the total system cost and system carbon emission characteristics.

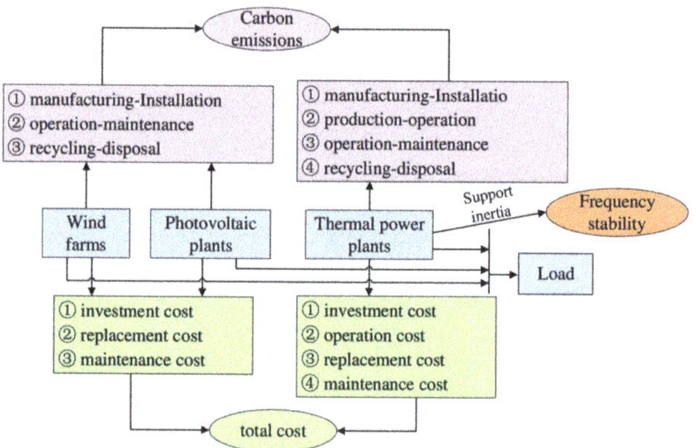

Figure 1. Wind-photovoltaic-thermal power system characteristics.

2.2. Relationship between Carbon Emission and Stability of Wind-Photovoltaic-Thermal Power System

For wind–photovoltaic–thermal power systems, good stability is the basis for the safe and stable operation of the power system, and reducing system carbon emissions is a realistic need to achieve the "double carbon" goal. However, in the case of wind–photovoltaic–thermal power systems, there is a contradiction between the goals of improving system stability and reducing carbon emissions.

Figure 2 shows that the thermal share of the system is positively correlated with carbon emissions; it is positively correlated with rotational inertia. When the share of thermal power increases, the carbon emission of the system accelerates and the rotational inertia of the system increases, which is conducive to improving stability. When the share of thermal power decreases, the carbon emission of the system decreases, the rotational inertia of the system decreases, and the stability of the system decreases.

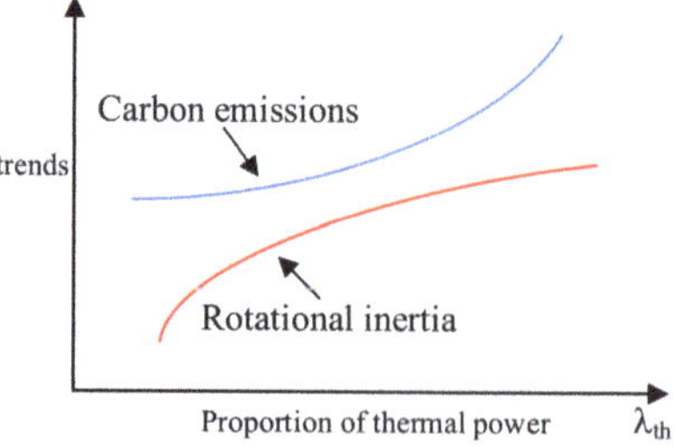

Figure 2. Trends in carbon emissions, rotational inertia, and proportion of thermal power.

2.3. Output Characteristics of Wind-Photovoltaic-Thermal Power Systems in the Context of New Power System

2.3.1. Wind Turbines Model

The WT output characteristics are related to the ambient wind speed and the power output characteristics of the unit.

$$
P_{wind} = \begin{cases}
0 & v < v_{in} \\
\frac{v^3 - v_{in}{}^3}{v_{out}{}^3 - v_{in}{}^3} P_{wind,N} & v_{in} < v < v_N \\
P_{wind,N} & v_N < v < v_{out} \\
0 & v > v_{out}
\end{cases}
\tag{1}
$$

where P_{wind} is the output power of the WT; v_{in} is the WT cut-in wind speed; v_N is the WT rated wind speed; v_{out} is the WT cut-out wind speed; $P_{wind,N}$ is the WT rated power.

2.3.2. Photovoltaic Model

The output of photovoltaic cells is related to the ambient temperature and the amount of solar radiation.

$$
P_{pv} = \frac{P_S}{G_S} G_R [1 - \gamma(T_R - T_\tau)]
\tag{2}
$$

where P_{pv} is the output power of the PV cell; P_S is the output power of the PV cell under standard conditions; G_S is the light intensity under standard conditions; G_R is the light intensity under actual conditions; γ is the power temperature coefficient, taken as $-0.5\%/°C$; T_R is the temperature of the PV cell under actual conditions; T_τ is the reference temperature value, taken as 25 °C.

2.4. Situation Awareness Model for Capacity Planning of Wind-Photovoltaic-Thermal Power System

Applying the situation awareness approach to the planning of multiple power sources in the context of new power systems [13], the situation awareness-based capacity planning model is divided into four stages: situation awareness, situation understanding, situation prediction, and assisted decision-making [20]. Using situation awareness → situation understanding → situation prediction → assisted decision making → situation awareness to form a closed loop to fully grasp the system state and improve the accuracy of capacity allocation. The wind-photovoltaic-thermal power system capacity planning situation awareness model is shown in Figure 3.

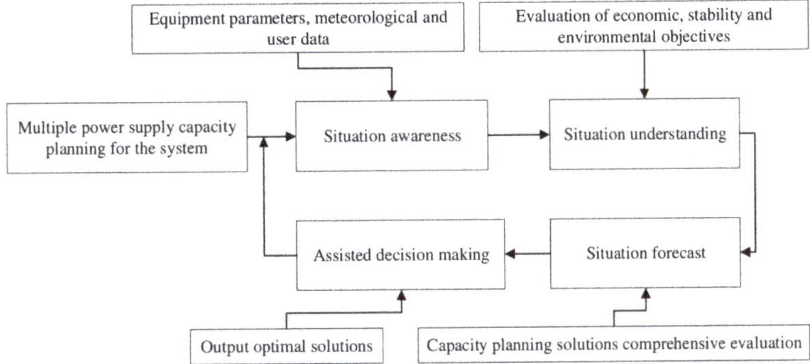

Figure 3. Situation awareness model for capacity planning of wind-photovoltaic-thermal power systems.

(1) Situation awareness: This is the stage of obtaining relevant data. This stage is mainly used to obtain equipment parameters, meteorological data, and user data in system

power capacity planning through system measurement techniques, meteorological information prediction techniques, and load-side data prediction techniques.

(2) Situation understanding: This is the stage of data analysis, which aims at understanding and mining the data obtained during the situation awareness stage, taking into account system stability, economy, environmental protection, etc., and analyzing the system operating dynamics of different capacity planning scenarios.

(3) Posture prediction: This is the state prediction phase. For the capacity planning of multiple power systems, posture prediction is a comprehensive evaluation and analysis of different capacity planning options.

(4) Assisted decision-making: Output the optimal solution, providing an effective basis for decision-making by relevant professionals.

3. Bi-Level Model for Optimal Capacity Allocation of Wind-Photovoltaic-Thermal Power Systems

3.1. Upper Level Model for Multi-Objective Optimal Configuration Considering the Environmental Friendliness and Economy of the System

The upper-level model is based on the objective of minimizing carbon emissions and total system cost over the whole life cycle of the system. The system capacity planning scheme is optimized based on the system power balance constraint, installed capacity constraint, generation unit output constraint, and thermal unit climbing constraint, and the decision variables are wind, PV, and thermal power output at each time.

3.1.1. Objective Functions

In this paper, the annual equivalent carbon emissions of wind power, photovoltaic power generation, and thermal power generation are calculated separately for the whole life cycle. The carbon emissions from wind, photovoltaic and thermal power plants are apportioned to the power generation process according to the carbon accounting model. The optimization objective is to minimize the annual carbon emissions of the wind-photovoltaic-thermal power system [21].

$$\min F_1 = \min \left\{ \begin{array}{l} \sum_{\substack{t=1 \\ i \in N_{wind}}}^{8760} P_{wind,i}(t) R_{wind} N_{wind} K_{wind,i} + \sum_{\substack{t=1 \\ j \in N_{pv}}}^{8760} P_{pv,j}(t) R_{pv} N_{pv} K_{pv,j} \\ \\ + \sum_{\substack{t=1 \\ k \in N_{SG}}}^{8760} P_{SG,k}(t) R_{SG} N_{SG} K_{SG,k} \end{array} \right\}$$

(3)

where R_{wind}, R_{pv}, and R_{SG} are the carbon emission factors of wind, photovoltaic and thermal power respectively for the whole life cycle of the system; $P_{wind,i}(t)$, $P_{pv,j}(t)$, $P_{SG,k}(t)$ represent the output power of the first wind turbine; N_{wind}, N_{pv}, and N_{SG} are the number of wind turbines, photovoltaic cells, and synchronous machines respectively; $K_{wind,i}$, $K_{pv,j}$, and $K_{SG,k}$ are the switching states of wind turbines, photovoltaic cells, and synchronous units respectively.

Carbon emission factors can be calculated based on the Carbon Accounting Model [22,23].

Considering the economy, the total cost of the wind-photovoltaic-thermal power system consists of four parts: investment cost, operation cost, replacement cost, and maintenance cost. In this paper, only the operating costs of thermal power units are considered, and the operating costs of wind power and photovoltaic are approximated to be zero. The optimization objective is to minimize the total cost of the system.

$$\min F_2 = \min\{P_{wind} + P_{pv} + P_{SG}\}$$

(4)

$$P_{wind} = \sum_{i=1}^{N_{wind}} \left(C_{wind,i} + \frac{C_{wind,r,i}}{(1+r_1)^{T_{wind,i}}} + \sum_{t=1}^{Tt} \frac{C_{wind,m,i}}{(1+r_1)^t} \right) \tag{5}$$

$$P_{pv} = \sum_{i=1}^{N_{pv}} \left(C_{pv,i} + \frac{C_{pv,r,1}}{(1+r_2)^{T_{pv}}} + \sum_{t=1}^{Tt} \frac{C_{pv,m,i}}{(1+r_2)^t} \right) \tag{6}$$

$$P_{SG} = \sum_{i=1}^{N_{SG}} \left(C_{SG,i} + \frac{C_{SG,r,i}}{(1+r_3)^{T_{SG}}} + \sum_{t=1}^{Tt} \frac{C_{SG,m,i}}{(1+r_3)^t} + T_{SG} \sum_{t=1}^{8760} \left[a_i P_{SG,i}(t)^2 + b_i P_{SG,i}(t) + c_i \right] \right) \tag{7}$$

where P_{wind}, P_{pv}, and P_{SG} are the total costs of wind, photovoltaic, and thermal power plants, respectively; $P_{SG,i}(t)$ is the output of thermal power unit i at time t; $C_{wind,i}$, $C_{pv,i}$, and $C_{SG,i}$ are the installed prices of a single wind turbine, a single photovoltaic cell unit, and a single synchronous machine, respectively; N_{wind}, N_{pv}, and N_{SG} are the number of wind turbines, photovoltaic cells, and synchronous machines, respectively; $C_{wind,r,i}$, $C_{wind,m,i}$ are the replacement and maintenance costs of wind turbines, respectively; $C_{pv,r,i}$, $C_{pv,m,i}$ are the replacement and maintenance costs of photovoltaic cells, respectively; $C_{SG,r,i}$, $C_{SG,m,i}$ are the replacement and maintenance costs of thermal power units, respectively. The cost of replacement and maintenance of thermal units. The life cycle of the wind turbine, PV cell, and thermal unit, respectively; the project life and discount rate, respectively. The consumption characteristics of the thermal units are shown in a_i, b_i and c_i respectively.

3.1.2. Conditions of Constraint

(1) Power balance constraints

Without considering the system network loss, the power generated by the system is equal to the power consumed by the load.

$$P_{LO}(t) = \sum_{i=1}^{N_{wind}} P_{wind,i}(t) K_{wind,i} + \sum_{j=1}^{N_{pv}} P_{pv,j}(t) K_{pv,j} + \sum_{k=1}^{N_{SG}} P_{SG,k}(t) K_{SG,k} \tag{8}$$

where $P_{LO}(t)$ represents the power consumed by the load; $P_{wind,i}(t)$, $P_{pv,j}(t)$, and $P_{SG,k}(t)$ represents the output power of the i-th wind turbine, the j-th photovoltaic cell, and the k-th synchronous machine, respectively; N_{wind}, N_{pv}, and N_{SG} is the number of wind turbines, photovoltaic cells, and synchronous machines, respectively; $K_{wind,i}$, $K_{pv,j}$, and $K_{SG,k}$ is the switching state of the wind turbine, photovoltaic cell, and synchronous unit, respectively.

(2) Installed capacity constraint

Wind-photovoltaic-thermal power systems should have a certain amount of spare capacity, taking into account the possibility of failure of turbines, photovoltaics, synchronous machines, or unknown sudden increases in load in the system.

$$\sum_{i=1}^{N_{wind}} P_{wind,i} + \sum_{j=1}^{N_{pv}} P_{pv,j} + \sum_{k=1}^{N_{SG}} P_{SG,k} \geq \lambda P_{LO,max} \tag{9}$$

where $P_{wind,i}$, $P_{pv,j}$, and $P_{SG,k}$ represent the rated power of the i-th wind turbine, the j-th PV cell, and the k-th synchronous machine respectively; $P_{LO,max}$ is the maximum load power; λ is the load power factor.

(3) Generator output constraints

The output of wind, photovoltaic and thermal power units should fluctuate within a certain range.

$$0 \leq P_{wind,i}(t) \leq P_{wind,i,m} \tag{10}$$

$$0 \leq P_{pv,j}(t) \leq P_{pv,j,m} \tag{11}$$

$$0 \leq P_{SG,k}(t) \leq P_{SG,k,m} \tag{12}$$

where $P_{wind,i}(t)$, $P_{pv,j}(t)$, and $P_{SG,k}(t)$ represent the output power of the i-th wind turbine, the j-the photovoltaic cell, and the k-th synchronous machine respectively; $P_{wind,i,m}$, $P_{pv,j,m}$, and $P_{SG,k,m}$ represent the maximum output power of the wind turbine, the photovoltaic cell, and the synchronous machine respectively.

(4)   Climbing constraints for thermal power units

Thermal power units are required to meet a creep constraint, where the rate of change in power cannot exceed the creep rate during normal operation and can break the creep rate limit during start-up and shut-down.

$$- V_L P_{SG,i,m} \leq P_{SG,i}(t) - P_{SG,i}(t-1) \leq V_h P_{SG,i,m} \tag{13}$$

where V_h, V_L are the maximum upward and downward climbing rates, respectively; $P_{SG,i,m}$ is the maximum output of thermal power unit i.

3.1.3. A Pareto Optimal Solution Set Decision Method Based on Grey Relativity Analysis

Grey relativity analysis (GRA) is a method of measuring the degree of association between factors based on the degree of similarity or dissimilarity of trends between them [24]. As the upper level optimization model is multi-objective optimization and it is difficult for the configuration solution to satisfy multiple objectives optimally at the same time [25]. The traditional method of using compromise weighting factors to transform into a single-objective function solution will inevitably affect the decision result of the configuration solution. To accurately evaluate the effect of multi-objective solution sets without destroying the integrity of the original solution set, this paper proposes a grey correlation method based on the Pareto optimal solution set evaluation method.

Firstly, GRA is used to calculate the correlation value between the Pareto optimal solution set and the ideal solution, and to establish a mapping between the Pareto optimal solution set and the correlation value to provide a basis for the lower level optimization model. The correlation value represents the degree of correlation between the solution set and the ideal solution. When evaluating the upper level model capacity configuration solution, the higher the correlation degree value, the greater the degree of correlation between the configuration solution and the ideal solution, and the better the configuration solution.

Assume that the set of optimal solutions of the upper level optimization model Pareto is $\{x_1, x_2, \ldots, x_n\}$, where $x_i = \{F_1, F_2\}, i \in n$, x_i denotes the set of objective values of configuration scheme i. Let the ideal solution $x_0 = \{\min\{F_1\}, \min\{F_2\}\}$, the correlation coefficient between the optimal solution set and the ideal solution can be solved using the GRA algorithm to construct the set of capacity configuration scheme-correlation mappings, $\{x_i, \gamma(x_0, x_i)\}$. x_i is the capacity allocation solution i, $\gamma(x_0, x_i)$ is the correlation of x_i based on the ideal solution, characterized as a label for the superiority or inferiority of solution x_i. The larger $\gamma(x_0, x_i)$, the better the solution.

3.2. Solving a Multi-Objective Configuration Upper Level Optimization Model for Systems Considering Environmental Friendliness and Economy

As the optimization upper-level optimization model is a multi-objective solution problem, the fast and elite mechanism of the non-dominated ranking multi-objective genetic algorithm (NSGA-II) has the advantages of efficiency and directness and is an effective method for solving multi-objective optimization problems [26].

The INSGA2-DS algorithm based on dominance strength uses (1) an improved fast sorting method based on dominance strength, (2) a novel distance algorithm that introduces the consideration of variance, and (3) a strategy of adaptive elite retention based on the NSGA-II algorithm. The improved algorithms can effectively improve the convergence and distribution problems of the NSGA-II algorithm. The introduction of the INSGA2-DS algorithm in this paper can effectively improve the distributivity and accuracy of the system capacity allocation scheme, avoid the flooding of good data, reduce the solution time, and improve the solution efficiency of the algorithm. Figure 4 shows the flow chart of the upper optimization model solution.

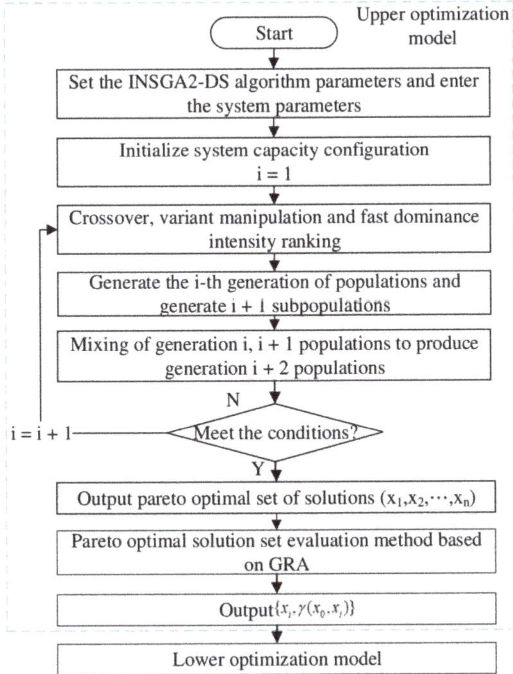

Figure 4. Flow chart of the upper optimization model solution.

3.3. Optimal Configuration of the Lower Level Model Considering System Stability

To improve the frequency stability of the system, the upper model configuration scheme is further optimized based on the upper optimized model, taking into account the level of inertia margin of the system. The power system time-series operation simulation method [27] is introduced to simulate the time-series operation of all the planning solutions derived from the upper model in turn to find the inertia security region for each solution. The optimal capacity planning scheme is then optimized with the objective of combining the best system stability and economic and environmental characteristics. The optimization variables are the different capacity planning solutions output by the upper model.

3.3.1. Inertia Security Region Model of the System

Inertia is the inherent ability of a power system to maintain frequency stability [28]. When the system is subjected to unpredictable power disturbances, frequency fluctuations occur within the system, when the rotational inertia present in the system helps to suppress rapid fluctuations in frequency and keep the frequency stable within a tolerable range [29,30]. Therefore, the inertia level of the system effectively reflects the frequency stability of the system.

According to the safety and stability standards proposed in the literature "Technical Guidelines for Safety and Stability Control of Power Systems", the minimum inertia required to ensure system frequency stability under N-2 faults in power systems relying only on primary frequency regulation and second line of defense safety and stability control measures is the safety critical inertia value under this fault scenario. When the actual inertia of the system is less than the safe critical inertia, the occurrence of a serious fault within the system will trigger the system's third line of defense safety device to act, and the system will be exposed to the risk of large area cut-off and load shedding. In this paper, the inertia value corresponding to the most severe failure scenario in the N-2 safety calibration of the system in the operating scenario is defined as the system safety critical inertia value M_{SIL} [6].

$$M_{SIL} = \max\left\{ M_{SIL,F_{12}}, \ldots, M_{SIL,F_{ij}} \right\} \tag{14}$$

where $M_{SIL,F_{ij}}$ is the safety critical inertia corresponding to the failure of component i, j; $i \neq j$ and $i, j \leq N_t$; $N_t = N_{SG} + N_w + N_p$; N_t is the total number of components; N_{SG} is the number of thermal power plants; N_w is the number of wind farms; N_p is the number of photovoltaic plants.

To quantify the system inertia level, the relative magnitude of the actual system inertia value to the safety critical system inertia value is defined as the system inertia margin.

$$K_m = \frac{M_{sys} - M_{SIL}}{M_{SIL}} \times 100\% \tag{15}$$

where K_m is the system inertia margin; M_{sys} is the actual system inertia value; M_{SIL} is the system safety critical inertia value K_m is the system inertia margin at a certain time. According to the operating characteristics of the power system, the system inertia margin varies at different times and is not continuous. The system Inertia security region is shown in Figure 5.

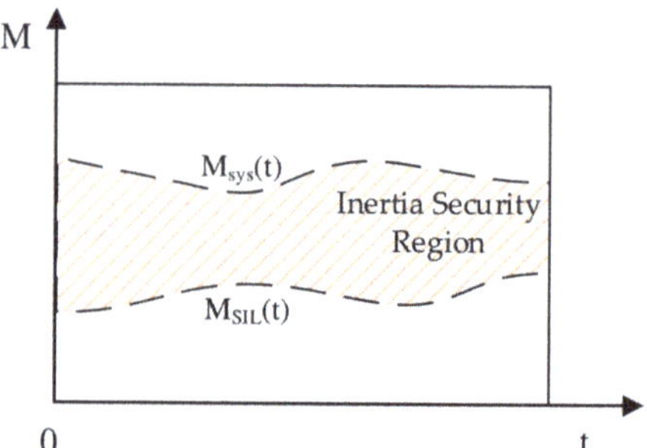

Figure 5. Inertia security region of the system.

Considering continuous operation periods, define the inertia security region of the system, which is the area of the system inertia margin over a length of time T.

$$K_M = \int_0^T \left[M_{sys}(t) - M_{SIL}(t) \right] dt \tag{16}$$

where $M_{sys}(t)$, $M_{SIL}(t)$ are the actual inertia and the safety critical inertia of the system at time t, respectively, and T is the length of time.

When the system inertia is the inertia security region, the system inertia can mitigate sudden changes in frequency caused by a potentially large disturbance fault in the system, avoiding large cuts in the system and load shedding.

In order to improve the calculation efficiency of the model, the rotational inertia of the system is considered in this paper. The rotational inertia of the turbine is ignored because the rotational kinetic energy provided by the turbine is related to the operating conditions, there are more variable factors, and it has less influence on the inertia of the system. Therefore, only the rotational inertia provided by the thermal power unit is considered.

The formula for calculating the actual inertia of the system:

$$M_{sys} = \sum_{i=1}^{N_{SG}} H_i P_i K_i \tag{17}$$

where M_{sys} is the actual inertia value of the generation system; H_i is the time constant of inertia of the thermal unit i; P_i is the rated power of the thermal unit i; K_i is the switching state of the thermal unit. When the thermal unit is on, $K_i = 1$, and when it is off, $K_i = 0$.

In the case of a power generation system, the actual inertia value of the system can be solved using the system operating scenario for the thermal power unit and its parameters.

The following equation for calculating the safe critical inertia value of the system is derived. Typically, when active disturbances occur in the system, ignoring damping effects, the equation of motion for the system equivalent rotor is:

$$2\frac{M_{sys}}{f_N}\frac{df(t)}{dt} = P_m(t) - P_e(t) \tag{18}$$

where f_N is the nominal frequency; $P_m(t)$ is the total mechanical power of the system at time t; $P_e(t)$ is the total electromagnetic power of the system at time t; $f(t)$ is the system frequency at time t.

In the event of an active disturbance in the system, $|RoCoF|$ reaches a maximum at the moment of the disturbance t_{0+} because frequency control measures act immediately to reduce the unbalanced power:

$$RoCoF(t_{0+}) = -\frac{f_N \Delta P}{2(M_{SIL} - M_{loss})} \tag{19}$$

where $RoCoF(t_{0+})$ is the rate of change of system frequency for the most severe fault in the system N-2 safety calibration fault set; f_N is the nominal frequency; ΔP is the active power disturbance from the limit expected fault; and M_{loss} is the loss of inertia due to the limit expected fault.

According to Equations (18) and (19), the system safety critical inertia is obtained:

$$M_{SIL} = \max \begin{cases} M_{loss} - \dfrac{f_N \Delta P}{2RoCoF_{min}} , \Delta P > 0 \\[3mm] M_{loss} - \dfrac{f_N \Delta P}{2RoCoF_{max}} , \Delta P < 0 \end{cases} \tag{20}$$

where $RoCoF_{max}$ is the upper limit of the rate of change of the system frequency; $RoCoF_{min}$ is the lower limit of the rate of change of the system frequency.

3.3.2. Objective Functions

To measure the degree of economy and environmental friendliness of the capacity allocation scheme of the upper level optimization model, a correlation factor μ_i is proposed. The correlation factor is calculated as:

$$\mu_i = \frac{\gamma_i}{\bar{\gamma}} \tag{21}$$

$$0 < \gamma_i \leq 1 \tag{22}$$

where γ_i is the correlation of option x_i; γ is the average of the correlation of all options γ_i characterizes the combined economic and environmental performance of option x_i. The larger γ_i is, the greater the correlation between option x_i and the ideal option, and the better the combined level of economy and environmental friendliness.

The correlation factor μ_i measures the degree of economy and environmental friendliness of the upper model capacity configuration. It is known that K_M is the system inertia security region. Obviously, the larger the system inertia safety domain is, the more beneficial to system stability. In order to improve the efficiency of the solution and make the decision scheme informative, the scenario of the maximum occurrence of the system limit expected failure day in one year is selected as a typical day, and the inertia safety domain of the typical day K_{M1} is defined:

$$K_{M1} = \int_0^{24} \left[M_{sys,i}(t) - M_{SIL,i}(t) \right] dt \tag{23}$$

where $M_{sys,i}(t)$ is the actual inertia of the system at time t for scenario x_i on a typical day; $M_{SIL,i}(t)$ is the critical inertia of the system at time t for scenario x_i on a typical day; and μ_i is the correlation factor for scenario x_i.

Based on the configuration scheme of the upper level optimization model, the objective function is established by considering the economy, environmental protection, and stability of the system:

$$\max F_3 = \max\{\mu_i K_{M1}\} \tag{24}$$

3.3.3. Conditions of Constraint

(1)   System inertia and rate of change of frequency constraints

The inertia and rate of change of frequency of the system shall be maintained within a range of:

$$M_{\min} \leq M_{SIL} \leq M_{\max} \tag{25}$$

$$M_{sys}(t) \geq M_{SIL}(t) \tag{26}$$

where $M_{\max}$ and $M_{\min}$ are the upper and lower limits of the system inertia, respectively.

(2)   System frequency rate of change constraint

$$RoCoF_{\min} \leq RoCoF \leq RoCoF_{\max} \tag{27}$$

where $RoCoF_{\max}$ and $RoCoF_{\min}$ are the upper and lower limits of the rate of change of the system frequency, respectively.

4. Bi-Level Model Solving for Optimal System Capacity Allocation

For system planning, the lack of actual output parameters for wind farms and photovoltaic power stations makes it difficult to perform direct calculations, so historical average meteorological data are used for output forecasting.

For the upper level model, the local historical average meteorological data information is combined with the predicted new energy output data based on the wind turbine and PV unit parameters, and the load data is predicted. As the upper optimization model is a multi-objective problem, INSGA2-DS is used to solve the algorithm and output the Pareto solution set for the upper model capacity configuration, and the Pareto optimal solution set evaluation method based on GRA is used to obtain the solution-correlation mapping set $\{x_i, \gamma(x_0, x_i)\}$ as input to the lower optimization model.

For the lower level model, first determine the various types of boundary conditions, introduce the power system time-series operation simulation method of literature [27], carry out year-round operation simulation based on the capacity configuration scheme of the upper level model, and then combine the system stability control strategy to generate N-2 safety check fault sets. Then, extract the system limit expected fault maximum occurrence day scenario, and solve the lower level optimization model [31,32]. Table 1 shows a Bi-level model for optimal system capacity planning. Figure 6 shows the framework for optimal capacity allocation of wind-photovoltaic-thermal power systems.

Table 1. Bi-level model for optimal system capacity planning.

Project	Upper Optimization Model	Lower Optimization Model
Decision variables	Wind, PV and thermal power output	Upper model solving solutions
Conditions of constraint	(1)(2)(3)(4)	(5)(6)
Objective functions	minF1, minF2	max F3
Solution algorithms	INSGA2-DS	Cplex solvers
Optimization objectives	Minimal carbon emission and lowest total cost	Combination of economy, environmental friendliness, and stability

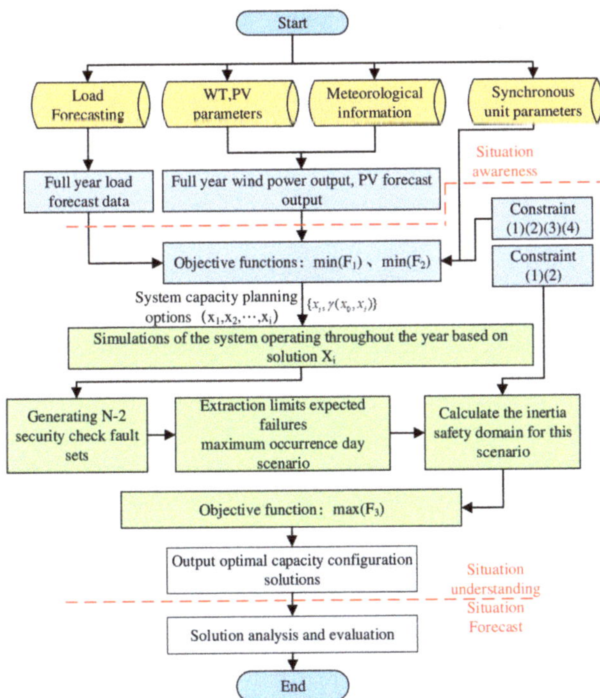

Figure 6. Framework for optimal capacity allocation of wind-photovoltaic-thermal power systems.

5. Case Study

5.1. Date Preprocessing

This paper uses a method for simulating the time-series operation of power systems based on meteorological data, as the planned wind-light-fire system lacks actual output data and is difficult to apply directly in simulation tools.

Based on the local average historical meteorological data information and historical load data for a region of the country, predictions are made including annual wind speed, temperature, light intensity, and annual load data, as shown in Figures 7–10. Suitable wind turbine, PV cell, and thermal power unit parameters are selected based on the load demand and meteorological data. The carbon emission factor parameters for the full life cycle of the system are shown in Tables A1–A3 in Appendix A. The cost and life cycle of each part of the system are shown in Table A4 in Appendix A. The INSGA2-DS algorithm was set to 100 iterations and run 10 times to obtain stable results for the algorithm. The Pareto optimal solution set for 60 sets of capacity allocation scenarios was obtained after the upper level model optimization solution, and the scenario-correlation mapping set was established and imported into the lower level optimization model for the solution.

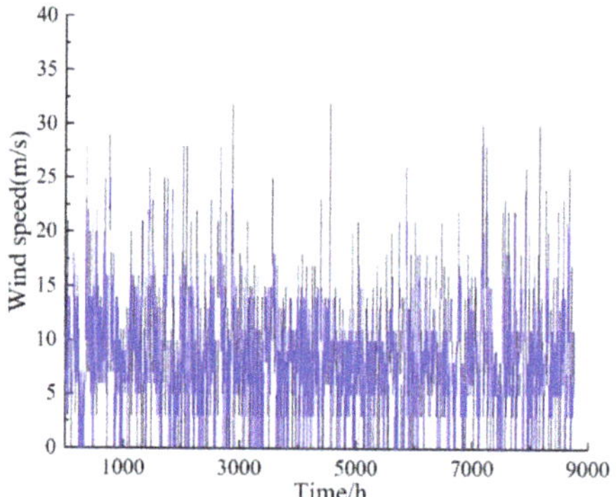

Figure 7. Full year wind speed forecast curve.

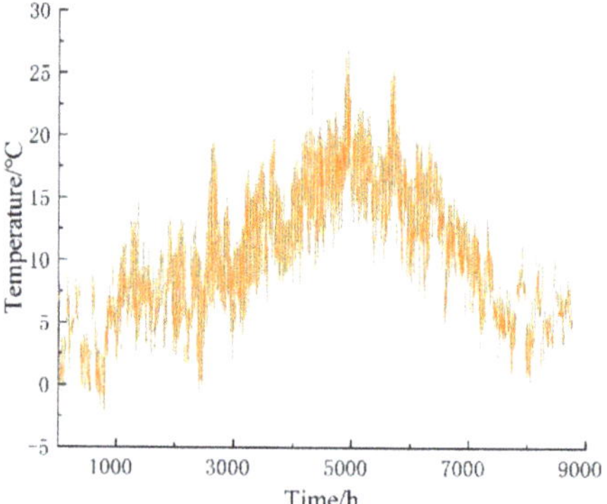

Figure 8. Full year temperature forecast curve.

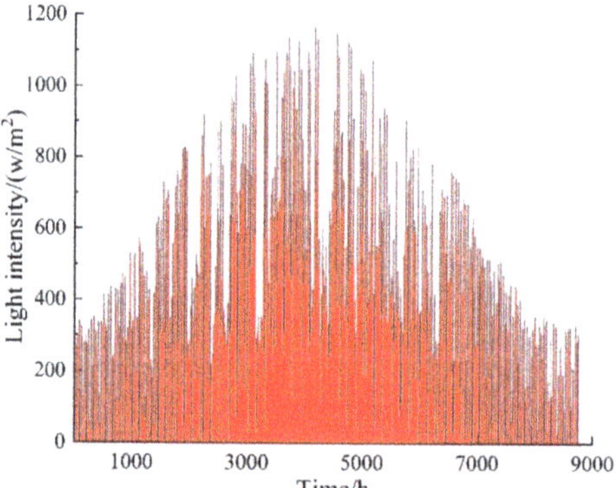

Figure 9. Full year light intensity prediction curve.

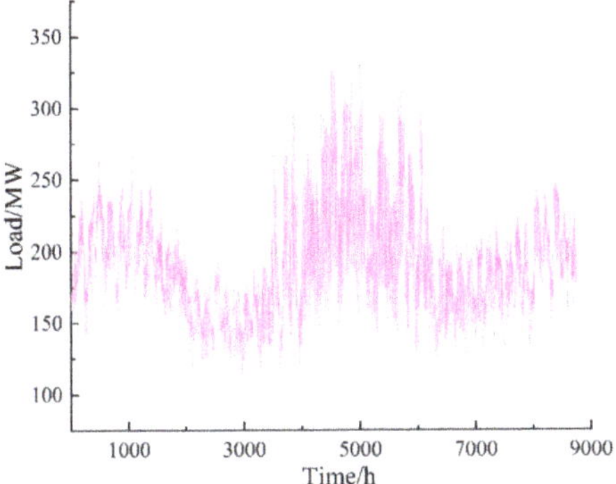

Figure 10. Full year load forecast curve.

5.2. Optimal Capacity Configuration Solution

The upper model is solved using the INSGA2-DS algorithm and the lower model is solved using the MATLAB software solver. Table 2 shows the optimal capacity configuration of the wind–photovoltaic–thermal power system that meets the requirements. Figure 11 shows the set of Pareto scenarios for the capacity configuration of the upper model.

Table 2. Optimal capacity configuration solution.

Configuration Solutions	Number
Number of WT	51
Number of PV cells	104,354
Number of thermal power units	8
Costs/¥	4.95×10^9
Carbon emission/kg	1.16×10^{10}
Inertia security region/MW·s^2	10,308.2
Correlation factor	1.38
Installed capacity of thermal power generation/MW	320
Installed capacity of wind power/MW	76.5
Installed capacity of photovoltaic power/MW	20.9

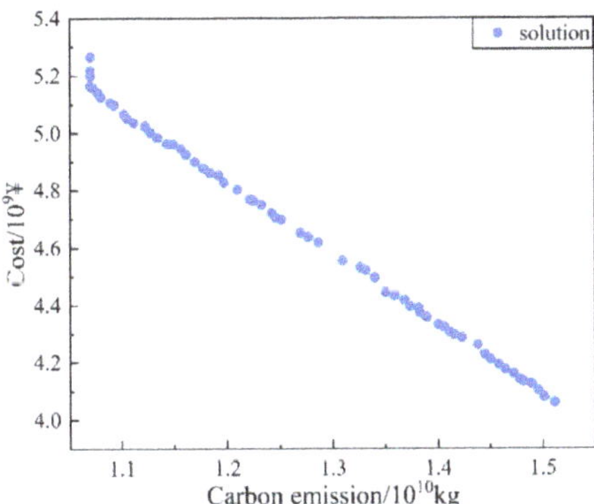

Figure 11. Pareto frontage diagram.

To verify the rationality of the optimal capacity allocation scheme, three different schemes are arbitrarily selected from the output scheme class of the upper optimization model for comparison and analysis. Option 1 is the optimal capacity allocation solution.

5.3. Simulation Analysis of the Timing Operation of Different Planning Scenarios

This paper uses a modified IEEE-39 node system as a research case. Figure 12 shows the modified IEEE-39 node network topology.

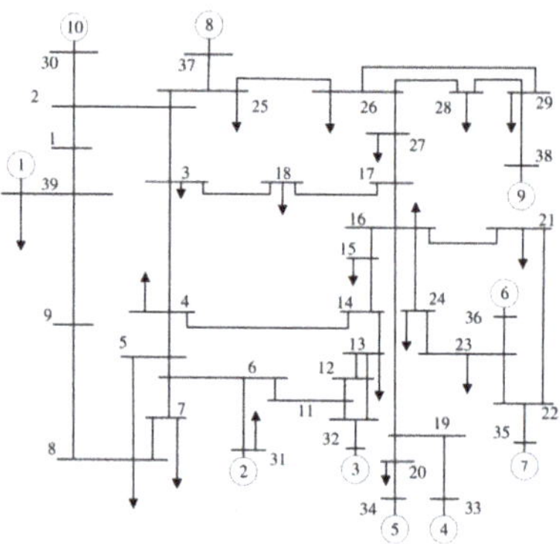

Figure 12. Modified IEEE-39 node network topology.

The IEEE-39 node arithmetic example is built on the MATLAB platform, with 1–8 connected to synchronous machines, 9 to an equivalent wind farm, and 10 to an equivalent PV plant. As the single generator output accounts for a relatively high total load, each limit expected fault type is either a synchronous unit tripping or a new energy field station going off-grid. The scenario of the maximum occurrence day of the limit scenario fault is selected for analysis. R_{max} and R_{min} are 2 Hz/s and −2 Hz/s respectively.

Scenario 1 is the optimal capacity allocation solution derived from the lower level model. During a typical day, 6 thermal units are expected to be on at moments 1–10; 8 thermal units are expected to be on at the remaining moments. As can be seen from Figure 13, during periods 1–3 and 12–24 on a typical day, the wind power output is less than the rated power because the actual wind speed is lower than the rated wind speed of the wind turbine, and the photovoltaic units can only generate power during the day, making it necessary for the thermal units to increase their power output to meet the power demand at the load side while satisfying the boundary conditions. During the 4–9 period, the wind power output reaches its maximum, and due to the low load demand at this time, the thermal power units must reduce their output by reducing the number of units on in order to reduce wind and light abandonment.

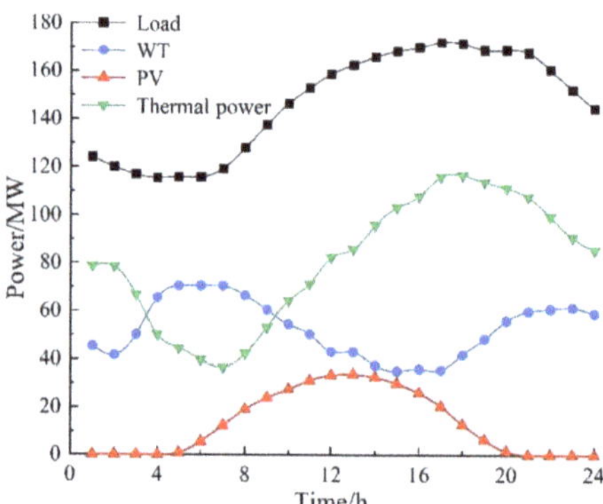

Figure 13. Typical daily power curve for scenario 1.

As can be seen from Figure 14, the moment of occurrence of the limit fault of the system in Scenario 1 is $t = 5$–9, the actual inertia of the system at all times during a typical day is greater than the value of the system safety critical inertia, and the inertia margins are all positive. The inertia of the system under the limit expected fault is sufficient to support the frequency fluctuation, which can effectively reduce the risk of a major outage accident.

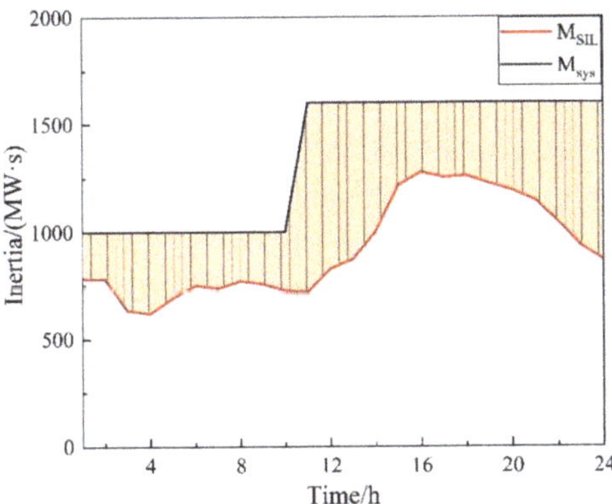

Figure 14. Typical daily inertia security region for scenario 1.

The typical daily inertia security region for scenario 1 is $K_{M1} = 10{,}308.2$ MW·s^2, $\mu_i = 1.38$ and $F_3 = 14{,}225.316$ MW·s^2. As the system operating state varies with time, the system limit failure varies from moment to moment, exhibiting the time-varying nature of the system safety critical inertia values. When $t = 3$, $t = 4$, and $t = 10$, the system corresponds to a smaller limit expected failure with a smaller inertia requirement, when the system inertia is more abundant. In contrast, when $t = 5$–9, the actual inertia of the system is close to the system's safe critical inertia, and the system is at a low inertia level at this time.

Scenario 2 is the capacity allocation option with the largest correlation factor, and the specific capacity allocation can be seen in Table 3.

Table 3. Different capacity configuration solutions.

Solution Configuration	Scenario 1	Scenario 2	Scenario 3	Scenario 4
Number of WT	51	71	15	74
Number of PV cells	104,354	104,525	103,790	104,739
Number of thermal power units	8	7	9	7

As can be seen in Figure 15, Scenario 2 has a different share of wind, PV and thermal power output due to the different number of installed thermal, wind and PV units compared to Scenario 1. In the periods 2–12 and 20–24 on a typical day, wind power output is higher due to the higher installed capacity of the turbines. Especially in the 3–12 period, wind power output is much higher than thermal power output. At this time, the number of thermal units must be reduced to meet the load demand.

As can be seen in Figure 16, the typical daily inertia security region for scenario 2 is $K_{M1} = -8614.6$ MW·s^2, $\mu_i = 1.55$, $F_3 = -13{,}352.63$ MW·s^2. At moments 3–14 and 19–24, the actual system inertia is lower than the system safety critical inertia value, with the system inertia deficit reaching a maximum of 1233 MW·s at moment 7, when the system faces a very high risk of frequency destabilization. This is due to the low thermal power output and the high proportion of new energy sources. When a major fault occurs in the system, such as a new energy source going off-grid, the system frequency will be destabilized due to the lack of sufficient inertia support, which will result in a large-scale power outage. The length of time that the actual inertia of the system is below the system safety critical inertia value is 517 h throughout a year.

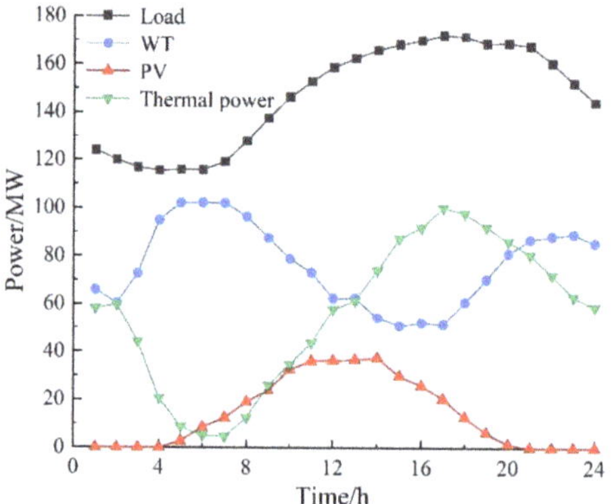

Figure 15. Typical daily power curve for scenario 2.

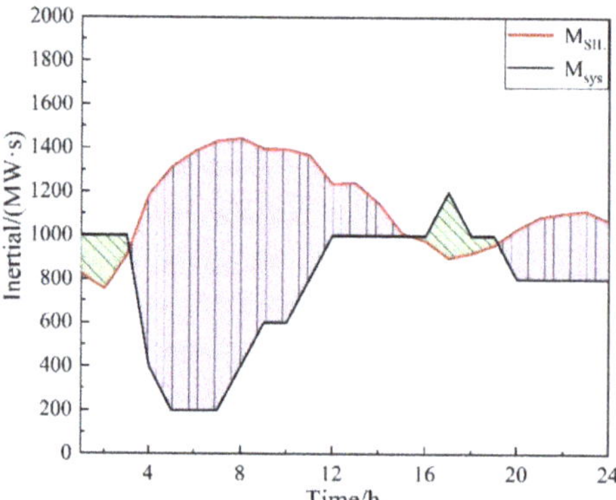

Figure 16. Typical daily inertia security region for scenario 2.

Scenario 3 is the least costly option to consider, with a typical daily inertia security region of K_{M1} = 16,802 MW·s^2, μ_i = 0.74, and F_3 = 12,433.48 MW·s^2. As can be seen from Figure 17, the system has a higher share of thermal power output and a smaller peak-to-valley differential. In the period 8–17, the combined share of wind and PV output is higher, peaking at around 26%. Figure 18 shows a typical daily inertia security region for Scenario 3.

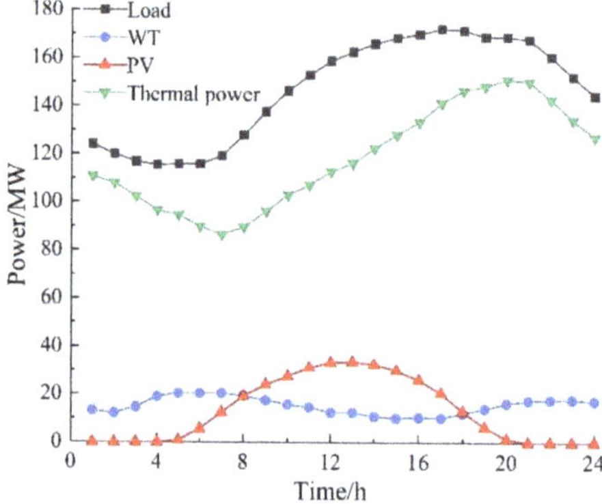

Figure 17. Typical daily power curve for scenario 3.

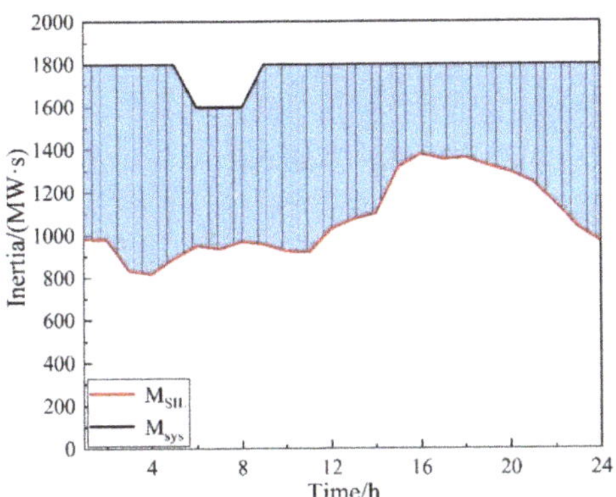

Figure 18. Typical daily inertia security region for scenario 3.

Scenario 4 is the case where carbon emissions are considered to be minimal. As shown in Figure 19, the typical daily output characteristics of scenario 4 are similar to those of scenario 2 due to the similarity between the capacity configuration scheme of scenario 2 and that of scenario 4. The typical daily inertia security region $K_{M1} = -8871.2$ MW·s^2, $\mu_i = 1.49$, and $F_3 = -13{,}218.09$ MW·s^2. The length of time during a year when the actual system inertia is below the system safety critical inertia value is greater than in Scenario 2, amounting to 780 h. Figure 20 shows a typical daily inertia security region for Scenario 4.

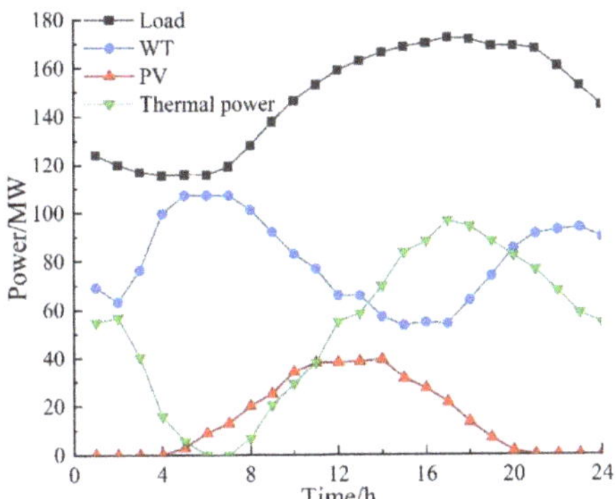

Figure 19. Typical daily power curve for scenario 4.

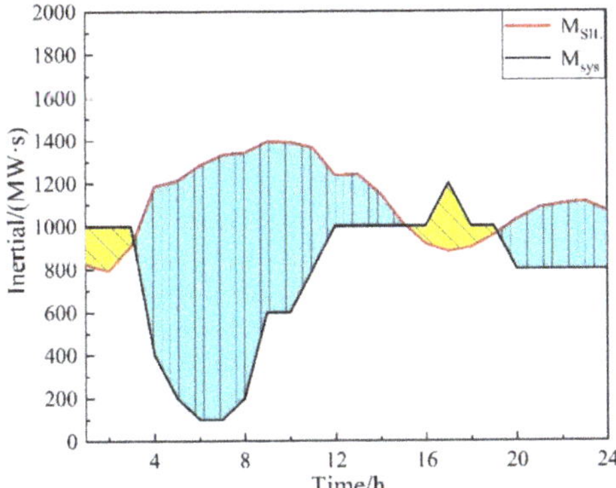

Figure 20. Typical daily inertia security region for scenario 4.

In summary, the system is able to ensure continuous and stable power supply on a typical day for all capacity configuration options, subject to constraints, such as meeting load demand and weather forecast data.

5.4. Comprehensive Characterisation of Different Configuration Scenarios

A comprehensive analysis of the economy, environmental friendliness, and stability of the four scenarios.

As can be seen from Figure 21, the inertia security region for scenarios 2 and 4 is negative, which is because the new energy output is higher during a typical day and the thermal units are in a lower output state, resulting in the actual inertia level of the system is lower than the system safety critical inertia value, and therefore scenarios 2 and 4 have a frequency stability risk. As the typical day is the maximum day scenario of the yearly limit expected fault occurrence, the inertia security region of scenarios 1 and 3 are 10,308.2 MW·s^2 and 16,802 MW·s^2 respectively, so there are no negative inertia margin operation scenarios for scenarios 1 and 3 during the yearly operation.

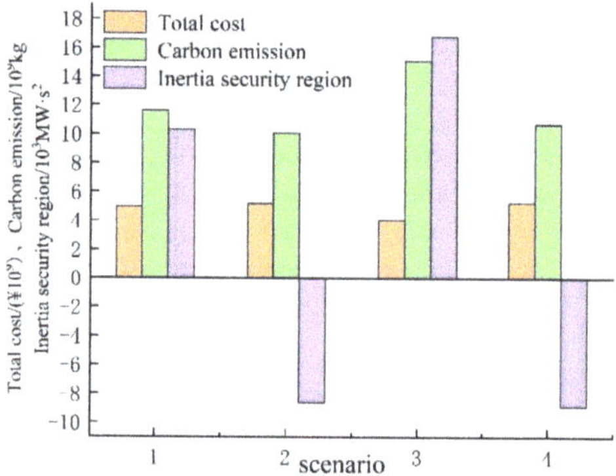

Figure 21. Characteristics of different scenarios.

The economic analysis of different scenarios shows that scenario 3 has the lowest total cost, which is because the unit price of new energy is higher than that of thermal units per unit of capacity, while scenario 3 has the smallest total amount of new energy installed and the system generation output is mainly borne by thermal units, thus the total cost of scenario 3 is low. In addition, scenario 1 also has a lower total cost of CNY 4.951 × 10^9, which is 4.18% and 6% lower than scenarios 2 and 4 respectively.

For the carbon emissions analysis, the carbon emissions from electricity generation are smaller for scenarios 2 and 4 due to their larger installed new energy capacity. The difference in carbon emissions between scenarios 1, 2, and 4 is not significant, within 7%. As the installed capacity of thermal power units is higher and the installed capacity of new energy is lower, the carbon emissions from scenario 3 are the largest, with scenario 1 emitting 3.514 × 10^9 kg less carbon than scenario 3, or approximately 23.25%.

An analysis of the characteristics of the different options shows that as the correlation factor increases, the carbon emissions of the system gradually decrease, while the total cost does not change much. Therefore, the correlation factor can be used to effectively evaluate the merits of the capacity allocation options. Scenario 2, with the highest correlation factor, is the best capacity allocation option if the system stability is not considered and only the system economy and environmental protection are taken into account. However, the actual inertia of the system will be lower than the safety critical inertia of the system during the operation of scenario 2, which will lead to low frequency load shedding or high cycle cut-off of the grid in case of serious failure. Therefore, the capacity configuration of scenario 2 requires an appropriate increase in the number of synchronous machines to increase the inertia of the system, while the number of turbines and PV units should be reduced to reduce wind and light abandonment and to enhance the economy of the system.

In summary, scenario 1 takes into account system economy, environmental friendliness, and stability. Scenario 1 can therefore be used as the best capacity configuration for the system.

5.5. Impact of Optimization Algorithms on Capacity Planning

To verify the superiority of INSGA2-DS in solving capacity planning problems, the NSGA-II algorithm was used for comparative analysis. As shown in Figure 22, the Pareto frontier solution is more widely distributed and can effectively avoid getting trapped in a local optimum.

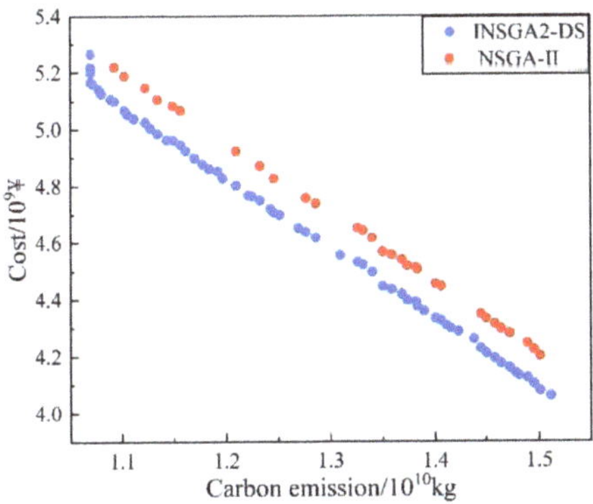

Figure 22. NSGA-II and INSGA2-DS optimization results.

A comparison of the operational characteristics of the two algorithms is shown in Table 4. For the same number of populations, INSGA2-DS has a shorter computation time than NSGA-II and converges at a faster rate, with a computational efficiency improvement of about 17%. Therefore, INSGA2-DS is more suitable for the problem of capacity optimization allocation of wind-light-fire systems.

Table 4. Comparison of NSGA-II and INSGA2-DS characteristics.

Algorithms	Population Size/Unit	Number of Convergence Iterations/Time	Calculation Efficiency/s
NSGA-II	300	120	101
INSGA2-DS	300	100	84

6. Conclusions

This paper proposes a situation awareness-based capacity optimization strategy for wind–photovoltaic–thermal power systems. A bi-level model is established for the optimal allocation of system capacity. The upper model takes into account the carbon emissions and total system cost of the whole life cycle of the system and ensures the effectiveness and practicality of the upper model through the system power balance constraint, installed capacity constraint, generator output constraint, and thermal unit climbing constraint. The Pareto-based capacity allocation scheme is solved using the INSGA2-DS algorithm, and the Pareto optimal solution set evaluation method based on GRA is used to establish the scheme-relation degree mapping set, which is used as the input of the lower model. The lower model integrates the maximum inertia security region of the system, the best economy, and environmental protection as the optimization objectives to optimize the capacity allocation scheme. Finally, the effectiveness of the proposed strategy and algorithm is verified by means of an arithmetic example.

This provides new practical ideas and methods for planning the capacity allocation of wind–photovoltaic–thermal power systems in the context of new power systems, and is a guide to the problem of planning the capacity of power sources in the context of the new power system. The wind–photovoltaic–thermal power system capacity optimization model developed in this paper can ensure the best system stability and minimize carbon emissions and total costs within a certain range.

The method proposed in the paper focuses on three forms of power sources, namely wind, light, and fire, and will be followed by subsequent studies to include multi-energy matching of systems, such as hydropower and energy storage.

Author Contributions: Methodology, D.L.; investigation and data curation, X.C.; resources, L.G.; formal analysis, W.H.; software, J.H.; writing—original draft preparation, X.C.; writing—review and editing, Z.H. All authors have read and agreed to the published version of the manuscript.

Funding: This research received no external funding.

Informed Consent Statement: Informed consent was obtained from all subjects involved in the study.

Data Availability Statement: Data are contained within the article. Data sharing is not applicable to this article.

Conflicts of Interest: The authors declare no conflict of interest.

Appendix A

Table A1. Thermal power unit parameters.

Parameters	Number
Power rating/MW	40
Maximum power/MW	40
Minimum power/MW	10
a/(¥/MWh)	0.024
b/(¥/MWh)	78
c/¥	960
Inertia time constant/s	5
R/(kg/kWh)	0.95

Table A2. Wind turbine parameters.

Parameters	Number
Power rating/MW	1.5
Cut-in wind speed V_{in}/(m/s)	3
Rated wind speed V_n/(m/s)	10
Cut-out wind speed V_{out}/(m/s)	30
R/(kg/kWh)	0.012

Table A3. Photovoltaic cell parameters.

Parameters	Number
P_s/(kWh)	0.2
G_s/lx	1000
γ/(%/°C)	−0.5
T_τ/°C	25
R/(kg/kWh)	0.035

Table A4. System component costs and life cycle.

Parameters	Thermal Power Units	Wind Turbines	Photovoltaic Cells
Investment cost/(¥)	1.125×10^8	1.097×10^7	1242
Replacement cost/(¥)	1.125×10^8	1.097×10^7	1242
Maintenance cost/(¥/yr)	1.125×10^6	1.097×10^5	12.42
Life cycle/(yr)	15	20	30

References

1. Bistline, J.E.T.; Brown, M.; Siddiqui, S.A.; Vaillancourt, K. Electric sector impacts of renewable policy coordination: A multi-model study of the North American energy system. *Energy Policy* **2020**, *145*, 111707. [CrossRef]
2. Helisto, N.; Kiviluoma, J.; Holttinen, H.; Lara, J.D.; Hodge, B.-M. Including operational aspects in the planning of power systems with large amounts of variable generation: A review of modeling approaches. *Wiley Interdiscip. Rev.-Energy Environ.* **2019**, *8*, 341. [CrossRef]
3. Xu, Y.; Tu, J.; Yin, Z. The capacity selection of wind photovoltaic power generations based on KELM method. *Electr. Meas. Instrum.* **2019**, *56*, 73–80.
4. Tang, H.; Yang, G.; Wang, P.; Li, Q.; Zhang, L.; Liu, S.; Qin, J. Capacity Optimal Configuration of Wind/PV Hybrid Power System Based on Carbon Dioxide Emission. *Electr. Power Constr.* **2017**, *38*, 108–114.
5. Xu, Y.; Lang, Y.; Wen, B.; Yang, X. An Innovative Planning Method for the Optimal Capacity Allocation of a Hybrid Wind-PV-Pumped Storage Power System. *Energies* **2019**, *12*, 2809. [CrossRef]
6. Liu, J.; He, D. Profit Allocation of Hybrid Power System Planning in Energy Internet: A Cooperative Game Study. *Sustainability* **2018**, *10*, 388. [CrossRef]
7. Ye, C.-J.; Liu, W.-D.; Fu, X.-H.; Wang, L.; Huang, M.-X. Capacity allocation of hybrid solar-wind energy system based on discrete probabilistic method. *Turk. J. Electr. Eng.* **2015**, *23*, 1913–1929. [CrossRef]
8. Xu, M.; Zhuan, X. Optimal planning for wind power capacity in an electric power system. *Renew. Energy* **2013**, *53*, 280–286. [CrossRef]
9. Emmanuel, M.; Doubleday, K.; Cakir, B.; Markovic, M.; Hodge, B.-M. A review of power system planning and operational models for flexibility assessment in high solar energy penetration scenarios. *Sol. Energy* **2020**, *210*, 169–180. [CrossRef]
10. Dhaliwal, N.K.; Bouffard, F.; O'Malley, M.J. A Fast Flexibility-Driven Generation Portfolio Planning Method for Sustainable Power Systems. *IEEE Trans. Sustain. Energy* **2021**, *12*, 368–377. [CrossRef]
11. Lin, X.; Wen, Y.; Yang, W. Inertia Security Region: Concept, Characteristics, and Assessment Method. *Proc. CSEE* **2021**, *41*, 3065–3079.
12. Wen, Y.; Lin, X. Minimum Inertia Requirement Assessment of Microgrids in Islanded and Grid-connected Modes. *Proc. CSEE* **2021**, *41*, 2040–2053.
13. Basu, C.; Padmanaban, M.; Guillon, S.; Cauchon, L.; De Montigny, M.; Kamwa, I. Situational awareness for the electrical power grid. *IBM J. Res. Dev.* **2016**, *60*, 7384562. [CrossRef]
14. Yang, J.; Zhang, P.; Xu, X.; Xu, T. Research Status of Power Grid Situation Awareness Technology in China and Abroad. *East China Electr. Power.* **2013**, *41*, 1575–1581.
15. Ge, L.; Li, Y.; Chen, Y. Key Technologies of Situation Awareness and Implementation Effectiveness Evaluation in Smart Distribution Network. *High Volt. Eng.* **2021**, *47*, 2269–2280.
16. Oree, V.; Hassen, S.Z.S.; Fleming, P.J. Generation expansion planning optimisation with renewable energy integration: A review. *Renew. Sustain. Energy Rev.* **2017**, *69*, 790–803. [CrossRef]
17. Wang, Z.; Shi, Y.; Tang, Y.; Men, X.; Cao, J.; Wang, H. Low Carbon Economy Operation and Energy Efficiency Analysis of Integrated Energy Systems Considering LCA Energy Chain and Carbon Trading Mechanism. *Proc. CSEE* **2019**, *39*, 1614–1626.
18. Ardente, F.; Beccali, M.; Cellura, M.; Lo Brano, V. Energy performances and life cycle assessment of an Italian wind farm. *Renew. Sustain. Energy Rev.* **2008**, *12*, 200–217. [CrossRef]
19. Luo, W.; Khoo, Y.S.; Kumar, A.; Low, J.S.C.; Li, Y.; Tan, Y.S.; Wang, Y.; Aberle, A.G.; Ramakrishna, S. A comparative life-cycle assessment of photovoltaic electricity generation in Singapore by multicrystalline silicon technologies. *Sol. Energy Mater. Sol. Cells* **2018**, *174*, 157–162. [CrossRef]
20. Oladeji, I.; Zamora, R.; Lie, T.T. An online security prediction and control framework for modern power grids. *Energies* **2021**, *14*, 6639. [CrossRef]
21. Yang, B.; Guo, Y.; Xiao, X.; Tian, P. Bi-level Capacity Planning of Wind-PV-Battery Hybrid Generation System Considering Return on Investment. *Energies* **2020**, *13*, 3046. [CrossRef]
22. Kumar, I.; Tyner, W.E.; Sinha, K.C. Input-output life cycle environmental assessment of greenhouse gas emissions from utility scale wind energy in the United States. *Energy Policy* **2016**, *89*, 294–301. [CrossRef]
23. Kadiyala, A.; Kommalapati, R.; Huque, Z. Characterization of the life cycle greenhouse gas emissions from wind electricity generation systems. *Int. J. Energy Environ. Eng.* **2017**, *8*, 55–64. [CrossRef]

24. Li, C.; Zhao, G.; Meng, J.; Zheng, Z.; Yu, S. Multi-Objective Optimization Strategy Based on Entropy Weight, Grey Correlation Theory, and Response Surface Method in Turning. *Int. J. Ind. Eng.-Theory* **2021**, *28*, 490–507.
25. Tan, Y.-Y.; Jiao, Y.-C.; Li, H.; Wang, X.-K. A modification to MOEA/D-DE for multiobjective optimization problems with complicated Pareto sets. *Inf. Sci.* **2012**, *213*, 14–38. [CrossRef]
26. Verma, S.; Pant, M.; Snasel, V. A Comprehensive Review on NSGA-II for Multi-Objective Combinatorial Optimization Problems. *IEEE Access* **2021**, *9*, 57757–57791. [CrossRef]
27. Liu, X.; Lu, X.; Lou, Y.; Zhao, Y.; Wei, L.; Zhao, W. Optimal setting of wind-thermal-bundled capacity ratio based on chronological operation simulation. *Power Syst. Prot. Control* **2021**, *49*, 53–62.
28. Rezkalla, M.; Pertl, M.; Marinelli, M. Electric power system inertia: Requirements, challenges and solutions. *Electr. Eng.* **2018**, *100*, 2677–2693. [CrossRef]
29. Gu, H.; Yan, R.; Saha, T.K.; Muljadi, E. System Strength and Inertia Constrained Optimal Generator Dispatch under High Renewable Penetration. *IEEE Trans. Sustain. Energy* **2020**, *11*, 2392–2406. [CrossRef]
30. Wang, B.; Yang, D.; Cai, G. Review of Research on Power System Inertia Related Issues in the Context of High Penetration of Renewable Power Generation. *Power Syst. Technol.* **2020**, *44*, 2998–3007.
31. Wang, J.; Dong, F.; Ma, Z.; Chen, H.; Yan, R.; Klemes, J.J. Multi-objective optimization with thermodynamic analysis of an integrated energy system based on biomass and solar energies. *J. Clean. Prod.* **2021**, *324*, 129257. [CrossRef]
32. Li, J.; Xu, W.; Cui, P.; Qiao, B.; Feng, X.; Xue, H.; Wang, X.; Xiao, L. Optimization configuration of regional integrated energy system based on standard module. *Energy Build.* **2020**, *229*, 110485. [CrossRef]

MDPI

St. Alban-Anlage 66

4052 Basel

Switzerland

Tel. +41 61 683 77 34

Fax +41 61 302 89 18

www.mdpi.com

Energies Editorial Office

E-mail: energies@mdpi.com

www.mdpi.com/journal/energies